THE GIANT BOOK OF EASY-TO-BUILD ELECTRONIC PROJECTS

BY THE EDITORS OF ELEMENTARY ELECTRONICS

FIRST EDITION

SECOND PRINTING

Printed in the United States of America

Library of Congress Cataloging in Publication Data

Main entry under title:

The Giant book of easy-to-build electronic projects.

Includes index.
1. Electronics—Amateurs' manuals.
I. Elementary electronics.
TK9965.G495 1984 621.381 83-4982
ISBN 0-8306-0199-6
ISBN 0-8306-0599-1 (pbk.)

Contents

Preface

Elementary Electronics and its sister electronics magazines by Davis Publications have been considered by electronics hobbyists as the best source of easy-to-build, inexpensive electronic projects. For this reason, we decided to compile some of the best projects ever published in these magazines and make a series of books from them. *The GIANT Book of Easy-to-Build Electronic Projects* is the first in this series. Watch for the rest of the volumes to follow shortly.

With very few exceptions, the recommended parts are easy to obtain. We have even kept in the parts lists the names of certain suppliers who are out of business or no longer stock those parts as a reference for you to use when shopping for parts. Aware that using components of a higher grade, tolerance, and price will add nothing to a final performance, we strongly urge you to follow each parts list as closely as possible. Expensive substitutions will only hurt your wallet, not help the project.

An excellent source of parts suppliers' names, addresses, and stock information is *Electronics Buyer's Guide—3rd Edition* by Edward A. Hall, TAB book No. 345. With these two books, in fact, we believe you will have no excuse whatsoever not to complete these unique projects. Good luck!

Chapter 1

Projects for the Home

SCRAMBLE PHONE

Demand ultimate privacy in your telephone conversations? If so, a secure phone link can be yours with these three easy and exciting steps.

☐ Place your call with Ma Bell's instrument.

☐ When your party answers, switch to your special scrambler phones.

☐ A soft buzz in the ear piece says your security phone link is operational. Your line is secure. Conduct your call in complete privacy no matter how many ears are party-line listening!

Refer to Figs. 1-1 through 1-4.

All it takes is a pair of Scramble Phones. And with this double-duty unit, you can try your hand at decoding scrambled conversations that are sometimes heard on radio receivers covering the VHF high band (148-176 MHz). The basic scrambler circuit can be simply modified for radio by removing two fixed resistors and replacing them with a dual-potentiometer.

Wait a minute! Before your soldering iron overheats, let us say that this scrambler will decode information that is encoded in the *single inversion* mode only. The highly sophisticated scramblers that are sometimes used today cannot be decoded with this decoder, but in many areas the single inversion system is still in use and may be decoded with our units.

How It Works. IC-1 and the associated circuitry form a stable audio tone generator which feeds a buffer amplifier, Q1 and Q2. The tone output is taken from the emitters of the transistor pair to supply a carrier voltage for a balanced modulator made up of four diodes—D1 through D4—and T1 and T2. If the two transformers and the four diodes are perfectly matched (which is almost impossible to achieve and not necessary in any case) no carrier will appear at the input or output of T1 or T2. In a practical circuit, a small amount of unbalance will occur and produce a low-level carrier tone at the input and output

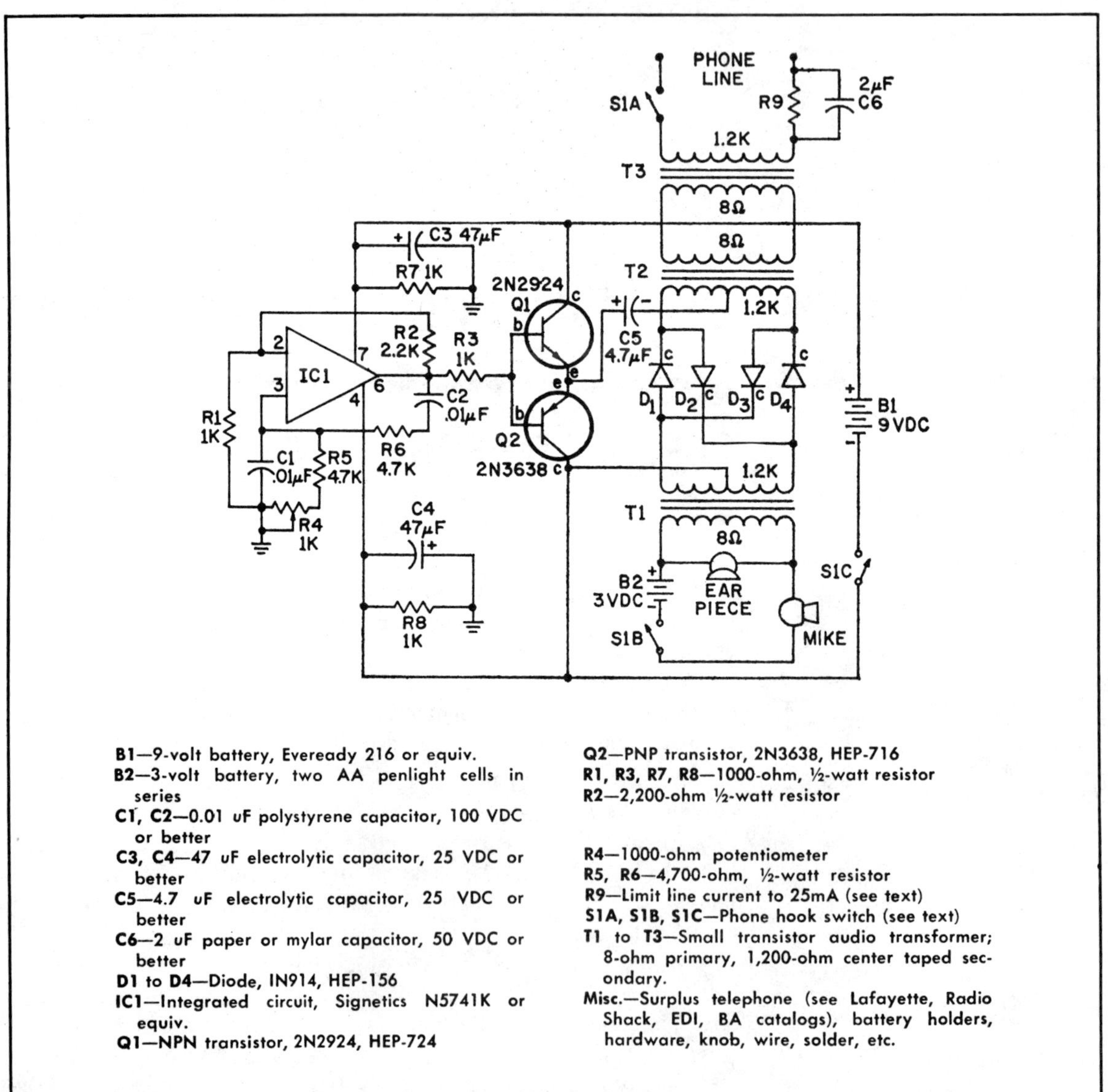

B1—9-volt battery, Eveready 216 or equiv.
B2—3-volt battery, two AA penlight cells in series
C1, C2—0.01 uF polystyrene capacitor, 100 VDC or better
C3, C4—47 uF electrolytic capacitor, 25 VDC or better
C5—4.7 uF electrolytic capacitor, 25 VDC or better
C6—2 uF paper or mylar capacitor, 50 VDC or better
D1 to D4—Diode, IN914, HEP-156
IC1—Integrated circuit, Signetics N5741K or equiv.
Q1—NPN transistor, 2N2924, HEP-724
Q2—PNP transistor, 2N3638, HEP-716
R1, R3, R7, R8—1000-ohm, ½-watt resistor
R2—2,200-ohm ½-watt resistor
R4—1000-ohm potentiometer
R5, R6—4,700-ohm, ½-watt resistor
R9—Limit line current to 25mA (see text)
S1A, S1B, S1C—Phone hook switch (see text)
T1 to T3—Small transistor audio transformer; 8-ohm primary, 1,200-ohm center taped secondary.
Misc.—Surplus telephone (see Lafayette, Radio Shack, EDI, BA catalogs), battery holders, hardware, knob, wire, solder, etc.

Fig. 1-1. Scramble Phone schematic and parts list.

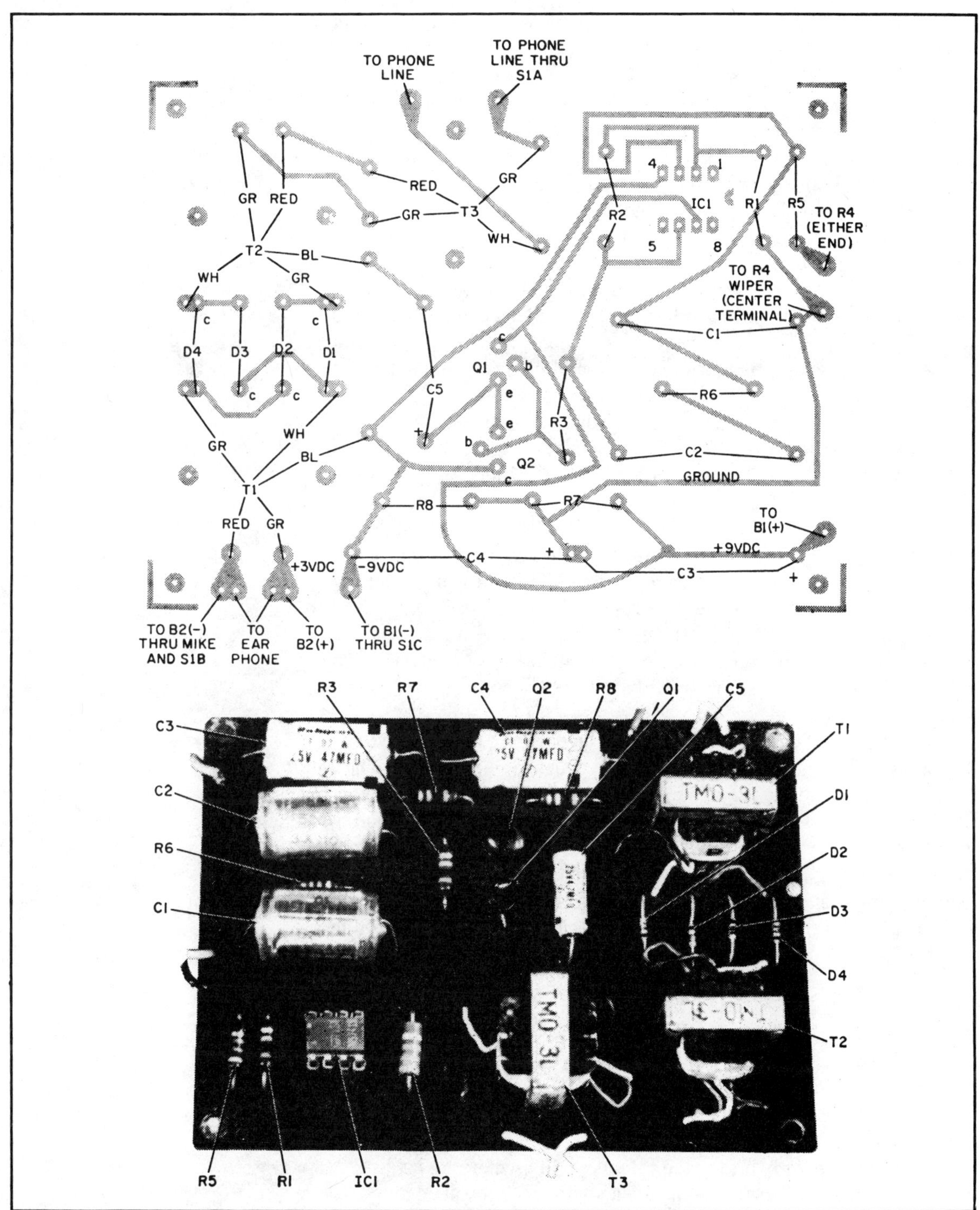

Fig. 1-2. Exact part placement on the PC board. View is from the component (top) side of board.

of the balanced modulator. This tells you your scramble phone is working.

A telephone carbon mike and ear piece are connected to the low impedance winding of T1, with a three volt battery supplying the necessary mike current. When the mike is spoken into, the carrier voltage is allowed to pass, in part, through transformers T2 and T3 and on to the telephone network. The only purpose of T3 is to match the impedance offered by most telephone lines.

Trim potentiometer R4 is used to make a fine frequency adjustment of the oscillator so that two scrambler units may be synchronized to the same carrier frequency. Both oscillators must be operating at the same frequency to produce the best decoded speech quality. This control is referred to as the speech *clarity* control. The best overall carrier frequency range to use for speech scrambling is between 2 kHz and 3.5 kHz.

Listening In. If the scramble phone is to be used for only receiver speech decoding, then only one unit is required. The operation is much the same as for telephone encoder/decoder purposes, with the exception that it is used only as a decoder. The carrier oscillator is made variable so the decoder may be synchronized to the same carrier frequency as is used in the encoder. The output of the receiver is connected to the 8-ohm winding of T2 (T3 is not required for this use), and the decoded information is developed across the 8-ohm winding of T1. A small speaker may be connected across this winding, or a low impedance earphone will do for monitoring the decoded speech. No mike or 3-volt battery is necessary for decoding operations.

Putting It Together. The circuit layout isn't critical, and any suitable scheme can be followed, but the layout shown for the PC board would be a good one to use. No matter what construction plan is used, PC board or bread board, extra care should be taken when connecting the IC, diodes, and transistors to the circuit. Care should also be taken when connecting the three transformers, so that the low-impedance and high-impedance windings are not reversed.

The size of the PC board allows the scrambler to be mounted in the base of a standard telephone. All parts located inside the phone, with the exception of the hook switch, can be removed to make the construction job an easy one. Check the pictures when mounting the board and batteries.

In some telephones, the hook switch contains enough switch contacts to function as the three switches, S1A, S1B, and S1C; however, if you have one that does not contain enough contacts, a separate switch must be added to switch the battery power. For the scramble phone to automatically bridge the telephone line when the handset is offhook, at least one section of the hook switch must be used for S1A.

If a dial telephone is selected, the dial may easily be removed and

Fig. 1-3. It is possible to decode some older scrambler systems heard on two-way radio bands. However, this single-inversion mode circuit will not work with multi-inversion systems also in use. Dual pot is Allen-Bradley CJK1N200P103U or equiv.

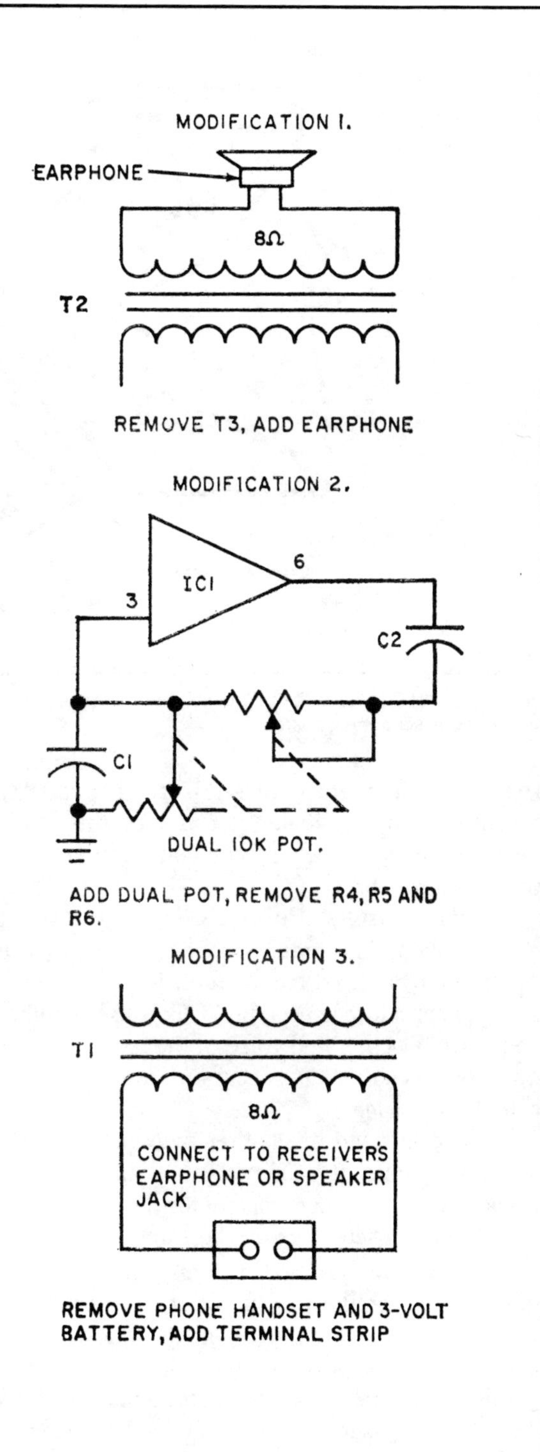

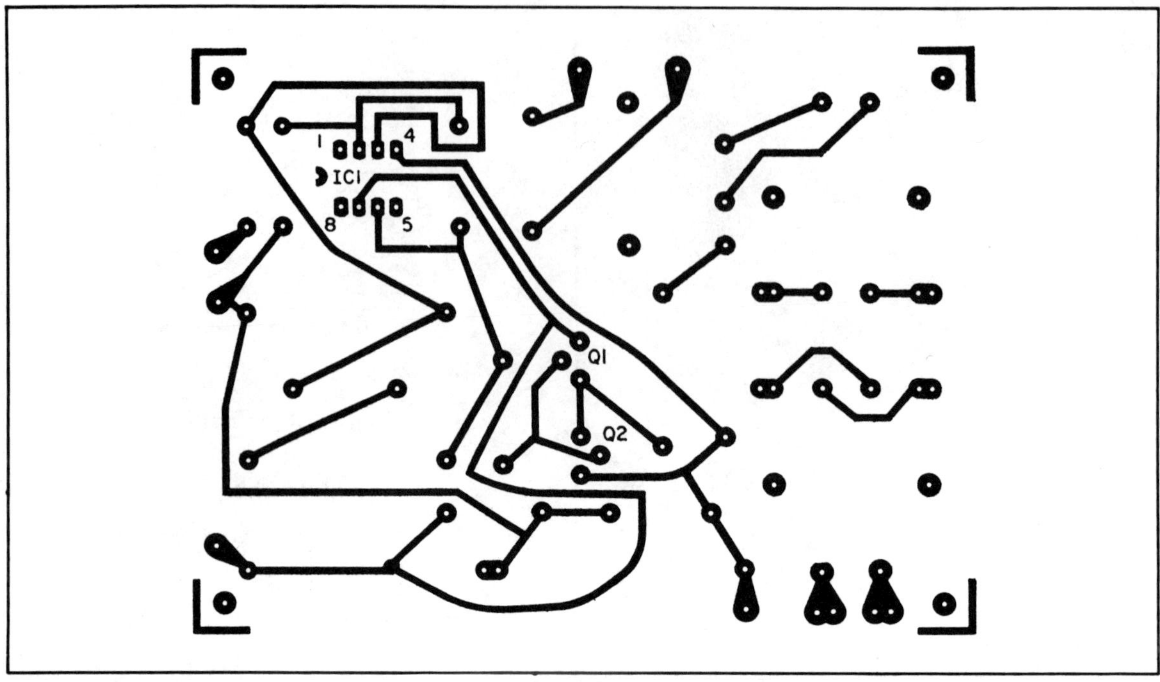

Fig. 1-4. Exact PC board size. Transfer image to copper-clad board using a piece of carbon paper. This is the bottom (copper) side of your scrambler board.

replaced with a metal or plastic plate. The *clarity* potentiometer can be mounted at any convenient location on the phone's base plate, but be very careful not to let any component interfere with the operation of the hook switch.

Scrambling A Phone. Connect the two scramble phones together (phone line outputs connected to each other) but keep them separated by at least twenty feet. Lift either of the hand sets and you should hear a low-level tone. Talk into the mike and you should hear your own unscrambled voice in the ear piece. This reception of your own voice is normal and occurs when using a standard telephone; it is called the *sidetone*.

Have a friend or another member of your family talk over the scramble phone. If your reception isn't clear or sounds like Donald Duck, adjust the *clarity* control for the best voice quality. This simply puts the two oscillators on the same frequency.

Scrambler Hook-Up. This job is a simple one. All that's required is to parallel the output of transformer T3 with the telephone lines. But before doing so, make the following tests. If you are in doubt about which two wires on the telephone terminal block are the telephone circuit, take a dc volt meter and check between pairs until 24 to 48 volts is measured. This test must be performed with the telephone on hook. The second important check to make before connecting the scrambler phone determines the line current. This test is made as follows. Set the VOM to measure dc current on the 50 to 100mA range, and place the meter in *series* with a lead from the high (1.2 k) impedance winding of T3. Pick up the phone. If the circuit current is greater than 25mA, then the resistor/capacitor network C6 and R9 must be added in series with the scrambler phone and the telephone circuit. This should reduce the circuit current to a value close to 25mA, but if not, adjust the value of R9 (start with a 1000-ohm, ½-watt resistor) until this current value is reached.

Security Link-Up. After connecting one of the scrambler phones at your location and another at the home of a friend, dial his number with your standard telephone. When the party answers and agrees to go to the scrambler mode, pick up your scramble phone, and have your friend do the same. You can now continue your conversation in complete secrecy. If either of the scrambler oscillators should drift in frequency, just set the *clarity* pot for the best voice quality.

Scrambled Signal Decoding. If you desire to use the scrambler for receiver speech decoding only, then make the modifications shown. Basically, resistors R4, R5, and R6 are removed and replaced with a 10K dual pot to allow the IC oscillator to be tuned over a wide frequency range. Connect the input of T1 to the output of your radio receiver and a small speaker or earphone to the output of T2. Transformer T3 is not required for speech decoding; it can be removed.

Tune the receiver to a station that is using the single inversion mode of speech scrambling, and adjust the oscillator frequency slowly until the speech begins to sound normal. Even when the scrambled information is decoded, the quality will not be up to hi-fi standards; however, if you are able to understand every word spoken, then you're right up town with the troops! Lots of luck—and remember to hang up all four phones when you're through.

ELECTRONIC METRONOME

A metronome is a useful device for the musical student. The classical metronome is a pendulum that swings back and forth, producing a clicking sound that is easily heard above the music being played.

The Electronic Metronome is similar to a classical metronome. It is self contained and produces an easily heard click. Further, the Electronic Metronome can be powered by two 9-volt transistor batteries, providing many hours of operation.

The Metronome uses a single 556 dual timer integrated circuit, an inexpensive buzzer and a handful of other components. Total parts cost should be under $10.

The buzzer is driven by U1. One side of the IC is wired as an oscillator, while the other side serves as a constant width pulse generator.

Square-Wave Output. Rate control R1 determines the operating frequency of the oscillator. The square-wave output of the oscillator is coupled to the constant width/pulse/generator. Each time the oscillator signal goes low, a positive, short-duration pulse is generated by the constant width pulse generator. Its output drives the buzzer that produces a click like 5-kHz tone.

R2 limits the upper frequency of the oscillator. The values of R5

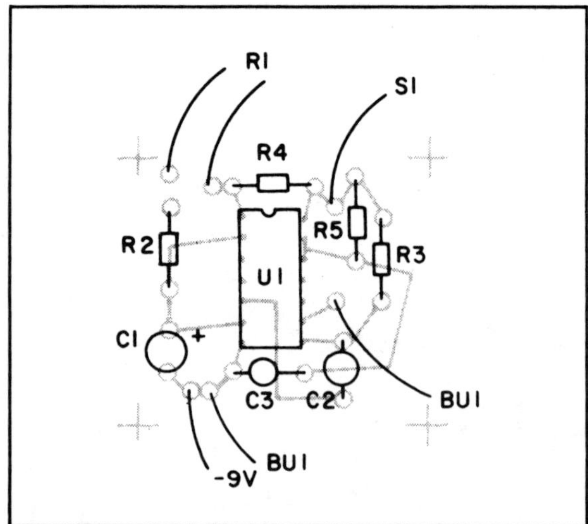

Fig. 1-5. Be careful when you solder the components. Watch for solder bridges.

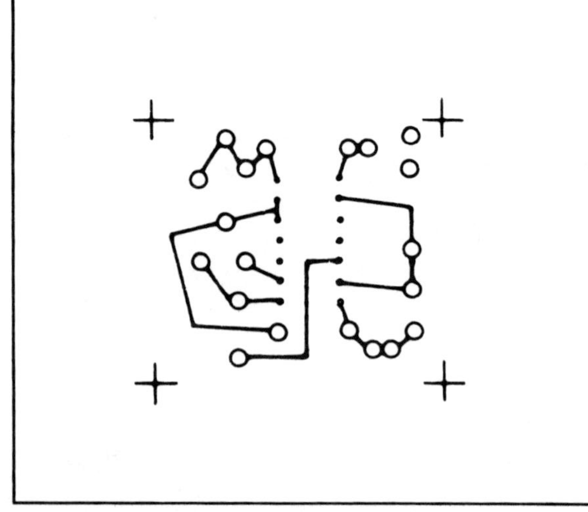

Fig. 1-6. This PC board is fairly easy to etch yourself. It may be fun and instructive.

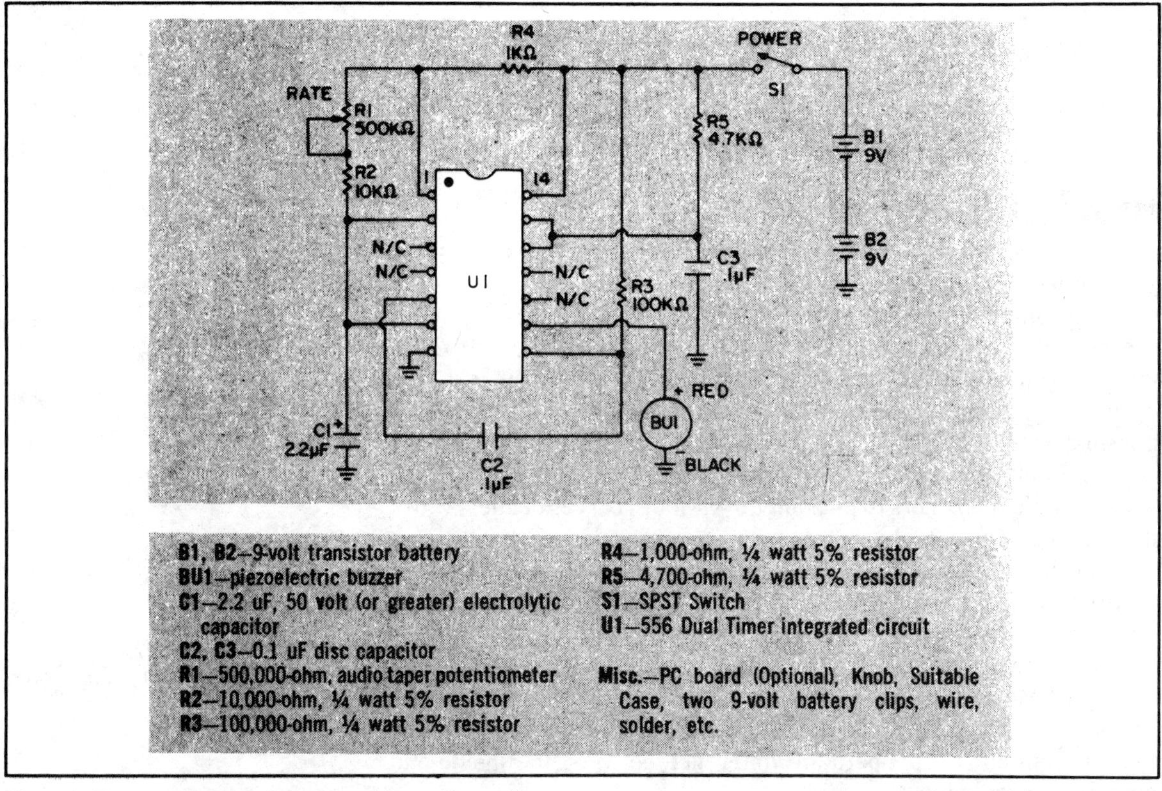

B1, B2—9-volt transistor battery
BU1—piezoelectric buzzer
C1—2.2 uF, 50 volt (or greater) electrolytic capacitor
C2, C3—0.1 uF disc capacitor
R1—500,000-ohm, audio taper potentiometer
R2—10,000-ohm, ¼ watt 5% resistor
R3—100,000-ohm, ¼ watt 5% resistor
R4—1,000-ohm, ¼ watt 5% resistor
R5—4,700-ohm, ¼ watt 5% resistor
S1—SPST Switch
U1—556 Dual Timer integrated circuit

Misc.—PC board (Optional), Knob, Suitable Case, two 9-volt battery clips, wire, solder, etc.

Fig. 1-7. Electronic Metronome schematic and parts list.

and C3 determine the width of the pulse that drives the buzzer.

Use The Buzzer. Once the PC board is completed, perform final wiring. When completed, the unit can be housed in any case large enough to hold the individual components. Using the buzzer as a template, mark the two mounting holes and the sound emitting center hole.

Finally, mount the PC board in the case. Determine a convenient location and mount the two 9-volt batteries. Power up the unit, placing the rate control to midposition. You should hear a repeating click from the buzzer.

The Electronic Metronome is ready to use. You may wish to calibrate the metronome in units of standard beats per minute. To do this, simply place the rate control to any setting and count how many clicks occur in a minute. Mark this setting. Do this at a number of rate control positions.

HEAT LOSS SENTRY

The money spent on heating and cooling your home represents your largest energy expenditure. As you are well aware, this cost can easily amount to over $1000 a year at today's prices for energy. With the dramatic increase in energy costs, it behooves everyone to do everything possible to reduce his energy consumption. This will help reduce oil imports, while keeping your personal expenses as low as possible.

Many of our utility companies are instituting a program of energy surveys for homeowners to pinpoint the various sources of energy loss in our homes. One way this is done is to pressurize the home under test with an air blower and use smoke generators to detect the passage of air from within the home to the outside. These passageways represent points of heat loss (or gain) in winter and summer.

With the help of the Heat Loss Sentry, you can perform the same tests for heat loss, using not smoke as the detecting mechanism but temperature change. These tests can be made in winter or summer. All that is required is a temperature difference between the inside and outside of your home.

Refer to Figs. 1-8 through 1-11.

The Heat Loss Sentry is a low-cost, high-quality instrument, sensitive enough to detect changes in temperature as low as one degree Fahrenheit. It is self-contained in a small cabinet and powered by a readily available 9-volt transistor radio battery that provides many hours of operation. An easy-to-construct probe contains a temperature sensing device used to locate sources of air leaks throughout the home. A built-in battery monitor circuit in the instrument alerts the user when the battery is near the end of its useful life. Although the Heat Loss Sentry has been designed as a heat loss detector, it is accurate enough for use as a thermometer over its range of 20 degrees Fahrenheit.

Circuit Theory. The Heat Loss Sentry has been made possible by the development of an accurate, low-cost temperature sensor integrated circuit, LM335. This is a three-terminal IC, designed to look like a 3-volt zener diode with an accurate temperature coefficient of 10 millivolts per degree Kelvin. (The Kelvin temperature scale is identical to the more familiar centigrade or Celsius scale, with zero degrees Kelvin equal to −273° C, or absolute zero.) The IC can be accurately calibrated to any desired temperature. Typically, the LM335 will provide one degree C accuracy over its entire operating range when it's calibrated at any temperature.

Refer to the schematic diagram. U1 and U2 are each an LM335 IC, connected in a differential amplifier circuit to detect a temperature difference between these two devices. U1 is mounted in a probe assembly, used to detect temperature changes, and U2 is contained in the instrument cabinet and acts as the reference. The adjustment lead of U2 is connected to a potentiometer (not panel mounted) so the meter reading can be set to center scale. In energy leak detection, center scale becomes the nominal or average temperature being measured.

When the Heat Loss Sentry is calibrated to center scale, the voltage across U2 is adjusted to be sufficiently below the voltage of U1 so that the output voltage of operational amplifier U3A drives the meter to center scale. Since U3A has an accurate gain of 18 determined by the ratio of resistors R6 and R5, the 10 millivolt per degree Kelvin sensitivity of U1 is amplified to 180 millivolts per degree Kelvin. This is equivalent to 100 millivolts per degree Fahrenheit. Resistors R7 and R8 are multiplier resistors which convert the 1mA meter movement to a voltmeter of 2 volts full scale. This provides a total meter range of 20 degrees Fahrenheit, or a relative scale of ±10 degrees with zero at center scale. Once calibrated to center scale, the sensor probe will produce an indication in any environment with a different temperature. A meter deflection downward occurs for colder temperatures, and an upward deflection occurs for warmer temperatures. If the total temperature change is 10 degrees or less, the actual differential can be read directly from the meter scale.

IC U3B is operated as a voltage comparator to constantly monitor battery voltage when the instrument is operating. This is accomplished by feeding a reference voltage across zener diode D1 to the positive input of U3B. A portion of the battery voltage is fed to the negative input of U3B. Voltage from a new battery is sufficient to develop a higher voltage at pin 9 of U3B than the D1 reference voltage. As a result, the U3B output is at zero potential and LED 1 is extinguished. As battery voltage decreases a point is reached when voltage at pin 10 of U3B exceeds pin 9 voltage. This results in U3B output rising to battery potential, and illuminating LED 1. The user is thus alerted that the battery is near the end of its useful life and should be replaced.

Construction. The entire circuit, with the exception of the sensing probe and front panel components, is contained on a printed circuit board. Shown is the foil layout as seen from the copper side of the board. Also shown is the component side, showing the parts layout. The printed circuit board has been designed to mount directly on the back of the meter, using the meter screws for both mechanical and electrical assembly. Before constructing your printed circuit board, take into account the center-to-center distance of the studs of the meter, if you decide to use a different 1mA movement than that specified in the parts list.

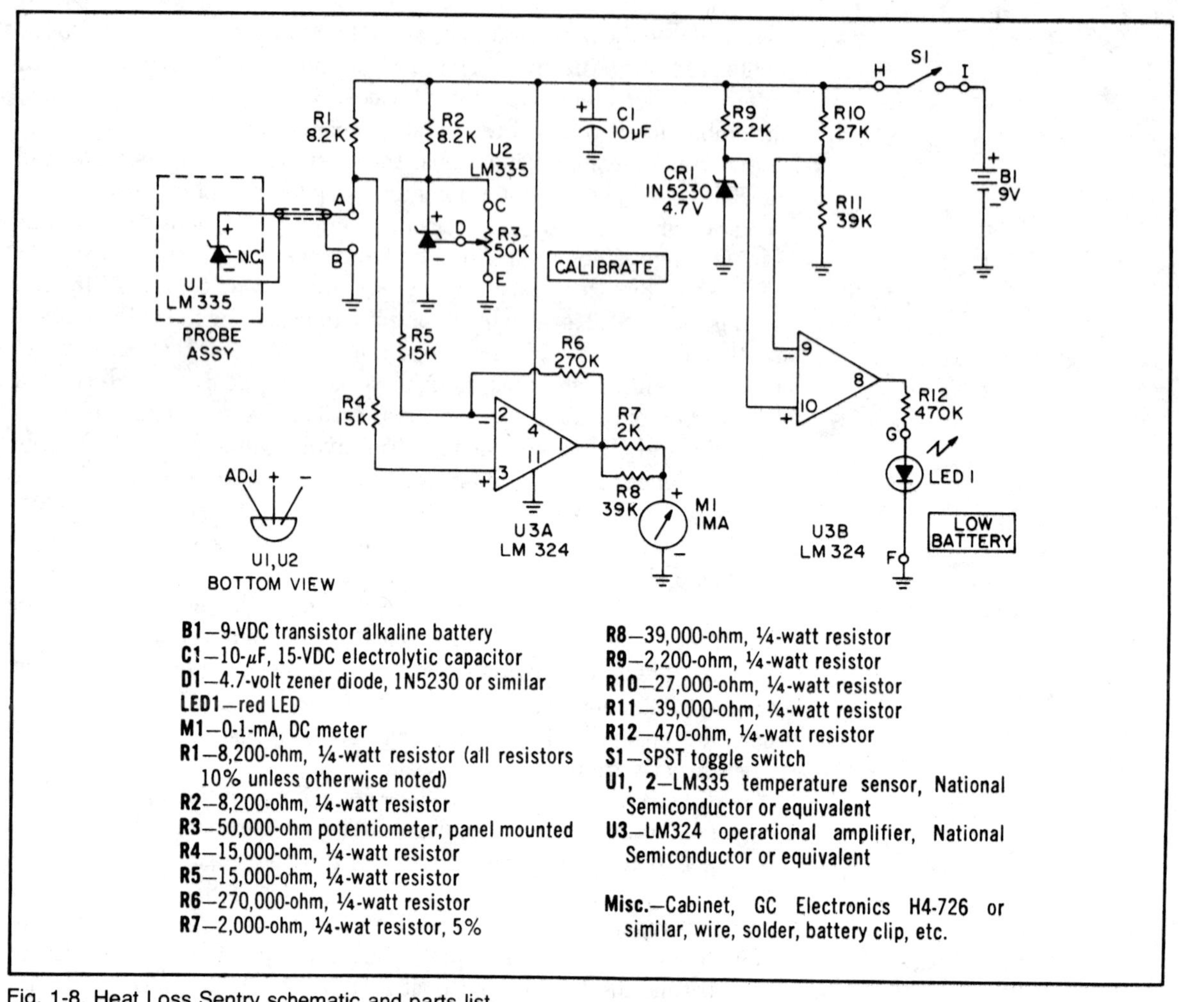

B1—9-VDC transistor alkaline battery
C1—10-μF, 15-VDC electrolytic capacitor
D1—4.7-volt zener diode, 1N5230 or similar
LED1—red LED
M1—0-1-mA, DC meter
R1—8,200-ohm, ¼-watt resistor (all resistors 10% unless otherwise noted)
R2—8,200-ohm, ¼-watt resistor
R3—50,000-ohm potentiometer, panel mounted
R4—15,000-ohm, ¼-watt resistor
R5—15,000-ohm, ¼-watt resistor
R6—270,000-ohm, ¼-watt resistor
R7—2,000-ohm, ¼-wat resistor, 5%
R8—39,000-ohm, ¼-watt resistor
R9—2,200-ohm, ¼-watt resistor
R10—27,000-ohm, ¼-watt resistor
R11—39,000-ohm, ¼-watt resistor
R12—470-ohm, ¼-watt resistor
S1—SPST toggle switch
U1, 2—LM335 temperature sensor, National Semiconductor or equivalent
U3—LM324 operational amplifier, National Semiconductor or equivalent

Misc.—Cabinet, GC Electronics H4-726 or similar, wire, solder, battery clip, etc.

Fig. 1-8. Heat Loss Sentry schematic and parts list.

It is recommended that you use a socket for U3, rather than soldering it directly into the printed circuit board. This will permit ease of service should it ever be required. Be sure that the orientation of U3 is correct. Pin 1 of U3 is clearly marked on the parts layout and foil layout by a small dot. The same precautions hold for U2, the diodes, and electrolytic capacitor. These parts are polarized and must be placed into the circuit in the proper direction. A bottom view of U1 and U2 is seen on the schematic diagram.

Connection between the printed circuit board and external components are made through a series of pads marked with letters A through I. These connections are clearly shown on the schematic diagram. It is best to use wires of different colors to help prevent wrong connections. The sensing probe is connected to terminals A and B of the printed circuit board. Make this connection with a convenient length of flexible shielded wire. Maintain the correct polarity when connecting U1. The shield connection of the cable should be tied to the negative lead of U1, and to terminal B of the printed circuit board. Feed the probe cable through a front panel grommet.

Power to operate the circuit is obtained from a 9-volt transistor radio battery, mounted directly to the printed circuit board. Connect the battery to the circuit with a battery clip made for this purpose. The layout easily provides room on the board for this. The battery can be secured to the board with a homemade clamp constructed from a piece of sheet copper, or by any other means you care to use. The parts list specifies a normally open, spring return, power switch. This was chosen to prevent the unit from being left on when not in use, and depleting the battery.

LED 1 is mounted on the front panel of the instrument using a small amount of epoxy. Use a pair of different colored wires to make the connections between the LED and printed circuit, and be careful not to bend the stiff leads of the LED where they enter the plastic body. This might render the LED defective.

Refer to the illustration of a typical probe assembly. If available, you may use a short piece of plastic or synthane tubing for the probe. You can even construct a probe from a piece of wood doweling. It is not recommended to use metal tubing for the probe, because the heat conduction from your hand may affect the temperature sensing performance of the sensor, U1.

Connect the shielded wire to U1, using the + and – terminals of the IC as shown on the schematic diagram. The adjustment terminal of U1 is not used. Insulate the connections carefully, and insert the IC and wire into the probe. Secure the IC and wire inside the probe with epoxy or silicon rubber compound. Allow part of the case of U1 to protrude outside the probe so that it is more sensitive to temperature change. Allow the assembly to harden before placing it in use.

For a professional-looking instrument, you can use the meter scale shown, which fits the meter specified in the parts list, as well as others. The existing meter scale can easily be removed by prying the plastic cover off the meter and removing two small screws. Be careful not to disturb the delicate needle. Paste the new scale on the back side of the meter scale, and reassemble it into the meter.

Checkout and Use. When the unit is fully wired, check for wiring errors. Then, connect a 9-volt transistor battery to the power input terminals. Activate the power switch and rotate the zero adjust control over its full range. You should be able to adjust the meter reading from zero to full scale, with some extra range left in the potentiometer. Set the control so that the meter reads half scale. While holding the power control on, place your fingers over the sensing tip of the probe. The meter reading should increase to beyond full scale. If the unit performs as specified, it is operating properly.

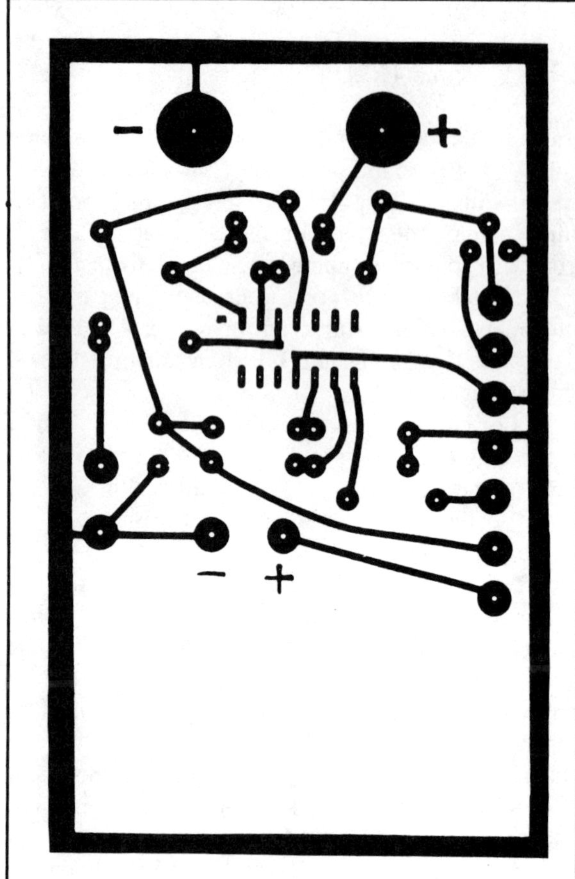

Fig. 1-9. This is the foil side down view of the Heat Loss Sentry PC board. Care must be exercised in etching board.

You may wish to check the Low Battery indicator circuit to determine if it is operating properly. To do this, you must substitute a variable voltage dc supply for the battery. Set the supply to 9 volts and connect it to the power input terminals observing correct polarity. Turn the power switch of the Heat Loss Sentry on, and observe the Low Battery indicator as the power supply voltage is reduced. The Low Battery indicator should become illuminated as the power supply voltage approaches approximately 6.5 volts. Due to variations in zener diodes, you may wish to change the value of R11, if necessary, so that the LED lights at approximately 6.5 volts battery voltage. Once this is done, the checkout of the instrument is complete. Reconnect the battery to the instrument.

When Heat Loss Sentry is operated, you may notice that the Low Battery indicator blinks as the power is turned on and off. This is a normal reaction, which occurs as the circuit voltage passes from zero to battery voltage then back to zero.

To operate, hold the power switch on and adjust the meter to

Fig. 1-10. The foil side up diagram illustrates parts placement on the top of the PC board. Heat Loss Sentry requires relatively few components.

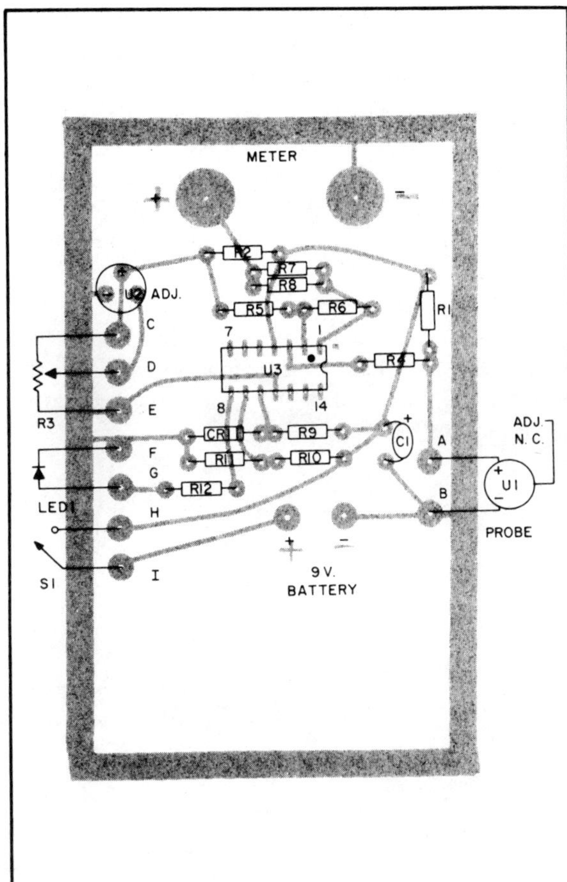

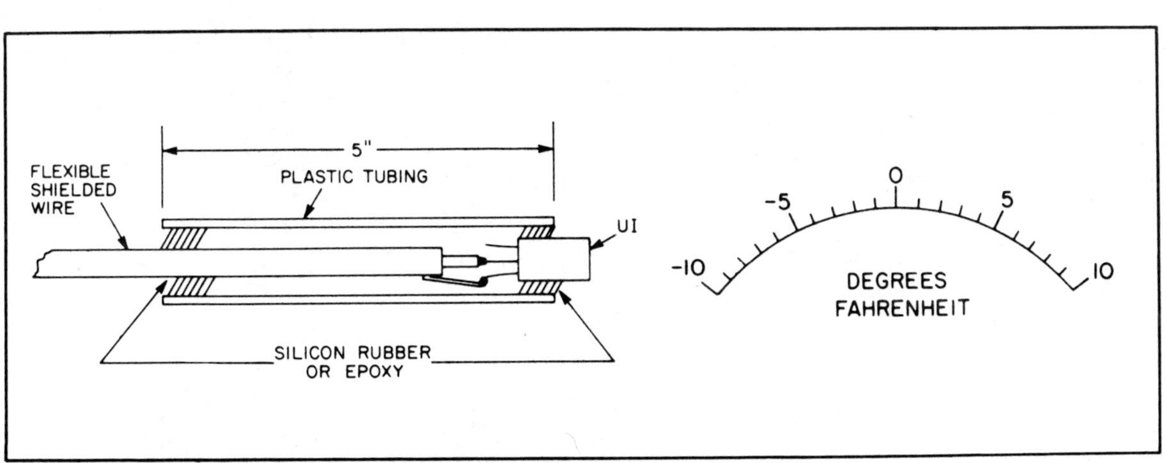

Fig. 1-11. To the left is an illustration of the heat sensing probe. Follow the set-up closely and use the glue. At right is an exact size drawing of the meter face. Cut it out and paste it right on.

center scale. Holding the probe, search out any area where you suspect an air leak between inside and outside of your home. The meter will give an immediate indication if there is a change in temperature. In the case of very small leaks, allow sufficient time for the unit to react. This may take several seconds. Once a change of temperature has been detected, it is best to remove the probe from the leak and allow its temperature to stabilize to room temperature before searching out another leak. It takes a few minutes to familiarize yourself with this instrument.

Another interesting use for this device is in troubleshooting defective electronic circuits. When the probe is held close to defective ICs, resistors, and other components, a higher than normal temperature will be indicated.

KITCHEN HELPER

Here is an inexpensive project—the cost of parts is under $10—which will make many of your kitchen appliances perform better than they were originally designed to. It is a motor speed control combined with an optional automatic on-off-on-off cycling pulser/interruptor. The pulsing feature is particularly important for kitchen blenders, mixers and food processors. The short pause, during which the appliance stops, allows you to see the progress of the food preparation. You can then stop before your food processor grinds and mashes everything to bits. Modern kitchen appliances frequently operate at such high speeds, that a few seconds difference in running time can transform an exquisite meal into meat loaf. Many appliances such as blenders also mix the food better when operating with an on-off cycle, which prevents the food from sticking to one side of the bowl. It should be noted that the duration of both the "on" cycle and the "off" cycle in the Kitchen Helper can be adjusted separately. The continuously variable speed/dimmer control used in conjunction with the pulser adds another desirable feature to many of your appliances. The pulsing/interrupting feature can also be bypassed by flipping a switch, and the Kitchen Helper becomes a regular variable-speed or light dimmer control for your power tools, Christmas lights, Halloween pumpkins, and other electrical devices. In fact, you may have trouble deciding between the kitchen and the work bench, as to where it should be kept. Don't fret too much—build a pair!

Refer to Fig. 1-12.

How It Works. The schematic diagram consists of two distinct sections: The low-voltage on-off pulser with a heart made out of our old friend/the 555 timer (IC1), and the high-voltage section, consisting of a 600-watt/110-volt Triac speed control. A small lamp (L1) and a photocell (PC) tied together, act as a light coupler by separating the high and low voltage sections. When the timing circuit puts a voltage across the lamp (L1), it lights up, the resistance of the photocell decreases, and the Triac conducts. Potentiometers R5 and R6 control independently the "on" and "off" cycles of the timer, with time on—$1.1 \times C4 \times R5$, and time off—$1.1 \times C6 \times R6$. With the values chosen for resistors and capacitors, the "on" and "off" cycles can be set between 0 and 5 seconds. Switch S1, when closed, bypasses the timing section of the circuit. Power for the low-voltage section is provided by transformer T1 with the associated rectifier bridge (Z1) and capacitor C3.

The high-voltage section of the circuit is a standard Triac motor speed/dimmer control for lights and appliances up to 600-watts.

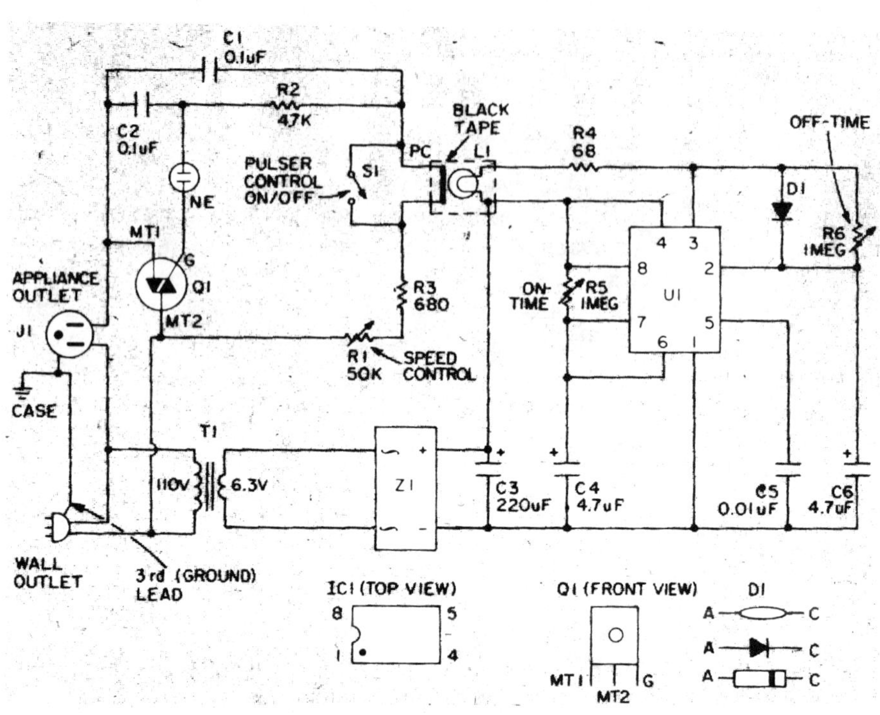

C1, C2—0.1-μF, 200-VDC tubular capacitor
C3—220-μF, 25-VDC electrolytic capacitor
C4, C6—4.7-μF, 35-VDC electrolytic capacitor
C5—0.01-μF, 35-VDC mylar capacitor
D1—1N4000 diode
J1—3-prong AC appliance receptacle
L1—6-volt pilot lamp, low current-type
NE—NE-2 neon lamp
PC—CdS photocell (Radio Shack #276-116 or equiv.)

Q1—Triac rated @ 200-volts @ 6-Amperes (GE-X12 or equiv.)
R1—50,000-ohm, linear taper potentiometer
R2—4,700-ohm, ½-watt resistor
R3—680-ohm, ½-watt resistor
R4—68-ohm, ½-watt resistor
R5, R6—1,000,000-ohm, linear taper potentiometer
S1—SPST switch
T1—transformer with primary rated @ 110-VAC/secondary @ 6.3-VAC @ 300 mA.
U1—555 timer
Z1—full-wave bridge rectifier; 200 PIV @ 4-Amperes
MISC.—cabinet, perfboard, hookup wire, solder, knob, AC plug and line cord combo., etc.

Fig. 1-12. Kitchen Helper schematic and parts list.

Capacitors C1 and C2, resistors R1, R2, R3 and the photocell resistance, set the firing point of the Triac, and vary its duty cycle for conduction. Potentiometer R1 is used as the speed/dimmer control, and neon light NE provides the hysteresis required by the speed control circuit for smooth operation.

Construction. The circuit can be built easily on a 2½ by 3½-inch perfboard, using point-to-point wiring. No special wiring precautions are necessary, except for the section of the circuit which carries ac voltage. It should be well-insulated, and kept away from the rest of the circuit and the cabinet. Use a 3-prong cable and jack, with the ground wire connected to the cabinet. The light coupler consists of the photocell and the lamp tied together with black electrical tape. Make sure that the active side of the photocell faces the lamp, and that the photocell pins do not touch the lamp wires.

The "on" and "off" controls, R5 and R6, can be mounted externally on the case, or internally on the perfboard. Mounting them inside makes for neater appearance, but changing the timing becomes a chore. We found that 3-seconds "on" and 2-seconds "off" were optimum for most applications.

Operation. Operation of the Kitchen Helper is very simple. Plug it into an ac outlet, and plug your appliance into jack J1 on the case of the Kitchen Helper. You can vary the speed or brightness/if you use it as a dimmer/with R1. If you don't need the pulsing feature flip switch S1. That's all there is to it!

REMOTE CONTROLLER

We all know how annoying TV commercials (and some programs) can be. Sometimes you would love to be able to flick off their loud, abrasive chatter until the show comes back on, but you are watching TV to relax, not to jump up and down every 10 minutes. With this simple remote control unit you can turn the sound on and off with the blink of a flashlight.

All you do is aim a flashlight at a small box sitting on top of the TV. When the unit receives the first flash of light the sound is turned off and an indicator light comes on (verifying your signal). When you want the sound back another flash of the light and the sound is restored.

There are a number of good reasons for using a flashlight as the transmitter. First, of course, it is simple and inexpensive. Also it allows the receiver to be simple, and hence easy to build and trouble-shoot. Naturally this system requires the operation to have a free line-of-sight to the TV, but a viewer always has that. Further the unit responds to any number of flashlights.

Refer to Figs. 1-13 through 1-15.

How It Works. The circuit operates as follows. The signal light is received by a photo-transistor (P1) recessed behind the front panel. A photo-transistor is a transistor where the base signal is effectively provided by a light source, the brighter the light the more it turns on. In this circuit, the photo-transistor is used as one leg of a voltage divider. When the light strikes P1 the voltage across it drops. This "falling edge" is amplified by Q1 and then used to trigger the monostable multivibrator (IC1).

This device (IC1) is often called a *one shot* because it outputs one pulse of uniform width each time the input goes high or low (depending on how it's connected up). The length of this pulse is set by C2 and R5, in this case about .5 sec. The one shot here is acting as a buffer, taking the rough signal from the flashlight and converting it into a nice, clean, clock pulse for the flip-flop (IC2).

The flip-flop is used here as a memory to keep track of whether the TV sound is on or off. One property of the J-K flip-flop is that if both J and K inputs are held high (5V), the output will *toggle* with each clock pulse. In other words if the output is low, it will go high with the first clock pulse and then low again with the next clock etc. This is just the kind of action we need. When the output of IC2 goes high it will turn the sound off and hold it off until it gets another clock pulse. To accomplish this the output of the flip-flop is used to drive the relay control transistor Q2. When IC2 goes high, the transistor turns on providing a

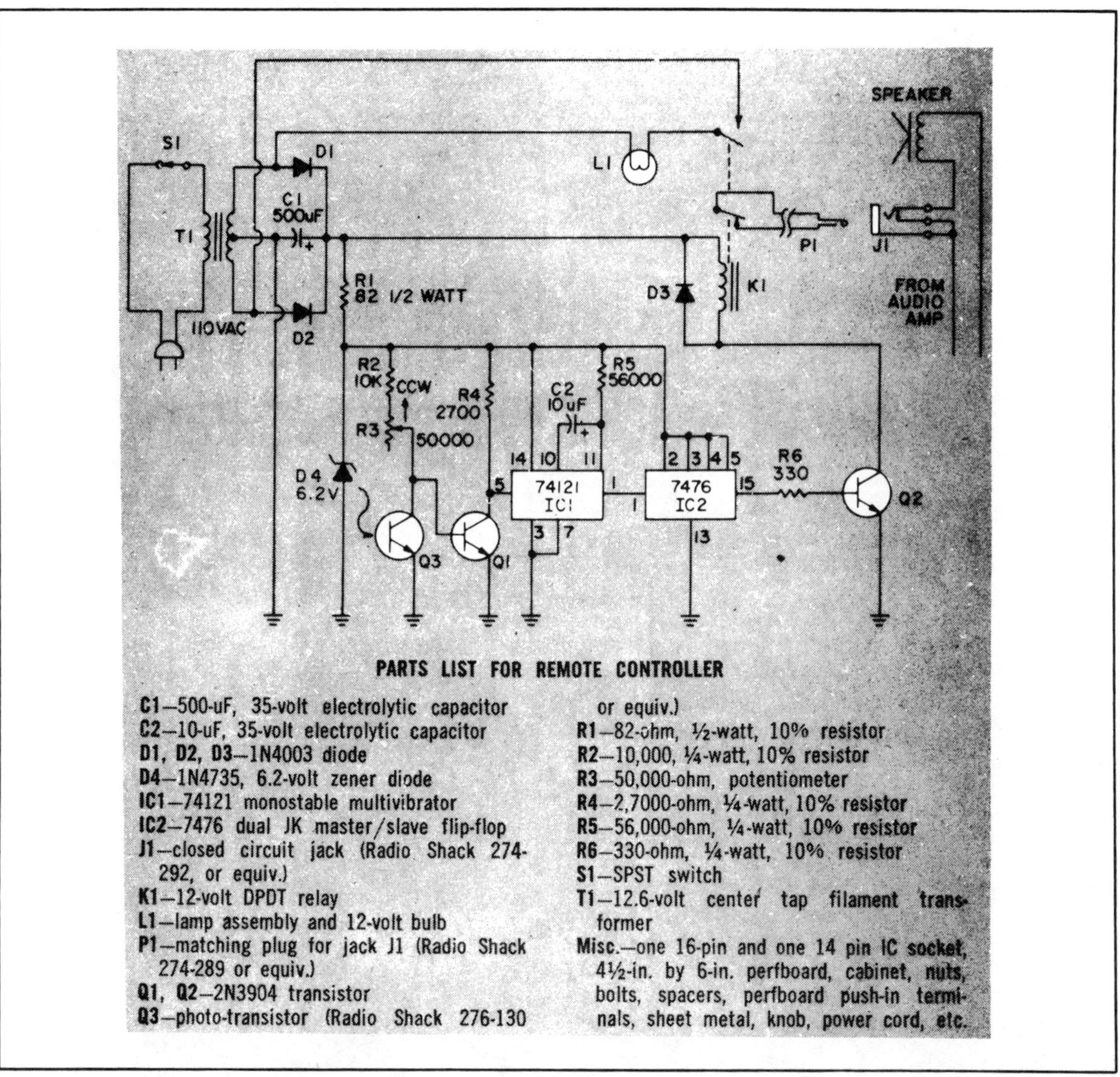

Fig. 1-13. Remote Control schematic.

path for the relay coil current to ground, thus energizing the relay.

Note diode D3 across the relay coil. This is necessary as it provides a safe path for the built-up energy in the coil to dissipate when the relay is deenergized. Without the diode high voltage spikes occur which quite probably would cause false triggering in other parts of the circuit. The relay performs two functions. The first set of contacts (normally closed) controls the sound by opening the TV speaker circuit. The other set of contacts (normally open) controls the indicator light which is powered by the 12.6V ac from the transformer. The function of this light is more important than you might think as it gives

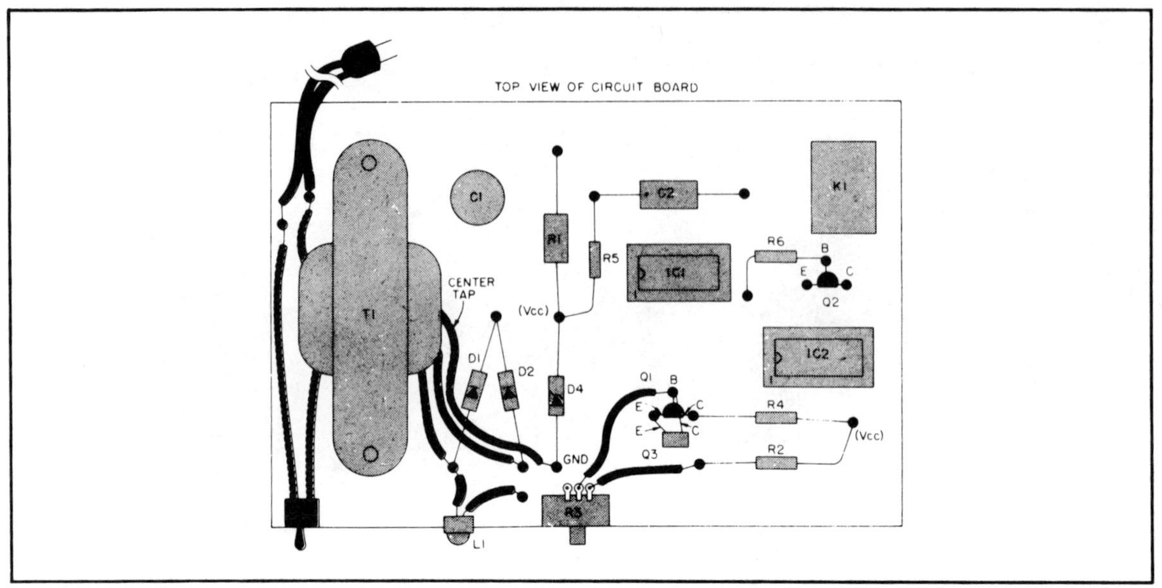

Fig. 1-14. Top view of the circuit board shows how the photo transistor is mounted directly on the base and emitter leads of transistor Q1.

the operator a positive indication of what state the unit is in. (It's not always obvious from program material.)

The power supply is simple and straightforward. The 110V ac is run through a power switch into a 12.6V filament transformer. The 12.6V ac is rectified by a full-wave rectifier using the center tap with

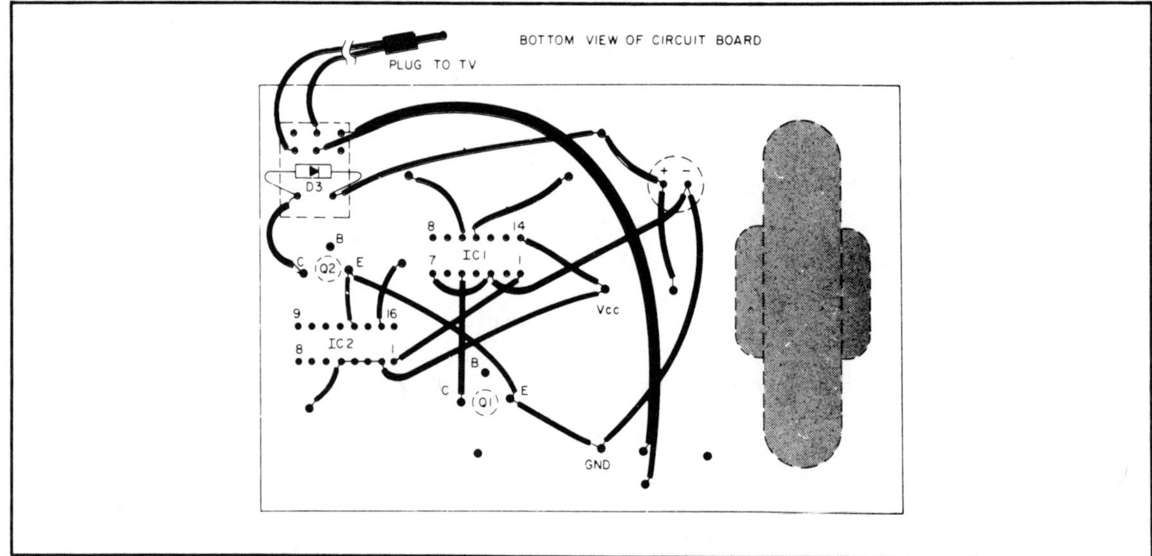

Fig. 1-15. Point-to-point wiring and perfboard construction make this project a snap to put together. Take care hooking up the relay.

D1 and D2, then smoothed by C1. This unregulated voltage is adequate for the relay coil, but not for the ICs, so the 6-volt zener is employed to provide a regulated 6-volts for these components.

Construction. Probably the easiest way to construct the circuit is on a 4½-inch by 6-inch perforated circuit board using push-in terminals for connection points. Start by laying in the transformer, relay the IC sockets. If you use a board with 0.1 inch centers the IC socket pins should stick right through and can be secured with two small screws. Mounting the transformer and relay will, of course, require enlarging some holes. Next, the push-in terminals may be inserted and the components soldered into place. Note that the photo-transistor is mounted directly on top of Q1, facing the front. It should be as far above Q1 as it's leads will allow. For a professional looking job turn the board over and solder the interconnecting wires on the backplane.

The circuit can now be mounted in a case using stand-offs. Mount the power switch, potentiometer, and indicator on the front panel. Cut a 1-inch hole directly in front of the photo-transistor to allow the flashlight beam in. Finally, it is best to enclose the whole unit to keep ambient light from activating the circuit.

Installation and Checkout. To complete the system a simple modification to the TV is required. Unplug the set, remove the back cover, and locate the two wires going to the speaker. As shown in the schematic, cut one of the wires and connect the two leads to a phone jack, which you mount in the side of the case. When the jack is correctly wired the sound will be normal when the control unit is not plugged in. But when it is plugged in the relay must be closed to complete the sound circuit.

Once the system is operational, some adjustment of the sensitivity potentiometer is necessary. Have someone operate the flashlight from the viewing area while you adjust the pot. Find the point where the unit responds to the flashlight but not to room lights etc. Now sit back, knowing you have at least some power over Madison Avenue.

RING-A-THING

Did you ever miss an important telephone call because you were out of range of sound of the telephone bell? It doesn't have to happen again if you build and install this inexpensive remote telephone bell. It is battery operated and can be located anywhere inside or outside your home. Since it is self-powered, it requires virtually no energy from the telephone line. The input impedance of the circuit, as seen by the telephone line, is almost 100,000 ohms, and the input resistance is infinite. When connected across the telephone line, it is undetectable and has no effect on telephone performance.

Refer to Figs. 1-16 through 1-18.

No Power Problems. The circuit derives its power from four rechargeable NiCad cells connected in series which provide 4.8 volts to drive an ordinary doorbell. Since the power demand on these cells occurs only when the telephone rings, the battery will operate the bell over 1000 times on one charge. This should last several months, depending upon how many calls you receive. A built-in battery charger is included in the circuit so that the cells may be conveniently charged from the ac power line at any time. Full recharge takes 14 hours, but the charger may be left in operation indefinitely, if desired, with no damage to the cells due to overcharge. This is possible since the charging circuit has been designed to deliver a limited current to the cells. The NiCad cells used in this circuit are size C, but other sizes may be used.

How It Works. When the telephone rings, a 20 Hz ac voltage of about 220 volts peak-to-peak is impressed across the telephone line. The series circuit composed of R1, R2, R3, C1, and C2 is connected across the line to provide isolation and act as a voltage divider for Q1. C1 and C2 provide dc isolation, since the line normally has a dc voltage of about 48 volts across it when the telephone is not in use. Q1 responds to the 20 Hz ringing signal by conducting current during each positive half-cycle applied to its base. CR1 prevents Q1 from being reverse biased during the negative half of the ringing signal. The emitter current of Q1 is applied to the base of Q2 causing it to saturate and act as a switch. This applies full battery voltage to the bell, causing it to ring. The voltage applied to the bell is essentially a 20 Hz square wave that produces a slightly different sound than that produced by pure dc. CR2 and C3 protects Q2 from any reverse voltage spikes produced by the collapsing magnetic field of the bell.

The battery charger circuit is composed of T1, a four-diode bridge rectifier, and R5. T1 provides isolation from the ac power line while

B1—Size "C" NiCad cell (Radio Shack 23-124 or equiv.)
C1, 2—0.22uF, 250-volt tubular or ceramic capacitor (Radio Shack 272-1070 or equiv.)
C3—22-uF, 16-volt tantalum capacitor (Radio Shack 272-1412 or equiv.)
D1, 2—General purpose silicon diode (Radio Shack 276-1103 or equiv.)
D2-6—0.5A, 100-volt or greater silicon diode (Radio Shack 276-1102 or equiv.)
Q1—NPN silicon transistor, 2N3904 (Radio Shack 276-2030 or equiv.)
Q2—NPN Silicon power transistor, 5-A, 2N4321 (Radio Shack 276-2020 or equiv.)
R1, 2—4700-ohm, ¼-watt, 10% resistor (Radio Shack 271-1300 or equiv.)
R3—10000-ohm, ¼-watt, 10% resistor (Radio Shack 271-1300 or equiv.)
R4—10-ohm, ¼-watt, 10% resistor (Radio Shack 271-1300 or equiv.)
R5—4.7-ohm, 1-watt, 10% resistor (Allied Electronics—address below—824-5049 or equiv., see text)
T1—6.3-volt, 1-A filament transformer (Radio Shack 273-050 or equiv.)
Misc.—Bell (Standard 6-volt doorbell), battery clips (Radio Shack 270-1436 or equiv.) wire, solder, etc.

Fig. 1-16. Ring-A-Thing parts list.

reducing the voltage to about 6 volts rms. The output of the bridge rectifier, a pulsating dc of about 9 volts peak, is applied to the four cells through a current limiting resistor, R5. This type of circuit is recommended for NiCad cells, and provides essentially a constant charge current regardless of the state of charge of the battery or power line voltage. By limiting the current to not more than one tenth of the ampere hour rating of the cells, the charger may be operated for any length of time without damage to the battery due to overcharge. When the cells attain full charge, the gases produced within the cell are recombined chemically, preserving electrolyte.

Construction. The entire circuit is built on a 6½-inch × 9½-inch printed circuit board. The foil layout is shown half size. The component layout is also shown. The cells are securely mounted to the printed circuit board using steel clips. This method of assembly is recommended since it would not be good practice to rely on the connecting wires of the cells to hold them in place. If the cells you are using do not have solder tabs, the wires can be soldered directly to the positive and negative metal parts of the cell. In this case do not use excessive heat when soldering so that the cells do not become damaged. When mounting the cells be sure to follow the exact polarity as shown in Fig. 1-18.

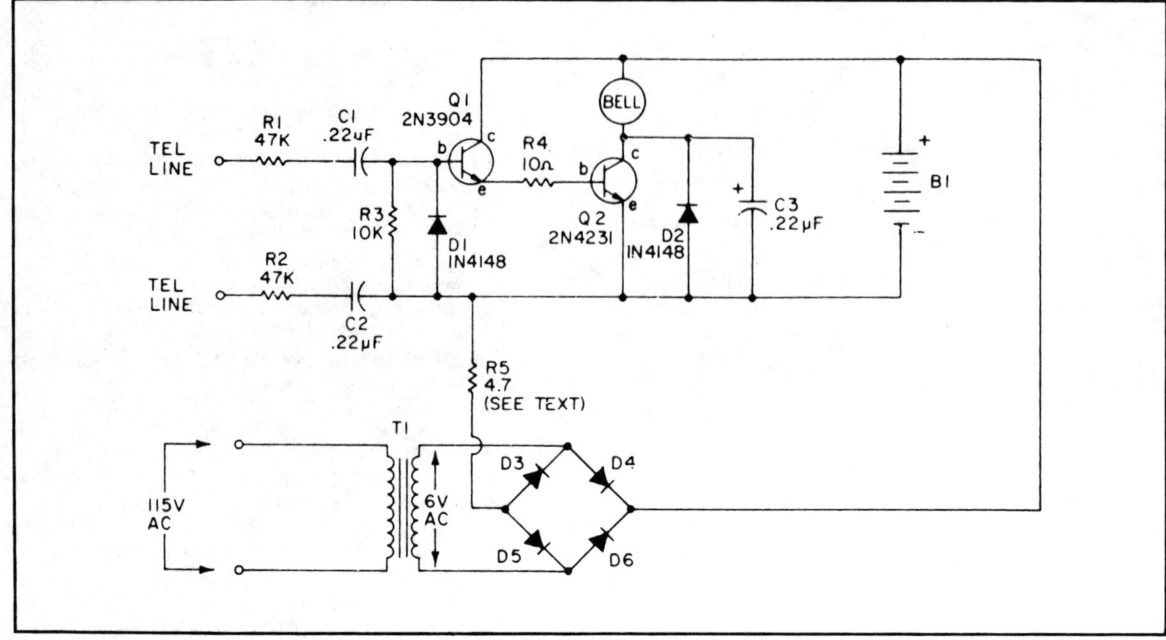

Fig. 1-17. Ring-A-Thing schematic.

The printed circuit layout for the bell connections may be changed to accommodate the type of bell you are going to use. Be sure to locate the mounting holes for the bell before laying out the printed circuit to avoid a conflict between the copper foil and mounting screws. Do not solder R5 into the printed circuit until instructed to do so in the test procedure outlined below. Q2 is mounted directly to the printed circuit board without a heatsink. None is required since the transistor operates as a switch, resulting in almost no power dissipation.

The method of connecting the telephone line and ac power cord to the printed circuit is left up to the builder. If the unit is to be used at a location not convenient to ac power you may want to use a small terminal strip or connector for the telephone line. This permits easy removal when battery charging is required.

Be Careful! We recommend that the cells be handled in a discharged state. NiCad cells are capable of delivering very large short circuit currents (50 amperes or more) even when only partially charged. Once the cells are mounted and wired to the printed circuit any accidental short circuit between the cells or other components on the board may cause a very large current flow. Such currents can easily burn out printed circuit wiring.

Checking the Circuit. After the unit is assembled and wired, measure the charger current so that the proper value of R5 can be placed in the circuit. Because different 6.3-volt filament transformers can vary considerably in output voltage and internal impedance, the current delivered by the charger should be checked and adjusted if necessary. The best method of measuring charging current is to insert a 0-1 ampere dc meter in series with R5. An alternate method is to measure the resistance of R5, connect it into the circuit, and measure the dc voltage across it. Current can then be calculated by dividing the voltage measurement by the resistance.

The recommended charge current for the C cells specified in the parts list is 120 milliamperes. This will charge the battery in 14 hours, and there would be no danger of overcharge if the line power was connected for several days. If you prefer to leave the charger permanently connected to the ac power line the charge current should be reduced by a factor of three, to 40 milliamperes. This will keep the cells at 100 percent charge without any danger of cell damage. If NiCad cells of other capacities are used the charge current should be set up to

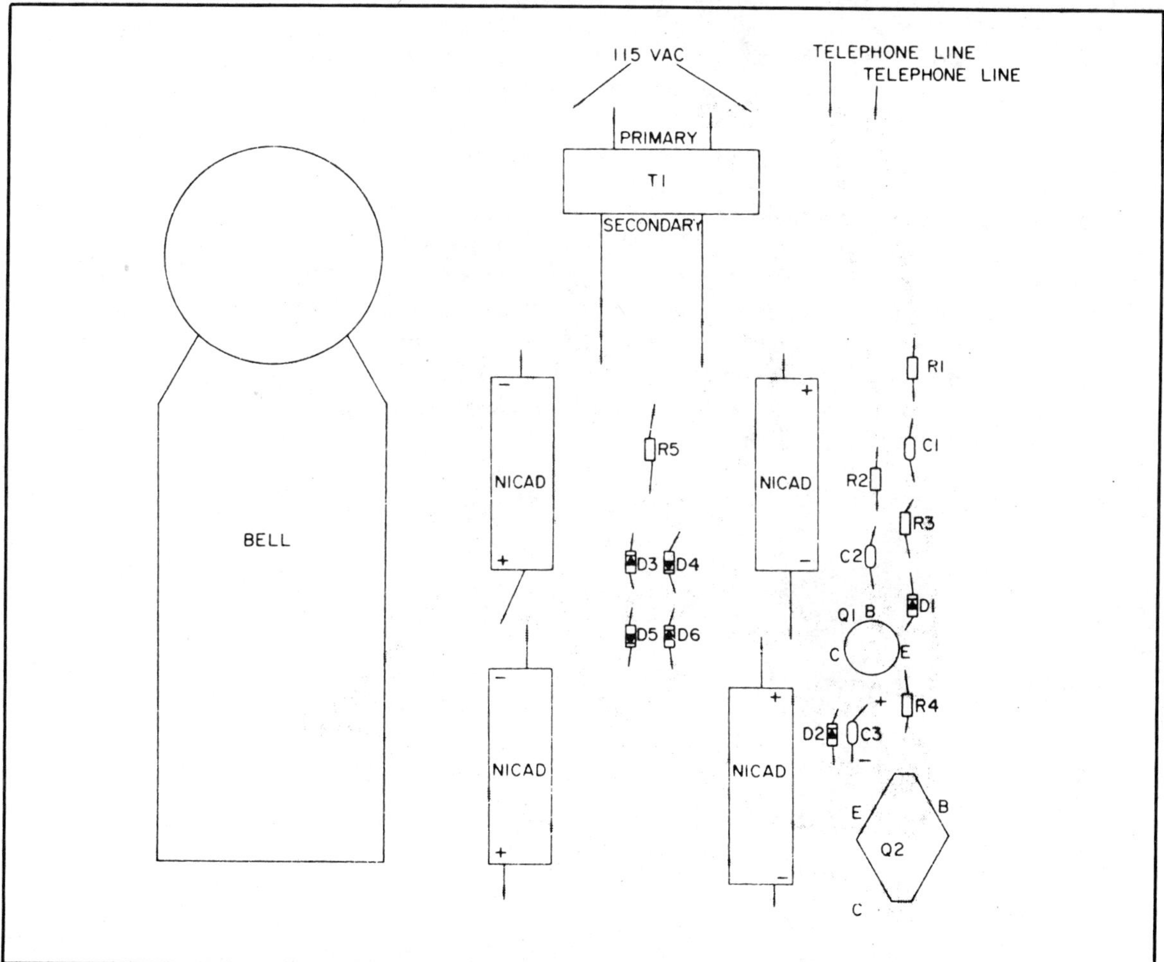

Fig. 1-18. The component layout of Ring-A-Thing, with the foil side down.

one one tenth of the ampere hour rating of the cells, or to 1/30 the rating if the charger is to be operated permanently. For example, if size D cells with a 3.5 ampere hour rating were used, the charge current would be set to 350 mA or 115 mA. The value of R5 should be changed, if necessary, to provide the desired current. Note that it is not necessary to set the current to exactly one tenth of the rating of the cells. Less current can be used with the disadvantage being that it would take longer than 14 hours to fully charge the battery. Do not use a larger current than specified. To do so will damage the cells on overcharge.

Installing Ring-A-Thing. The unit is connected across the telephone line as shown in the schematic. The only exception to this will be for two-party telephones. In this case, the telephone ringing signal is impressed between one of the telephone lines and ground. (The other party's ringing signal is impressed across the other line and ground). You will have to experiment to determine which of the two lines has your ringing signal. If you inadvertently connect the circuit to the wrong line, you will be answering the other party's calls! For the ground connection you may use any convenient ground point such as a BX ground.

Before installing the unit, you should operate the charger at least 14 hours to fully charge the battery, unless you plan to leave the charger connected permanently to the power line. With a fully charged battery, the unit will operate several months before a recharge is necessary. The power demand of the charger is about 2 watts and will have little effect on your electric bill if left operating. In this case, the unit would need no further attention, and you can forget it.

When You Start. You'll need a template to build Ring-A-Thing's printed circuit board. Take a look at the parts list to find out how to get it. It's as easy as answering the phone.

SCR LAMP DIMMER

Refer to Figs. 1-19 and 1-20.

The lamp dimmer that we are about to describe is a second-generation device. It uses no expensive diodes, iron core coils, or variable wire wound controls. It can be built in one evening, using the main component of all dimmers, the SCR. An inexpensive neon bulb is used to trigger the SCR into conduction. The device will dim up to 600 watts to incandescent light.

Choice of Switches. Actually the photographs show two different, although alike, units that were built. The unit shown with the student lamp was built first. The variable control in this unit has a reverse action switch, in other words the switch is off until the control reaches the clockwise end of rotation and then snaps on. The net effect of this action is that the lamp is dim at the start of CCW rotation and gets brighter as the control is rotated. At full rotation, the lamp will light to about 70 percent of full brightness, then the switch closes and puts full line voltage on the bulb. This is a fine arrangement and one used by most commercial lamp dimmers.

But a conventional variable control with a s.p.d.t. switch can also be used. You can have a full brightness at the CCW (counterclockwise) end of rotation and dim light at the CW (clockwise) end of rotation. Choice of switch depends on what you have in your "junk box," perhaps, or which type is readily available. A third type is listed in the parts list since it is conveniently ordered with control R2. It is a push-on, stay-on switch that can be activated regardless of the setting of R2.

Choice of Packaging. This dimmer is housed in a metal cover from a can of spray paint. The cover was painted with satin black

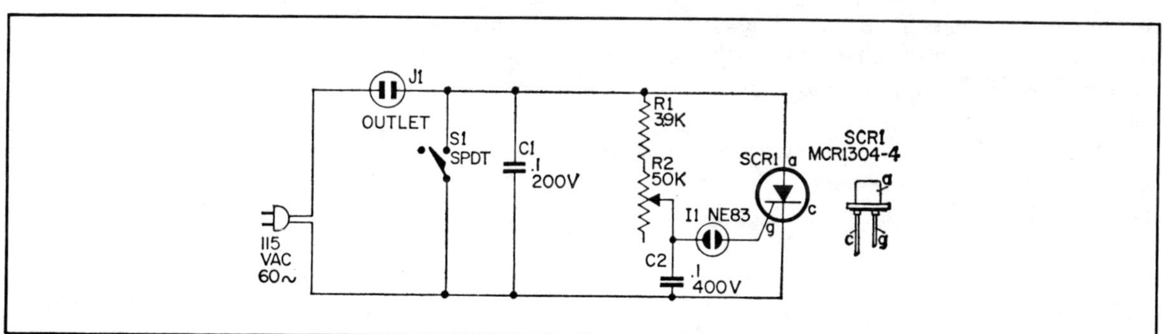

Fig. 1-19. Schematic diagram of lamp dimmer shows how NE83 in SCR gate circuit triggers device into conduction. Outlet J1 connects to lamp.

C1—.1-mfd., 200-volt tubular capacitor
C2—.1-mfd., 400-volt tubular capacitor
I1—Neon circuit control lamp, 60-100vdc start voltage, 10ma (G.E. NE-83 or equiv.)
J1—AC outlet (part of line cord)
R1—3900-ohm, ½-watt resistor
R2—50,000-ohm linear taper potentiometer with s.p.d.t. switch (Mallory U-35 and US-28, respectively or equiv.)
S1—S.p.d.t switch (see R2)
SCR1—Silicon controlled rectifier (Motorola MCR1304-4 or equiv. Order from Allied Radio)
Misc.—Spray can cover or equivalent housing for the dimmer, plywood, 6-foot line cord complete with plug and outlet, hardware, solder, etc.

Fig. 1-20. SCR Lamp Dimmer parts list.

lacquer and decals applied. All the components were wired and soldered in place outside the housing. Three holes were punched in the housing, before painting, one at top center for the control and one on each side for the line cord. A round disc was cut from ½-inch plywood, sprayed with black paint, and used as a bottom for the housing. It is held in place by two wood screws.

Of course, any number of housing arrangements are possible. Perhaps you want something that will blend with your room decor. Or, maybe you have an eye to mounting the components behind a console panel that controls all the electric and electronic equipment in your den. Either way, the choice is yours.

Wiring the Dimmer. Be careful when soldering the lead to the SCR that you do not overheat the case. The SCR is rated at 8 amperes, so no heatsink will be needed unless you intend to dim more than 300 watts. All components are supported by their own leads, and wiring is quite simple and straight forward.

The schematic diagram shows an additional 0.1-mF, 200-volt capacitor across the line to suppress radiation from the neon bulb.

The line cord is cut into two pieces, and each end is put through a rubber grommet inserted into the side holes in the housing. A knot is tied in the wire to keep it from pulling out of the housing. One end of each cord goes to the switch terminals (the two terminals that are *on* at the extreme CCW end of rotation) and the other end of each cord is spliced together.

The lamp that the dimmer will be used with will have its own off-on switch so none is needed on the dimmer housing. The dimmer itself will draw no current if the lamp is turned off so it can be left connected.

TOUCHOMATIC

Ever get tired using heavy, bulky switches for equipment that takes a lot of power? Well, snap on the Touchomatic and see that there is, indeed, a better way!

Refer to Figs. 1-21 through 1-24.

Ever get tired of having to leave the room to turn out the lights or having to get up from your chair to switch on the TV? Then give in and build the Touchomatic for a touch that will really tell. Its applications are limitless. You can even control the power input to your ham or CB shack from one central point with this snappy control box.

Anyone can have magic fingers with this do-all switch. And with the heavy-duty relay that's included, just under a 1000 watts of resistive power can be controlled. No modifications to your present equipment need be made, and operation is safe and reliable. But if you're the kind of fellow who touches all bases, you can even add an isolation transformer to make sure that the ac line voltage stays in place.

A Working Model. Taking a look at the schematic, you can see that R1, the touchplate (TP), R2, C1, and NE1 would form a basic relaxation oscillator if it were not for 1000-ohm resistor R3. With the addition of R3, a pulse-generating circuit is formed.

When someone touches the touchplate (TP), the resistance of his finger across points A and B is added in series to the combination of R1 and R2, the capacitor C2 begins to charge. When the voltage across C1 is finally sufficient to fire NE1, C1 will begin to discharge.

When NE1 fires, it produces a short between its terminals. Since R3 is connected across C1, they are effectively in series after NE1 fires. A voltage spike will then be passed by C2 and this will act as a positive triggering pulse. The pulse is fed to both SCR gates: SCR2 conducts, thereby closing relay K1. With a finger no longer on the

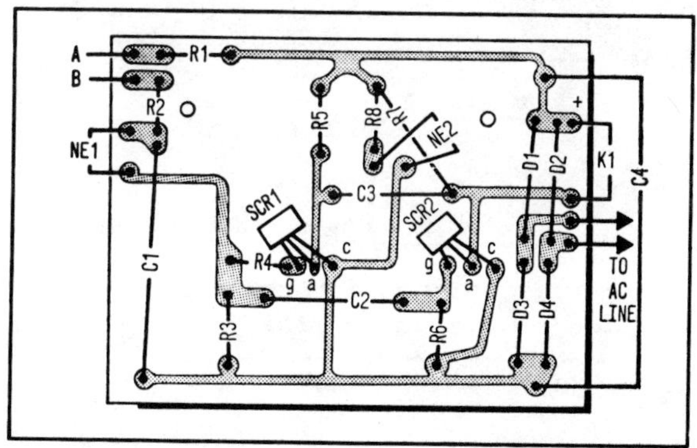

Fig. 1-21. Heavy lines show how components mount on top of PC board. Relay K1 fits between two holes at top of board, covering R5, R7, and R8. A perfboard and push-in terminals will also work fine.

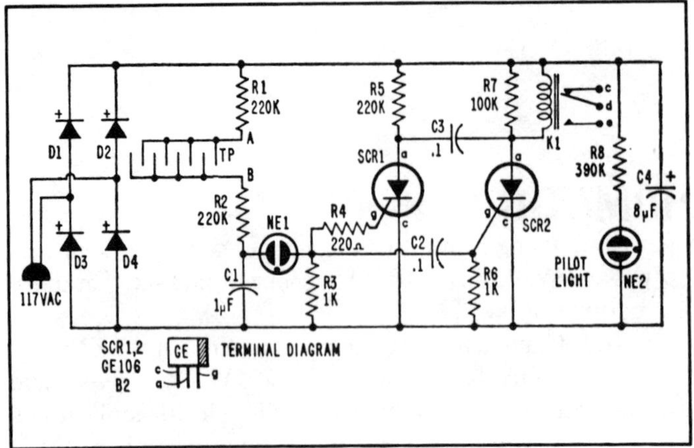

Fig. 1-22. Note that touchplate (TP) is connected to circuit with 300-ohm twinlead at points A and B. These two points are located on terminal strip TS1. With this design, an external input (i.e., from a switch) would connect to screw terminals of TS1. Twinlead is also used to connect points A and B at TS1 to PC board. Since R1 and R2 offer only minimal isolation from ac line, and diodes D1-D4 could short out, an isolation transformer is recommended for beginners.

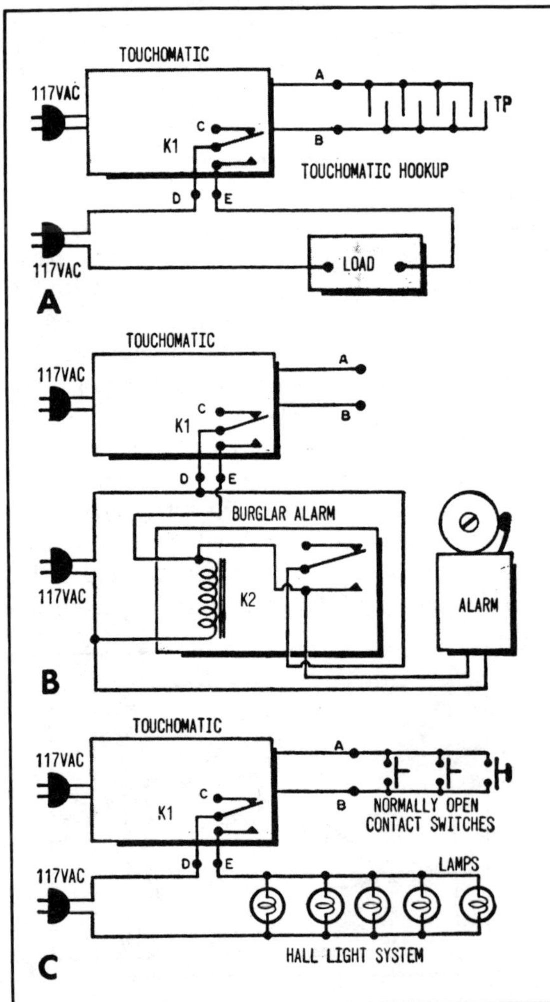

Fig. 1-23. Basic Touchomatic hookup is shown to the left in Fig. A. While relay K1 is a SPDT model, only two contacts (d and e) were used by author. However, experimenters may wish to use contact c for some kind of indicator to show that switch is off. Figure B is design for a burglar alarm and contains an external relay K2 to control current to alarm and to provide a latching circuit so that alarm cannot be switched off. Builder may want to add a switch so that relay may be de-energized at his command. Figure C illustrates typical lighting circuit where external switches connected in parallel can control lights at any one point.

C1—1-uF, 200-VDC capacitor
C2, C3—0.1-uF, 400-VDC capacitor
C4—8-uF, 450-VDC electrolytic capacitor
D1, D2, D3, D4—10B2 silicon rectifier
K1—110-VDC, spdt, 8000-ohm power relay (P&B MR5D, Allied 41 E 6760 or equiv.)
NE1, NE2—#51 neon bulb
R1, R2, R5—220,000-ohm, ½-watt resistor
R3, R6—1000-ohm, ½-watt resistor
R4—220-ohm, ½-watt resistor
R7—100,000-ohm, ½-watt resistor
R8—390,000-ohm, ½-watt resistor
SCR1, SCR2—GE 106B2 silicon controlled rectifier
Misc.—Plastic box and cover, touchplate (see text), panel light assembly for #51 neon bulb, 2-screw terminal strip, optional chassis-mounting AC outlet (Allied 47B0830 or equiv.), zip cord, 300-ohm twinlead, PC board or perf board (optional), rubber feet, spaghetti, wire, solder, hardware, etc.

Fig. 1-24. Touchomatic parts list.

touchplate, no more pulses are forthcoming because the C1 charge path is open.

The next contact with the touchplate will produce a pulse which triggers SCR1. SCR2 is now turned off by capacitor C3 which was charged by current passing through R6 and a SCR2. The firing of SCR1 in this way places a negative voltage across SCR2 which momentarily drops the relay current to a point below the holding current value of SCR2. (Holding current is the minimum current an SCR requires to remain in a conducting state once its gate voltage is removed.)

With SCR2 turned off, the relay will open and SCR1 will turn off due to the large resistance in series with its anode. Starved in this way, SCR1 turns off because of a forced lack of holding current.

Construction. Because of such high sensitivity, you might expect the parts layout to be critical. This just isn't so. If you follow the figure, you shouldn't have any trouble.

Together, R1 and R2 offer excellent isolation from the ac line. A transformerless power supply is used. But as said before, an isolation (power) transformer might be a good idea, particularly if you want to be certain that you're always on the safe side.

The actual form of the touchplate is pretty much up to you. The model shown was constructed by drilling holes in the box lid and placing solid wire in alternate positions so that touching any combination of two different leads (A and B) will close the relay. The wire should be rugged and spaghetti should be used to insulate one group of wires from the other (A from B).

Applications. Your touchplate needn't be the end of the line for the Touchomatic. While the basic Touchomatic hookup is shown, other hookups can be made to points A and B via terminal strip TS1. Circuits for a burglar alarm and hall light system are also shown.

The hall light system—as well as similar circuits—requires only low-current wires and switches, and the whole system is switched *on* or *off* at any one control point. These are just starting points for people who want to experiment—just use your imagination.

VISULERT

Are there times when you'd like to turn down the telephone bell so that baby or grandma can nap, and yet you need to know when that important call comes in? Because high platform noise overrides the normal telephone bell, and you're skeptical of the effectiveness of so-called loud ringers, do you have need for another means of alerting the shipping clerk to take a telephone call? Or, perhaps you know a deaf person who can't hear the phone bell at all.

Refer to Fig. 1-25.

Visulert, a small, self-contained, easily constructed telephone accessory, solves all these problems. And the beauty of it is that you don't have to connect it directly to the telephone lines.

An inductive pick-up coil ordinarily used for recording phone messages, placed on or under a telephone, picks up just the ringing pulses by magnetic induction and feeds them to an amplifier in the Visulert. This amplifier triggers an SCR that switches a lamp *on* and *off* in step with the pulsing of the ringing signal.

How It Works. Provided magnetic pickup MP1 is properly located within the ringer's magnetic field an electrical voltage is induced in the coil of MP1 whenever the ringer of a telephone is energized. This voltage is fed via jack J1 to the base of transistor Q1. The resulting amplified signal output on the collector of Q1 is coupled to the gate of silicon controlled rectifier Q2 and triggers it *on* whenever the signal appears on its gate. Lamp I1 is turned *on* each time Q2 is triggered *on* and remains *on* until Q2 is triggered *off* by a drop in the induced signal level. Since the ringer voltage is pulsating, the Visulert will flash its lamp *on* and *off*, following the ringer pulses.

Building Visulert. Our model is housed in a standard 4-inch × 2¼-inch × 2¼-inch aluminum minibox. Though the layout isn't critical, you will speed up your construction time by following our layout as shown in our photos.

All of the components are mounted either directly on the minibox or to tie strips, which support them away from the metal to prevent shorts. Before soldering electrolytic capacitors and diodes, check to be sure that you have them properly polarized. Also, doublecheck that connections to Q1 and Q2 are correct before soldering to avoid application of too much heat, if you must unsolder and resolder them, since excessive heat can damage solid-state devices. In fact, use an alligator clip as a heatsink by temporarily clipping it to each lead being soldered.

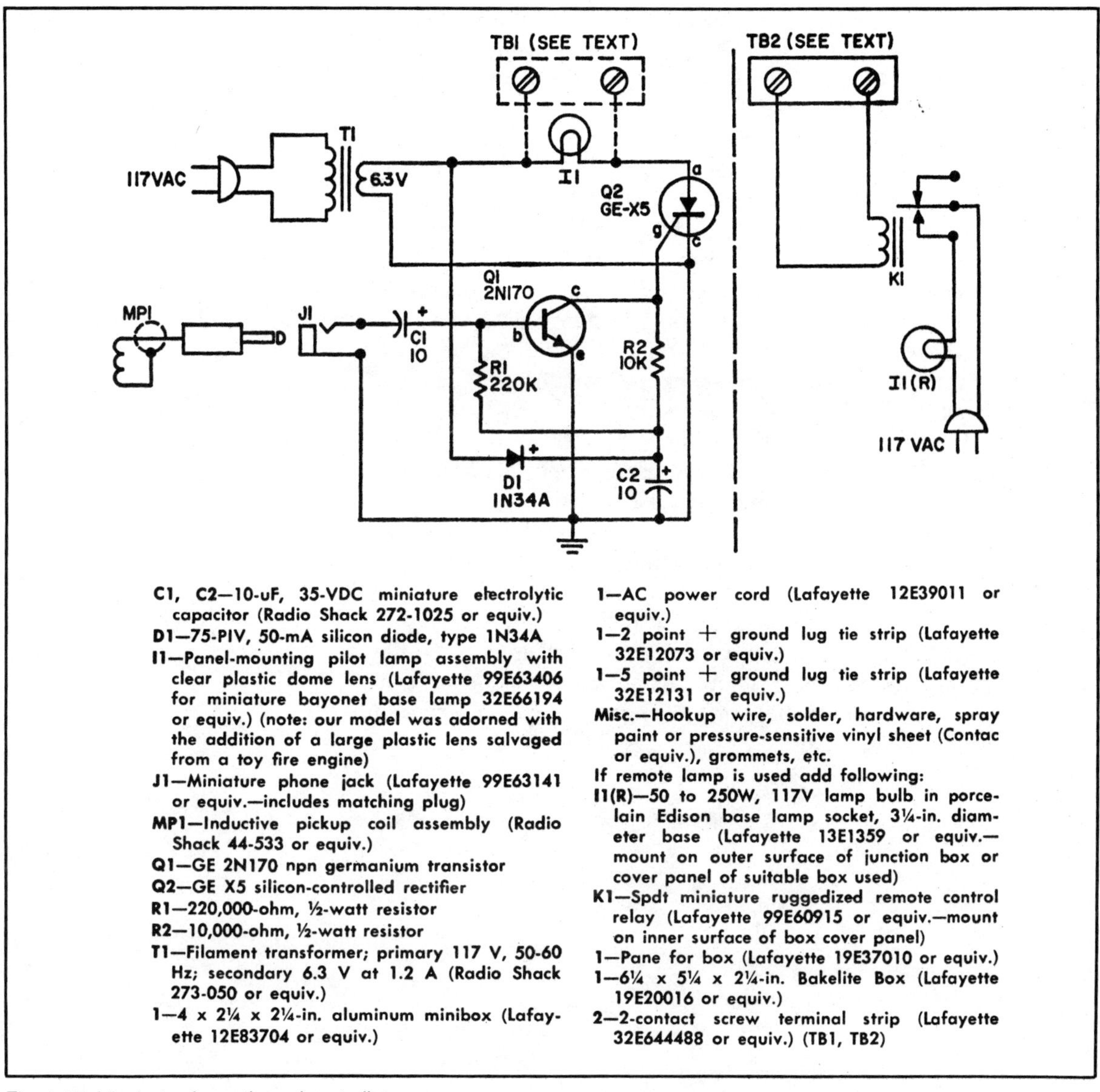

C1, C2—10-uF, 35-VDC miniature electrolytic capacitor (Radio Shack 272-1025 or equiv.)
D1—75-PIV, 50-mA silicon diode, type 1N34A
I1—Panel-mounting pilot lamp assembly with clear plastic dome lens (Lafayette 99E63406 for miniature bayonet base lamp 32E66194 or equiv.) (note: our model was adorned with the addition of a large plastic lens salvaged from a toy fire engine)
J1—Miniature phone jack (Lafayette 99E63141 or equiv.—includes matching plug)
MP1—Inductive pickup coil assembly (Radio Shack 44-533 or equiv.)
Q1—GE 2N170 npn germanium transistor
Q2—GE X5 silicon-controlled rectifier
R1—220,000-ohm, ½-watt resistor
R2—10,000-ohm, ½-watt resistor
T1—Filament transformer; primary 117 V, 50-60 Hz; secondary 6.3 V at 1.2 A (Radio Shack 273-050 or equiv.)
1—4 x 2¼ x 2¼-in. aluminum minibox (Lafayette 12E83704 or equiv.)
1—AC power cord (Lafayette 12E39011 or equiv.)
1—2 point + ground lug tie strip (Lafayette 32E12073 or equiv.)
1—5 point + ground lug tie strip (Lafayette 32E12131 or equiv.)
Misc.—Hookup wire, solder, hardware, spray paint or pressure-sensitive vinyl sheet (Contac or equiv.), grommets, etc.
If remote lamp is used add following:
I1(R)—50 to 250W, 117V lamp bulb in porcelain Edison base lamp socket, 3¼-in. diameter base (Lafayette 13E1359 or equiv.—mount on outer surface of junction box or cover panel of suitable box used)
K1—Spdt miniature ruggedized remote control relay (Lafayette 99E60915 or equiv.—mount on inner surface of box cover panel)
1—Pane for box (Lafayette 19E37010 or equiv.)
1—6¼ x 5¼ x 2¼-in. Bakelite Box (Lafayette 19E20016 or equiv.)
2—2-contact screw terminal strip (Lafayette 32E644488 or equiv.) (TB1, TB2)

Fig. 1-25. Visulert schematic and parts list.

Remote Lamp. In the event you require a brighter lamp than the standard bulb listed, or want the lamp located on a wall or site outside the area of the telephone—where it can be universally observed—make the following modification. Remove the bulb and connect the leads to terminal strip TB1 for connecting the remote lamp control leads. Mount a 6.3-V ac relay, a standard 110-V lamp socket, and TB2 in a container suitable for the remote location. Wire it as shown in the schematic. By using low voltage (6.3V ac) the interconnecting remote control leads can be small-sized insulated wire. The 6.3V that is switched by the SQR (Q2) to turn the low voltage lamp *on* and *off* will now be used to operate the remote relay, which will, in turn, control 117V ac to the larger lamp bulb.

Checking Out Visulert. After doublechecking your hookup for possible errors, shorts, or cold soldered connections, plug the power cord into an ac outlet, and plug magnetic pickup MP1 into J1. Now bring the pickup near power transformer T1. If the unit is working correctly the radiated ac field around the transformer will produce a signal in the magnetic pickup device, triggering the SQR (Q2) to turn *on* lamp I1. Each time you move the pickup close to the transformer, the lamp will be lit; as you move MP1 away from T1's magnetic field, the lamp will go out. When this checkup has been completed you can close up the minibox and place Visulert in service.

Using Visulert. The suction cup on the pickup coil we used serves a dual purpose. It permits you to easily orient MP1 into the magnetic field of the telephone ringer and also holds it in position once the ideal location is found. If the pickup you use is one of the flat types, place it under the phone near the exit of the handset cable.

Regardless of the type, you'll have to move the pickup around the base of the phone to locate the magnetic field of the ringer. Remember, of course, that the only time you can locate the pickup is when the phone is ringing, because the Visulert's operation is dependent upon the relatively high magnetic field of the ringer to develop a control signal to fire the SCR.

FRIDGALARM

Every creature in the world has its natural enemies, and the refrigerator is no different. Perhaps the most dangerous of the ice box invaders are dieters and children. Either is likely to lodge in front of the door and stare longingly inside, feverishly calculating which item would be least likely missed. Inevitably, as the hours of openness pass, a layer of permafrost grows inside that requires a chisel and a contingent of National Guardsmen to remove, and the electric bill rises ever the higher heavenward.

In our household, frugality is the mother of invention, so we followed our pursestrings to the workbench, grabbed a handful of parts, and created the Fridgalarm. Now, should one of the pantry predators decide to camp out in a lean-to made from the ice-box door for longer than our preset interval, the alarm lets out a piercing squeal until all the cold is again locked safely within.

Refer to Fig. 1-26.

The Circuit. Our circuit is based on a pair of versatile 555 timer chips and a photoresistor, and offers not only a useful project, but also a quick and fun lesson on how each part works. One of the 555s is used to time the period before the alarm goes off (the filching interval). The other generates a tone that serves as the alarm proper.

Let's start at the beginning and see exactly what makes the alarm work. The photoresistor is used to detect the lamp that lights inside the refrigerator to let you see how good the pickings are. The circuit is sensitive enough, though, to trigger even if the bulb has burned out. Once the door is closed, the inside of the average refrigerator is dark, really dark, and opening it changes the light level enough that it can easily be electronically detected.

Photoresistor R6 is a light-sensitive resistor. The more light it sees, the less it wants to conduct electricity and the higher its resistance becomes. In total darkness, its resistance is low enough that it effectively shorts the base of transistor Q1 to ground so that Q1 will not conduct. (Q1 is actually operating as an inverting amplifier.)

Resistor R1 limits the current through R6 and is effectively the only current-consuming element in the circuit when no light is present. The light sensitivity of the alarm can be adjusted to some degree by varying the value of R1.

When the light goes on, the resistance of R6 increases, and there is a corresponding voltage drop across it. When this voltage becomes great enough, Q1 begins to conduct and supplies current to the circuitry.

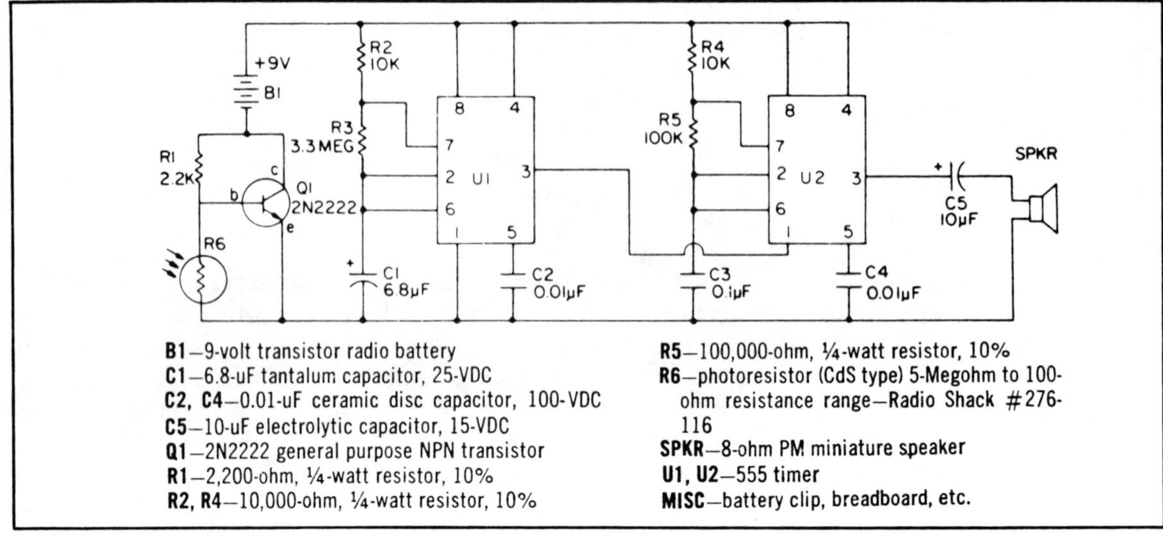

Fig. 1-26. Fridgalarm schematic and parts list.

U1 is the timer, determining how much time must pass after the light goes on and Q1 turns on, before the alarm is triggered. U1 is set up for astable operation to conserve battery life once the alarm sounds. In other words, it functions as a long-period oscillator, turning the sound on and off.

The initial timing period is determined by the time it takes to charge C1 two-thirds of the way up through R2 and R3. (The regular on and off periods of the alarm are one-half this value, because the 555 only discharges C1 to one-third of its capacity, hence C1 oscillates in charge between one-third and two-thirds, its most linear region.) As C1 reaches a charge of two-thirds of its capacity, the output of U1 (at pin 3) goes low, near ground, and thereby completes the voltage supply to U2 which also operates as an astable oscillator but with a much shorter period. As U2's output swings between high and low, it creates the sound that is coupled through C5.

Operation. With the values shown, the alarm will trigger about thirty seconds after the refrigerator door is opened, and sound for about fifteen seconds, then cycle on and off every fifteen seconds until the door is closed. Should you prefer the alarm to sound continuously until the door closes, remove the connection between pin 2 on U1 and the rest of the circuitry. This prevents the timer from resetting until supply voltage is removed by closing the door.

Should the tone we've chosen not be noxious enough for you, you can change its pitch by varying R5 (which changes the frequency of U2's oscillations), to a higher value producing a *lower* frequency, or by changing C3 to a lower value, producing a *higher* frequency. Similarly, the timing period before the alarm sounds can be varied by changing either C1 or R3.

TELEPHONE TRIGGER

How would you like to be able to energize any electrical device in your home, from anywhere in the country, and do it without paying for a phone call? You can do it, and it's perfectly legal. The remote control described here is a simple digital circuit that responds to the sound of the telephone bell. There is no need to make any connections to the telephone line.

The remote control circuit is protected against accidental operation through normal telephone calls by means of an automatic reset feature that cancels out the effects of any telephone calls made by others. This is accomplished by incorporating a time-delay circuit that allows the circuit to receive a valid code for a period of ninety seconds. When the ninety second period is completed, the circuit resets itself and waits for the next telephone call.

Refer to Figs. 1-27 through 1-30.

The proper code to activate the circuit consists of two rings of the telephone, a 25-second to 40-second delay, and two more rings. If, and only if this code is received before the 90-second delay is terminated will the device be activated. Any other combination of rings will not operate the circuit. Since it is very unlikely that anyone would ring your telephone with such a sequence, accidental operation is virtually eliminated. Included in the circuit is a group of LED indicators which monitor the control pulses and indicate the status of the circuit at all times.

How It Works. The best way to understand circuit operation is to refer to the figures showing several pertinent waveshapes in the circuit and the schematic. A crystal or ceramic microphone is used as the sensing element that detects the sound of the telephone bell. The output of the microphone is fed to the negative input of a comparator, IC 1. The positive input of the comparator is set to a positive dc voltage by means of a potentiometer which acts as a sensitivity control. This forces the output of IC 1 to +5 volts. During periods of silence there is insufficient output from the microphone to exceed the voltage setting of the sensitivity control, and the output of the comparator remains at 5 volts. When the telephone rings, the increase in sound energy causes the output of the microphone to exceed the setting of the sensitivity control. The output of the comparator oscillates between zero and 5 volts as the bell continues to ring. This is shown in the figure as waveform A.

The output of IC 1 is fed to the trigger input terminals of IC 2 and IC 3, pin 2. Each of these ICs is a 555 timer connected to operate as a

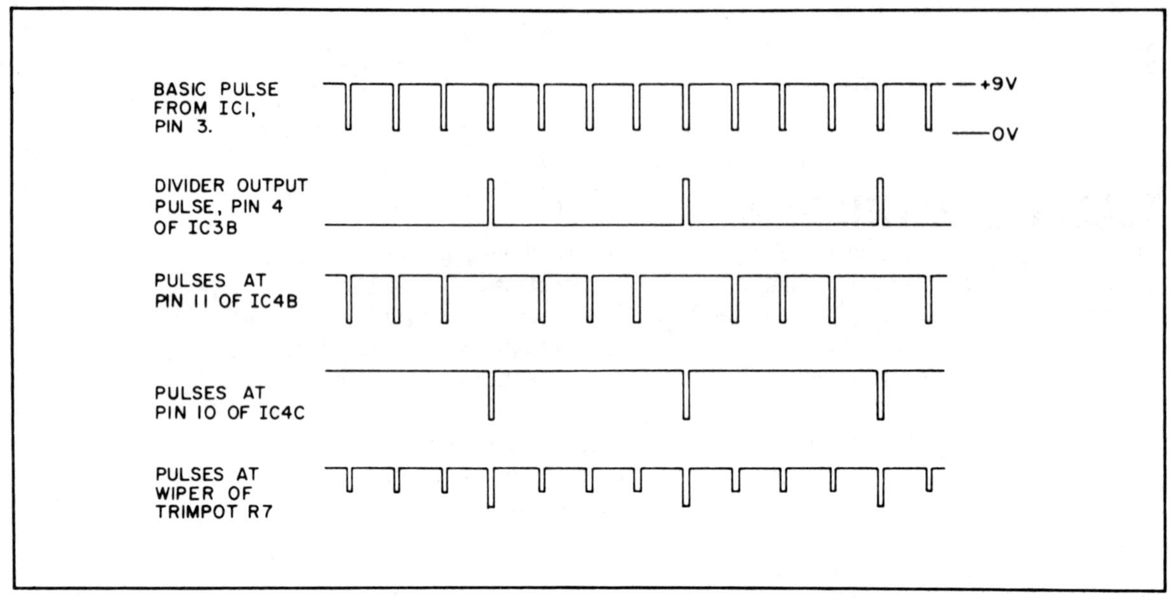

Fig. 1-27. The best way to understand the operation of a circuit such as the Telephone Trigger is by studying these waveshapes in conjunction with the schematic and the explanation presented in the text. Read the explanation over a few times, referring to the diagram and schematic as you do. Soon you'll be a logic-circuit, solid-state maven.

monostable or one shot multivibrator. IC 2 produces an output pulse of about 3 seconds duration, and IC 3 produces an output pulse of about 90 seconds duration. These output pulses are shown as waveforms B and C.

The purpose of IC 2 is to convert the rapidly oscillating output of IC 1 to a rectangular pulse of known duration, about 3 seconds. Note that the pulse time of IC 2 is greater than that of one ring, but stops before the start of the next ring. IC 2 provides an accurate waveform which can be counted by IC 4.

The output of IC 2 is inverted by IC 7D, which in turn drives LED "R." This provides a visual indication of the circuit response to the sound of the telephone.

IC 3 is used as the control of the binary counter IC 4. When the circuit is in a standby condition the output of IC 3 is at a logic level of zero. This is fed to IC 4 reset terminals 2 and 3 through NAND gate IC 6C and forces IC 4 to be set to a count of zero. When the first telephone ring is received, the output of IC 3 goes to a logic one state, thus allowing IC 4 to count for a period of 90 seconds.

The output of IC 2 feeds the clock input, pin 14, of IC 4 which clocks on the trailing or falling edge of the pulses. IC 5 is a 4 bit decoder which provides a zero logic level at any one of its output terminals as determined by the binary information fed from IC 4 to its input terminals. The output of IC 5 is used to drive a set of LEDs and also to control a 25 second timer, IC 8.

This is a 555 timer chip which operates as a one shot multivibrator in a similar manner as IC 2 and IC 3, except that its period of operation is 25 seconds. The purpose of IC 8 is to provide a 25 second time interval, starting after the second ring, which will cause the circuit to reset itself should a third ring be received before IC 8 resets itself. IC 8 is also retriggered after the fourth ring to prevent the appliance from being turned on if a fifth telephone ring is received. When IC 4 reaches a count of 2, the 25 second time is activated. This is shown in the figure as waveform D. If IC 4 receives a third clock pulse from IC 2 during this interval, it is reset to zero through inverter IC 7A and NAND gates IC 6D and IC 6C. Once IC 8 returns to its normal state after the period of 25 seconds, IC 4 is ready to receive additional clock pulses without being reset to zero.

When IC 4 reaches a count of 4, the 25 second timer is reactivated as shown in waveform D. The output of IC 5, pin 5, is prevented from reaching Q1 base until the 25-second period is over. Should IC 4 receive any more input pulses Q1 would not be activated at the end of the 25-second period, since the counter would then be at a count of 5 or more.

The outputs of IC 5 are shown as waveforms E. When IC 5 is fed a binary number from 0 to 9, the corresponding output terminal assumes a logic level of zero. All other output terminals remain at a logic level of one. By connecting LEDs to outputs 0, 1, 2, 3, and 4 (pins 1, 2, 3, 4, and 5), the status of binary counter IC 5 is visually indicated.

The coil of a 12-volt relay is connected in the collector circuit of Q1 so that it is activated whenever Q1 conducts current in response to the zero logic level of IC 5, pin 5. The output of IC 5 is not permanent, so one section of the relay contacts is connected to the coil circuit so that the relay latches and remains activated even though IC 4 is reset to zero at the end of the 90 second pulse time of IC 3. The other set of contacts of the relay is used as a single pole switch.

Power to operate the circuit is obtained through a 12 volt transformer feeding a full-wave bridge rectifier. The output of the rectifier is fed to IC 9 which is a fixed 5 volt regulator. The entire circuit, with the exception of the relay is returned to the unregulated output of the bridge rectifier, 12 volts.

Construction. The entire circuit is constructed on a printed circuit board. The crystal microphone cartridge was mounted directly on the board with adhesive, but also may be connected by means of a shielded wire to some remote location by the telephone bell. The foil pattern of the printed circuit board is shown full size, as is the component layout. Although this is a relatively simple digital circuit, you should use sockets for the IC and Q1. The initial checkout of the circuit will be simplified if IC 3 can be temporarily removed from the circuit. In the event that the unit ever requires service you will find that it is well worth the added cost of sockets. It is extremely difficult to remove

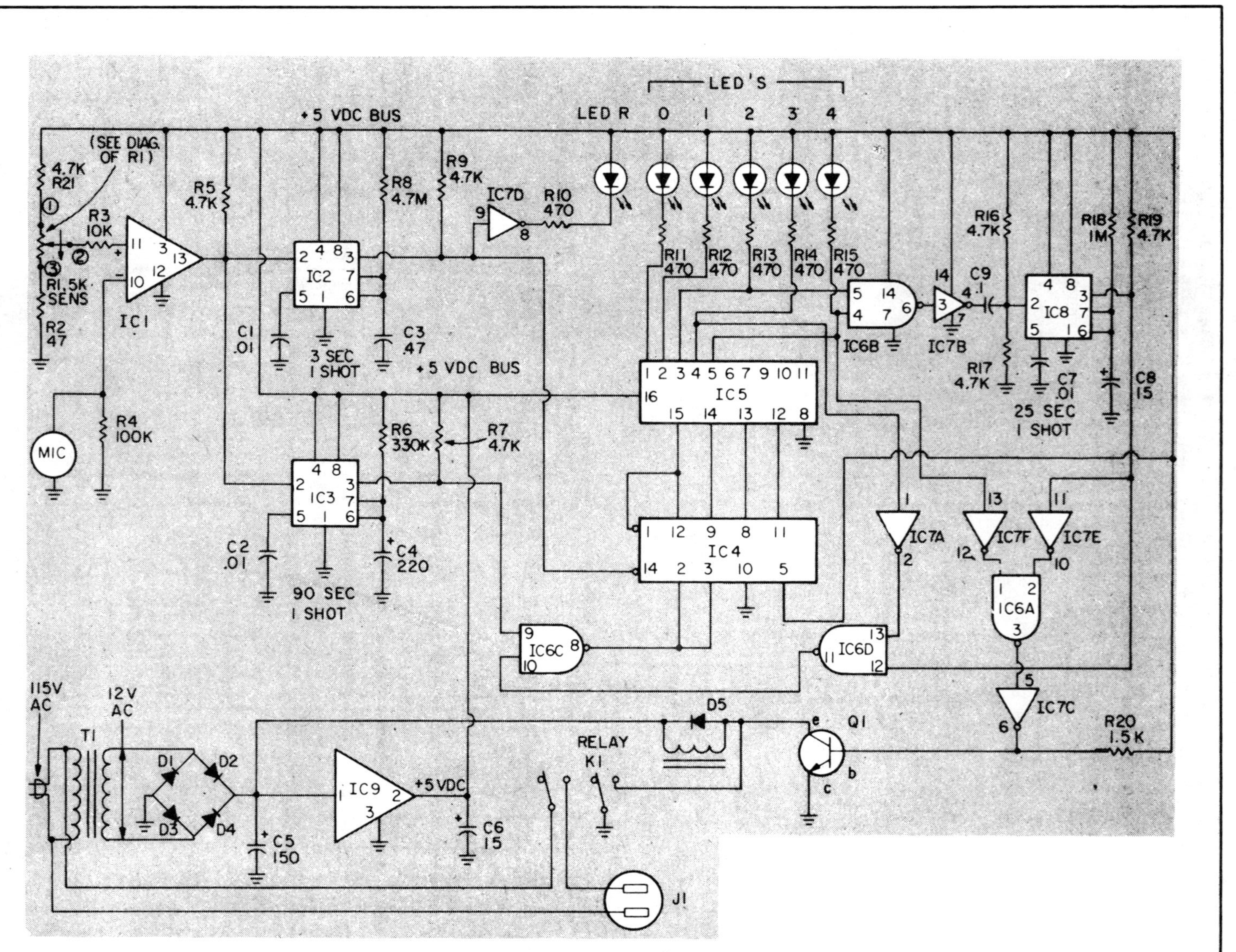

+5 VDC BUS
LED'S
LED R
0
1
2
3
4
(SEE DIAG. OF R1)
4.7K R21
R3 10K
R1 5K SENS
R2 47
IC1
R5 4.7K
IC2
3 SEC 1 SHOT
C1 .01
R8 4.7M
C3 .47
R9 4.7K
IC7D
R10 470
R11 470
R12 470
R13 470
R14 470
R15 470
+5 VDC BUS
MIC
R4 100K
IC3
90 SEC 1 SHOT
C2 .01
R6 330K
R7 4.7K
C4 220
IC5
IC4
IC6B
IC7B
C9 .1
R16 4.7K
R17 4.7K
IC8
C7 .01
25 SEC 1 SHOT
R18 1M
R19 4.7K
C8 15
IC7A
IC7F
IC7E
IC6A
IC6C
IC6D
IC7C
R20 1.5K
Q1
D5
RELAY K1
115V AC
12V AC
T1
D1
D2
D3
D4
C5 150
IC9
+5VDC
C6 15
J1

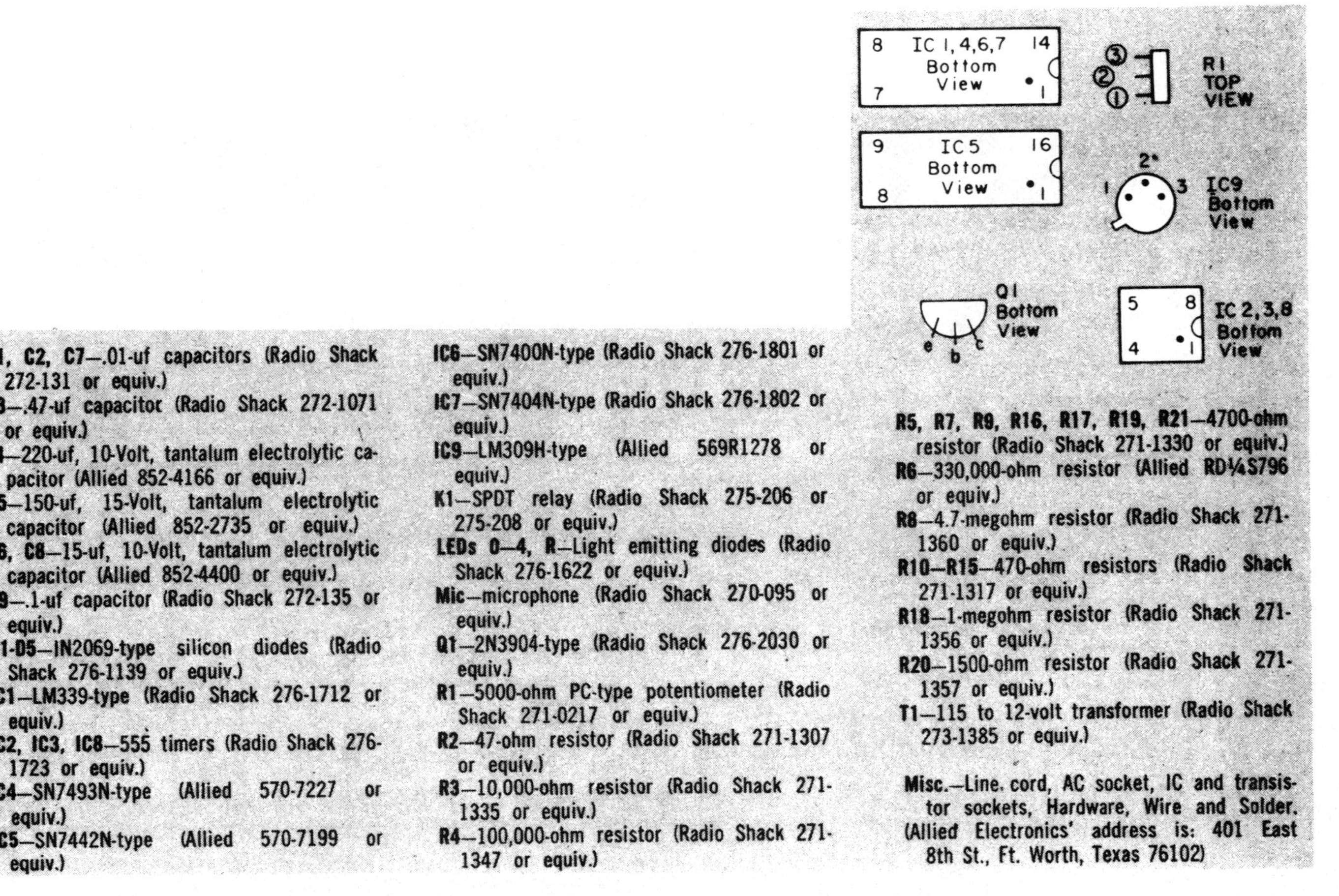

C1, C2, C7—.01-uf capacitors (Radio Shack 272-131 or equiv.)
C3—.47-uf capacitor (Radio Shack 272-1071 or equiv.)
C4—220-uf, 10-Volt, tantalum electrolytic capacitor (Allied 852-4166 or equiv.)
C5—150-uf, 15-Volt, tantalum electrolytic capacitor (Allied 852-2735 or equiv.)
C6, C8—15-uf, 10-Volt, tantalum electrolytic capacitor (Allied 852-4400 or equiv.)
C9—.1-uf capacitor (Radio Shack 272-135 or equiv.)
D1-D5—IN2069-type silicon diodes (Radio Shack 276-1139 or equiv.)
IC1—LM339-type (Radio Shack 276-1712 or equiv.)
IC2, IC3, IC8—555 timers (Radio Shack 276-1723 or equiv.)
IC4—SN7493N-type (Allied 570-7227 or equiv.)
IC5—SN7442N-type (Allied 570-7199 or equiv.)
IC6—SN7400N-type (Radio Shack 276-1801 or equiv.)
IC7—SN7404N-type (Radio Shack 276-1802 or equiv.)
IC9—LM309H-type (Allied 569R1278 or equiv.)
K1—SPDT relay (Radio Shack 275-206 or 275-208 or equiv.)
LEDs 0—4, R—Light emitting diodes (Radio Shack 276-1622 or equiv.)
Mic—microphone (Radio Shack 270-095 or equiv.)
Q1—2N3904-type (Radio Shack 276-2030 or equiv.)
R1—5000-ohm PC-type potentiometer (Radio Shack 271-0217 or equiv.)
R2—47-ohm resistor (Radio Shack 271-1307 or equiv.)
R3—10,000-ohm resistor (Radio Shack 271-1335 or equiv.)
R4—100,000-ohm resistor (Radio Shack 271-1347 or equiv.)
R5, R7, R9, R16, R17, R19, R21—4700-ohm resistor (Radio Shack 271-1330 or equiv.)
R6—330,000-ohm resistor (Allied RD¼S796 or equiv.)
R8—4.7-megohm resistor (Radio Shack 271-1360 or equiv.)
R10—R15—470-ohm resistors (Radio Shack 271-1317 or equiv.)
R18—1-megohm resistor (Radio Shack 271-1356 or equiv.)
R20—1500-ohm resistor (Radio Shack 271-1357 or equiv.)
T1—115 to 12-volt transformer (Radio Shack 273-1385 or equiv.)

Misc.—Line. cord, AC socket, IC and transistor sockets, Hardware, Wire and Solder. (Allied Electronics' address is: 401 East 8th St., Ft. Worth, Texas 76102)

Fig. 1-28. The schematic diagram for the Telephone Trigger, the automatic telephone relay-tripping circuit. There are plenty of components in the circuit, so be careful as you build this exciting project.

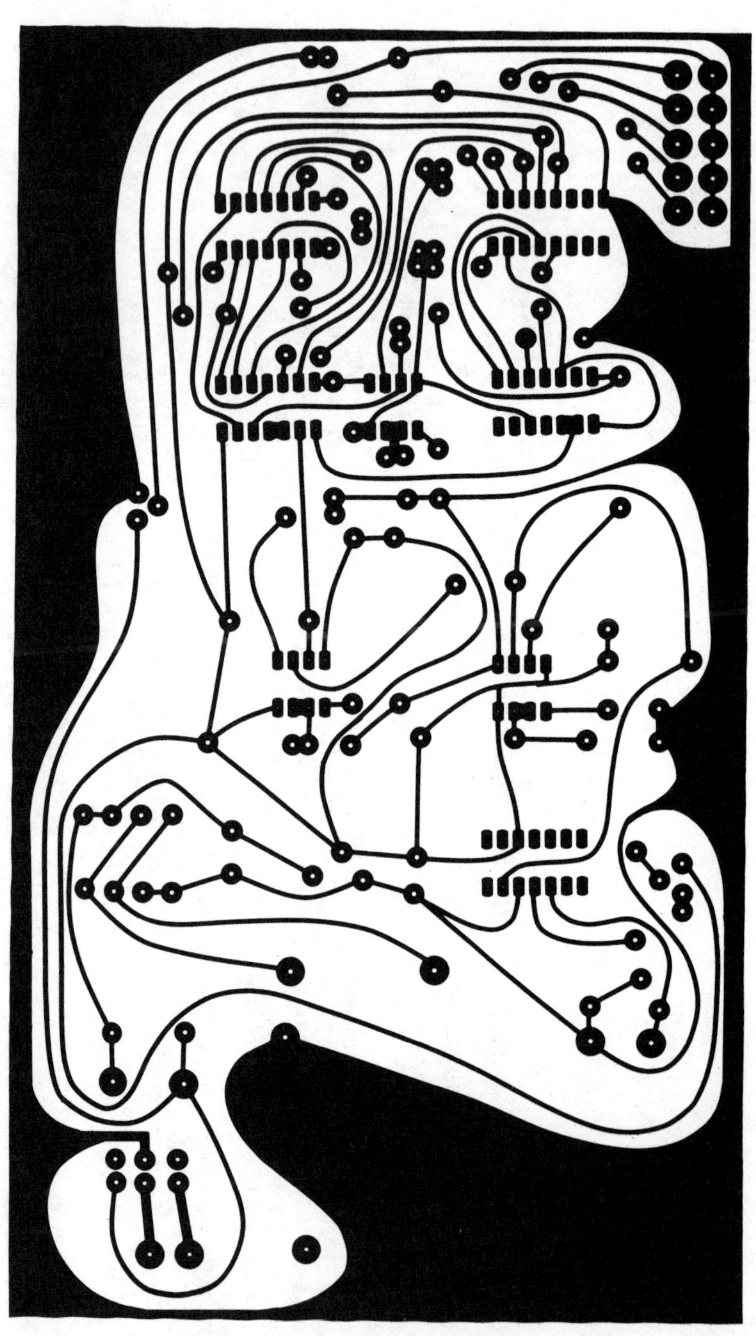

Fig. 1-29. Exact-size PC board layout to make your own Telephone Trigger. If this is your first time etching a complex board, avail yourself of some of the excellent layout aids available at many large electronics stores. You might also look into some of the photographic kits used to make boards by copying images such as this right off the magazine page, a definite help to those of us who have trouble drawing a straight line with a ruler. Doing your own PC board is an additional part of the enjoyment.

a multipin IC which has been soldered directly into a printed circuit board without destroying the IC. When mounting the electrolytic capacitors, diodes and LED indicators, be sure to observe the correct polarity as shown in the schematic.

Mount and solder all parts to the printed circuit board as shown in Fig. 1-30. After this is done you will be able to locate and insert the proper jumpers into the printed circuit board.

Be sure to use a relay which is capable of carrying the current of the appliance which is to be turned on. The parts list shows a choice of two relays. Part number 275-206 has a current rating of 10 amperes. You will have to use the higher current relay if you plan to operate a high current appliance such as an air conditioner or coffeemaker. If the remote control is to be used to operate a heating system, relay contacts can be connected into the thermostat circuit. Such a connection permits the use of the lighter duty relay. The relay coil driver transistor, Q1, can safely carry up to 150 milliamperes to drive the relay coil.

A receptacle for plug-in appliances is mounted on the circuit board and is wired directly to the line cord and relay as shown in the schematic. Be sure to use a line cord and wire which will safely carry the desired current. For 10-ampere operation use at least a No. 16 gauge wire.

Testing And Adjusting. Checks are made with only the built-in LED indicators and a dc voltmeter. The first check is on the timing of IC 2. It will be helpful if you temporarily remove IC 3 from the circuit to prevent it from resetting the counter while you perform the first part of the check.

Apply power to the unit and measure the voltage at pin 1 and pin 2 of IC 9. The voltage at pin 1 should be about 12 volts and the voltage at pin 2 should be 5± 0.25 volts, measured with respect to ground. Set the sensitivity control about three-fourths maximum clockwise. Gently tap the microphone while watching LED R. It should light when the microphone is tapped, and remain lit for about three seconds. Each time the microphone is tapped LED R should light for at least two seconds and not more than four seconds. It is important that the timing of IC 2 falls into this range so that it will be able to sense each telephone ring separately. You may change the value of R8, if necessary, to bring the timing of IC 2 within the range of two to four seconds.

To check the operation of IC 4 and IC 5 momentarily short pin 9 of IC 6 to ground to clear the counter and set it to zero. LED 0 should be lit. Gently tap the microphone and wait for LED R to be extinguished. When this occurs, LED 1 should light, indicating that the counter has advanced one count. Connect a voltmeter to pin 3 of IC 8. Tap the microphone while watching the voltmeter. At the end of the 3 seconds the counter should advance to a count of 2, and the voltage at pin 3 of IC 8 should rise from zero to about 5 volts. This voltage should hold for at

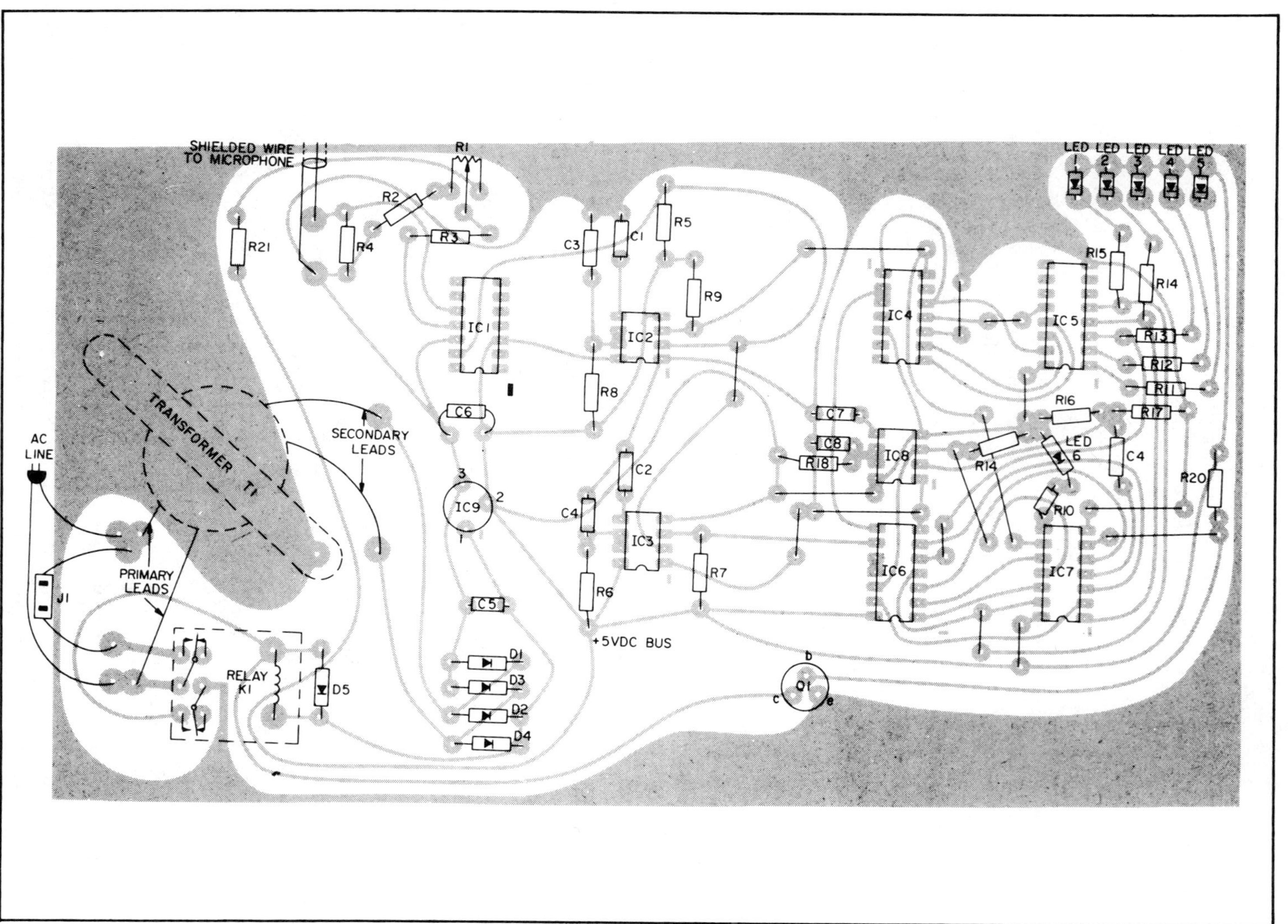

Fig. 1-30. Following this full-size component layout will help make construction of your Telephone Trigger as simple as dialing a phone. Make certain to observe the polarity of diodes and LEDs; the LEDs will have a notch on the on the negative lead. Some of the foil runs are close together so, if this is one of your first projects or even if you're a seasoned construction veteran, check for any solder bridges. If you find one, repair it carefully with a sharp knife blade.

least 25 and not more than 30 seconds. You may adjust R 18 to bring the timing of IC 8 into this range, if necessary.

After IC 8 has completed its operation, tap the microphone two more times to advance the counter to 4. At this time, IC 8 should again be activated. When IC 8 completes its second cycle, the relay should be activated.

You may check the operation of the reset circuitry by tapping the microphone three times to simulate three telephone rings without the required 25-second delay. When this is done, the counter should not advance to a count of 3, but should reset to zero after a count of 2.

Replace IC 3 in its socket. Check the timing of IC 3 by connecting a voltmeter to pin 3. The voltage should be zero before the circuit is triggered by tapping the microphone, and should rise to about 5 volts and hold for at least 80 and not more than 95 seconds. You may adjust R6, if necessary, to bring the timing into this range. The operation of IC 3 may be visually checked by advancing the counter to a count of 1 and watching the LED indicators. After IC 3 completes its 90-second time, the counter should be reset to zero.

The final adjustment is the sensitivity control. If possible have a friend call you up and let the telephone ring for about a minute. Locate the remote control microphone as close to the telephone bell as possible. Set the sensitivity control the maximum counter-clockwise (least sensitivity) position. Slowly turn the sensitivity control while watching LED R, and leave it set to the least sensitive position which gives a reliable detection of the sound. It is best to avoid excessive sensitivity so that the circuit does not respond to random noises in the house.

The remote control is now ready to be placed in operation by connecting the appliance to the receptacle on the circuit board and turning its power switch on. Once the hookup is made, the appliance will automatically be turned on when you call your own telephone number with the proper code. The 25-second delay time between the second and third rings is not critical, but should not be less than 25 seconds and not more than 40 seconds. The circuit will operate properly if you happen to get a partial ring when you call, but if there is any doubt that the phone has rung, it is best to hang up, wait two minutes, and try again. Also, since it is possible that you may call during a time that the circuit is in an activated state because of any calls made by others, it would be good practice to ring the code two times, spaced several minutes apart.

SIMPLE TOUCH SWITCH

Looking for a way to add a touch of class to your digital process? Try this touch switch. Not only does it add a note of distinction to a project, but it's bounce-free as well. Whenever a finger touches the contact plate, stray 60-Hz powerline interference is coupled into the circuit due to the antenna effect of your body. The 60-Hz pickup is rectified and filtered to provide a negative bias on Q1's gate, thus causing Q1 to turn off and Q2 to turn on. As a result, Q2's collector drops to ground potential. When the touch plate is released, the potential at Q2's collector terminal once again jumps high. You can use the output to drive either CMOS or TTL with ease.

Refer to Fig. 1-31.

Note that if you do your experimenting in a place devoid of 60-Hz powerline radiation—in the middle of a field of wheat, for example—the circuit will not work. The average home is full of 60-Hz radiation, however, so the switch should function well. If you have some difficulty, connect your system's electrical ground to an earth ground (the screw on your ac outlet's cover plate). This will boost the signal pickup.

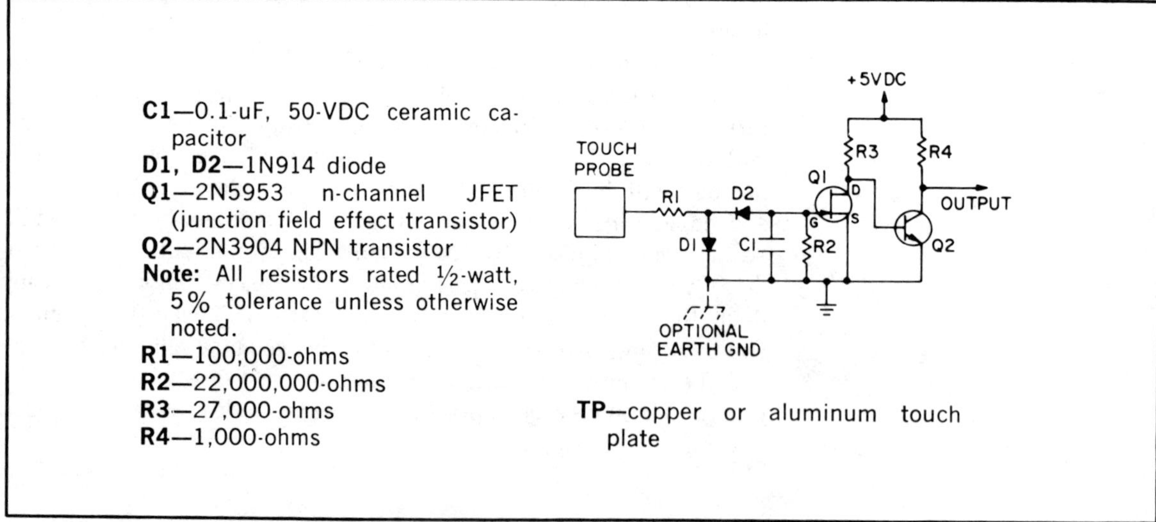

Fig. 1-31. Simple Touch Switch schematic and parts list.

POWER TOOL TORQUE CONTROL

As the speed of an electric drill is decreased by loading, its torque also drops. A compensating speed control like this one puts the oomph back into the motor.

Refer to Fig. 1-32.

When the drill slows down, a back voltage developed across the motor—in series with the SCR cathode and gate—decreases. The SCR gate voltage therefore increases relatively as the back voltage is reduced. The "extra" gate voltage causes the SCR to conduct over a larger angle and more current is driven into the drill, even as speed falls under load.

The only construction precaution is an extra-heavy heatsink for the SCR. The SCR should be mounted in a ¼-inch thick block of aluminum or copper at least 1-inch square; 2-inches if you drill for extended periods.

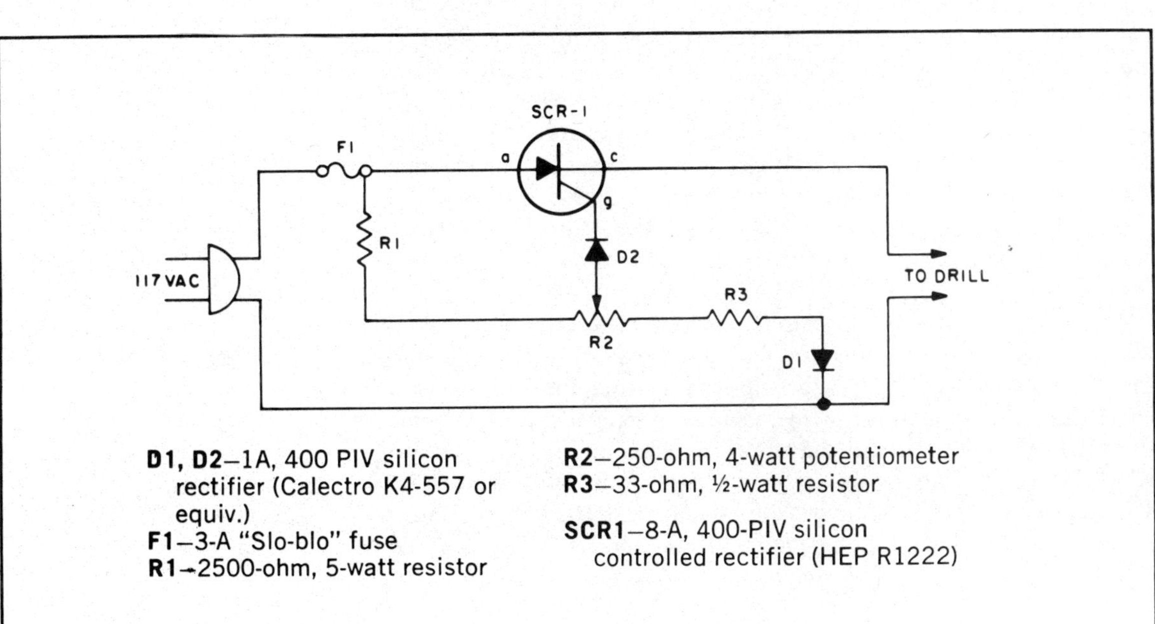

Fig. 1-32. Power Tool Torque Control schematic and parts list.

STEREO SPEAKER PROTECTOR

The advent of the superamplifier, capable of supplying 100 to 200 watts per channel on a continuous basis, has been both a blessing and a curse to the audiophile. The blessing is that a recording's dynamic range can now be more faithfully reproduced, even with inefficient loudspeakers. Unfortunately, these amps are so powerful that loudspeakers can often be overdriven, and eventually destroyed, if sufficient care is not exercised. If your amp lacks provisions for speaker protection, you may want to build the speaker protector diagrammed here.

Refer to Fig. 1-33.

The contacts of relay K1 are hooked in series with your right-hand and left-hand speakers in such a way that, when K1 is unenergized, its contacts close and complete the circuit to each loudspeaker.

Inputs to the protection circuit come from your amplifier's outputs (the same outputs that drive the speakers). If the signal feeding the *right* input is sufficiently large to charge C1 to a potential greater than the breakdown voltage of Q1's emitter, a voltage pulse will appear across R7. Similarly, excessive inputs to the 'left' channel will also produce a pulse across R7, this time due to the discharge of C2 by Q2. The pulse across R7 triggers SCR Q3, which latches in a conducting stage and energizes K1. This interrupts both speaker circuits, and the resulting silence should alert you to a problem. Cut back on your amplifier's volume; then, press and release S1 to reset the circuit and restore normal operation.

The circuit can be adjusted to trip at lower levels from 15 to 150 watts rms. To calibrate, feed a deliberately excessive signal to the 'right' input, and raise R3's wiper up from ground until K1 pulls in. Disconnect the signal from the *right* input, and apply it to the *left* input. Press S1 to reset the circuit, and raise R4's wiper up from ground until K1 pulls in again. The circuit is now calibrated. Your calibration signal should preferably be a continuous tone, but a musical passage of fairly constant loudness will probably suffice. K1's contacts should be rated to carry a 3-amp to 5-amp load.

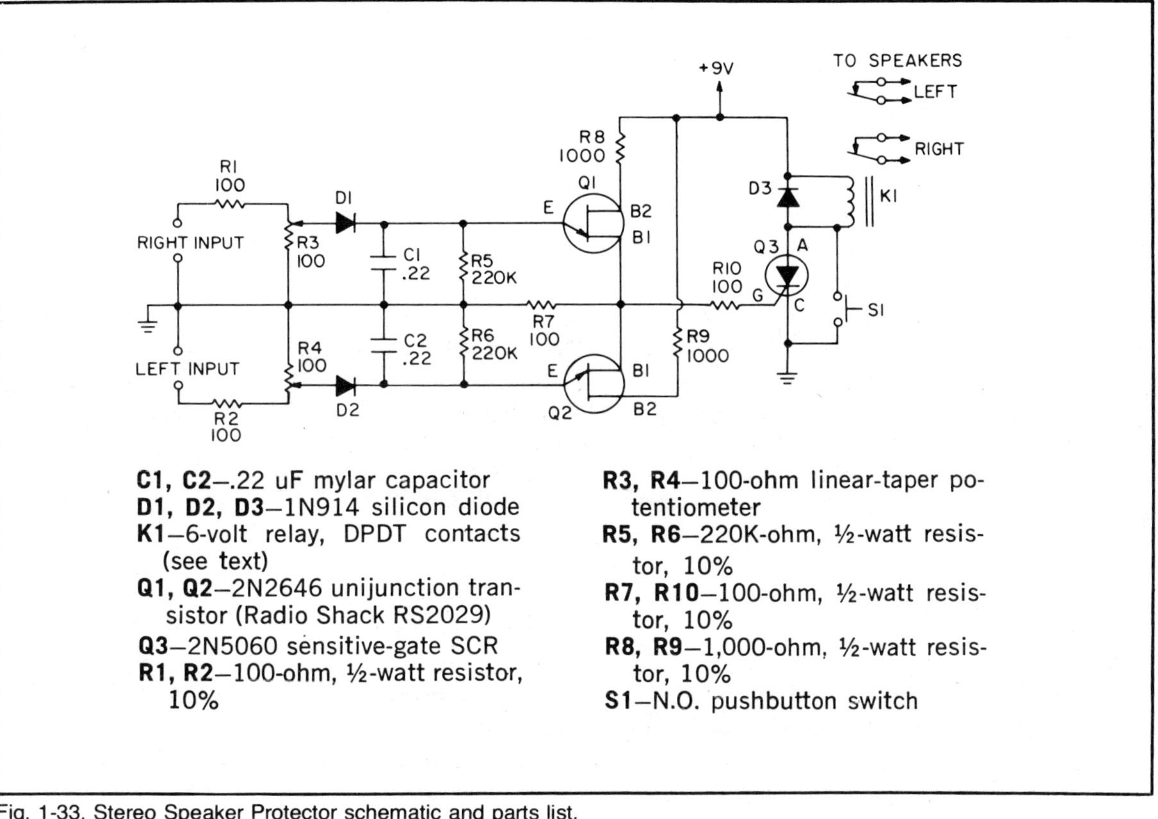

C1, C2—.22 uF mylar capacitor
D1, D2, D3—1N914 silicon diode
K1—6-volt relay, DPDT contacts (see text)
Q1, Q2—2N2646 unijunction transistor (Radio Shack RS2029)
Q3—2N5060 sensitive-gate SCR
R1, R2—100-ohm, ½-watt resistor, 10%
R3, R4—100-ohm linear-taper potentiometer
R5, R6—220K-ohm, ½-watt resistor, 10%
R7, R10—100-ohm, ½-watt resistor, 10%
R8, R9—1,000-ohm, ½-watt resistor, 10%
S1—N.O. pushbutton switch

Fig. 1-33. Stereo Speaker Protector schematic and parts list.

POOR MAN'S HOLD SWITCH

This is just one step more sophisticated than holding your hand over the telephone mouthpiece. We all find occasions when we would like to discuss something with the people in the room without sharing it with the party on the phone. This circuits provides dc continuity for the phone line to keep from losing a call when you hang the phone up. There is some danger, though, of putting the phone on "terminal hold," if you forget, because as long as you are switched to hold, it's just like leaving a phone off the hook: No one can call in, you can't call out.

Refer to Fig. 1-34.

Only two of the lines that reach your telephone are really part of the phone line, and these are most often the red and green wires that are in the cable between your phone and the wall. Other wires in the cable may carry power for lighting your phone, or may carry nothing. Check carefully. Also understand that if you make a connection to the phone line that inhibits the phone company's ability to provide service, they have the right to disconnect you for as long as they like. This is a proven, simple circuit that should cause no difficulty, but be careful.

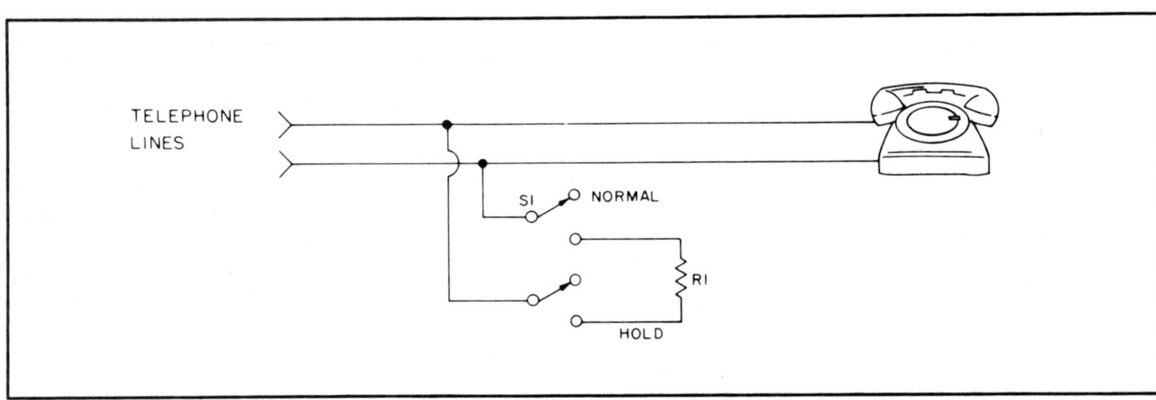

Fig. 1-34. Poor Man's Hold Switch schematic and parts list.

THREE-DIAL COMBINATION LOCK

Refer to Fig. 1-35.

Here's an effective little combination lock that you can put together in one evening's time. To open the lock, simply dial in the correct combination on the three rotary or thumbwheel switches. With the correct combination entered, current flows through R1 into Q1's gate terminal, causing the SCR to latch in a conductive state. This sends a current through relay K1, which responds by closing its contacts and actuating whatever load is attached. After opening the lock, twirl the dials of S1 through S3 away from the correct combination so that nobody gets a look at it. The lock will remain open and your load will remain on because the SCR is latched on. To lock things up, it's only necessary to interrupt the flow of anode current through the SCR by pressing pushbutton S4.

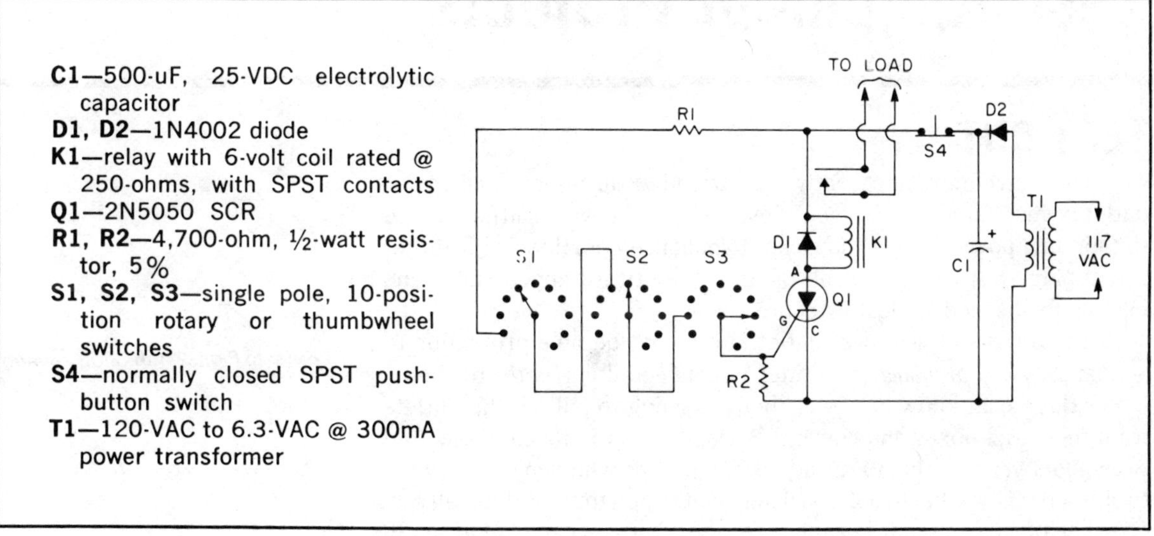

C1—500-uF, 25-VDC electrolytic capacitor
D1, D2—1N4002 diode
K1—relay with 6-volt coil rated @ 250-ohms, with SPST contacts
Q1—2N5050 SCR
R1, R2—4,700-ohm, ½-watt resistor, 5%
S1, S2, S3—single pole, 10-position rotary or thumbwheel switches
S4—normally closed SPST pushbutton switch
T1—120-VAC to 6.3-VAC @ 300mA power transformer

Fig. 1-35. Three-Dial Combination Lock schematic and parts list.

Chapter 2

Test Equipment Projects

TEST OUT

The convenient, apparently very friendly, three-prong ac power outlet in your home may *kill* you! Yes, it sits in the wall waiting for you to plug in a power tool or household appliance complete with three-prong plug, you trusting to all of its safe outward appearances and ending up shocked to death's door.

The three-slot ac power outlet offers considerable protection to appliance users *provided* the outlet is connected correctly to the ac lines. But we all know hardly anybody is going to pull all the outlets from their wall boxes and check the wiring; it's too much work. And what about your neighbors, relatives and friends who don't know what to check or know what to do! You don't want to pull their outlets, also.

Refer to Figs. 2-1 through 2-4 and Table 2-1.

The obvious answer is a quick test set that you can plug in safely to a wall outlet to give you a visual indication that the outlet is wired correctly. That's what Test-Out, a handy self-contained visual indicator, does in seconds and you can build it cheaply.

What It Does. Test-Out is a neon bulb indicating device that is plugged into the wall outlet. When the indication is normal, the outlet is wired correctly and you can so unplug it and go to the next outlet. When the indication is other than normal, the color-coded neon indication lets you know what is wrong and tells you what to do to make it safe.

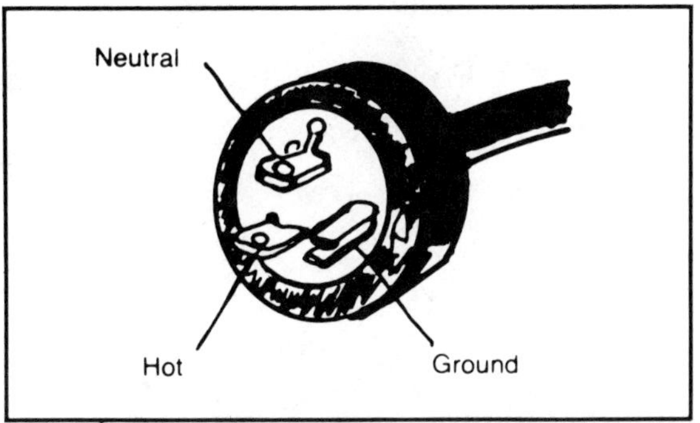

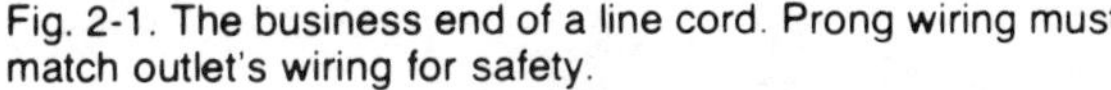

Fig. 2-1. The business end of a line cord. Prong wiring must match outlet's wiring for safety.

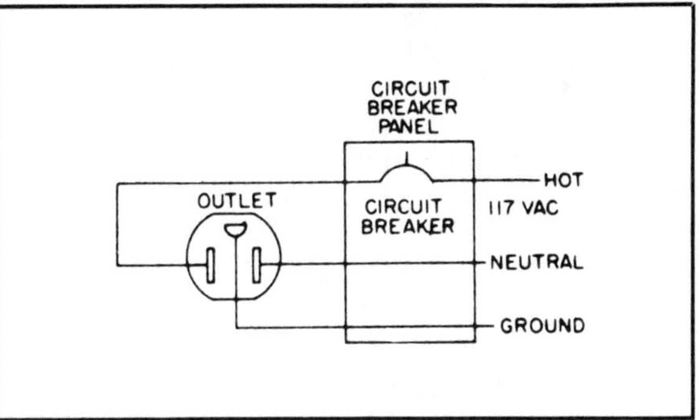

Fig. 2-2. Here's how your house's wiring diagram would look if it had only one outlet.

The ac outlet is where it's all at, so you have to know about its wiring hookup before you proceed. This drawing shows the wiring of typical outlet. In your home, almost all of the outlets are in-wall installations with the wall plate flush against the wall and duplex outlet plastic mold protruding slightly. The three wires in the box connect to the outlet—the black (hot) wire to the brass screw, the white (neutral) wire to the chrome-plated screw, and the green or bare (ground) wire to the green-painted screw. When wired in this fashion, the outlet is connected as shown here.

A lamp connected to the hot terminal and to either remaining slot, neutral or ground, will be illuminated. This is exactly what happens in Test-Out. When Test-Out is plugged into an outlet that is correctly wired, both the green and orange lights come on. See the schematic diagram. Trace the circuit for yourself. Now imagine that the outlet into which Test-Out is plugged has the *hot* and *neutral* wires reserved. The *green* and *red* lights will come on. This is a common wiring fault and should be corrected whenever it occurs. Some other less common, but still dangerous wiring faults or bad connections that Test-out can detect in a wall outlet are *open-ground* circuit, *hot* and *ground* connec-

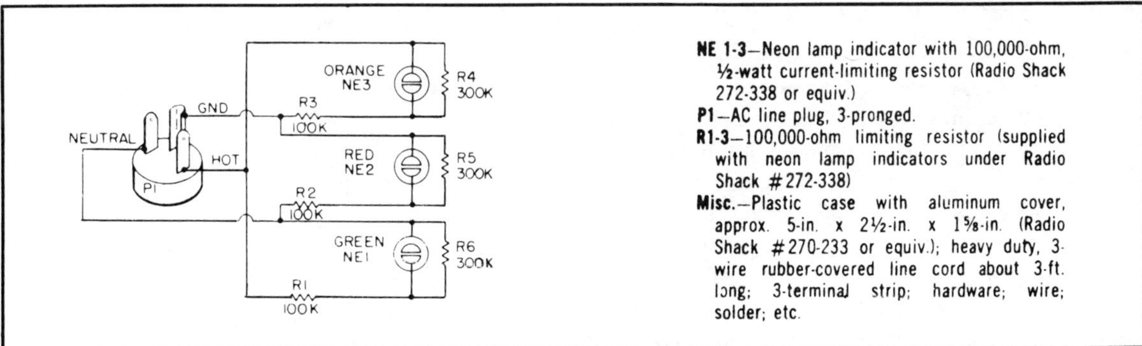

Fig. 2-3 Ac Outlet Tester schematic and parts list.

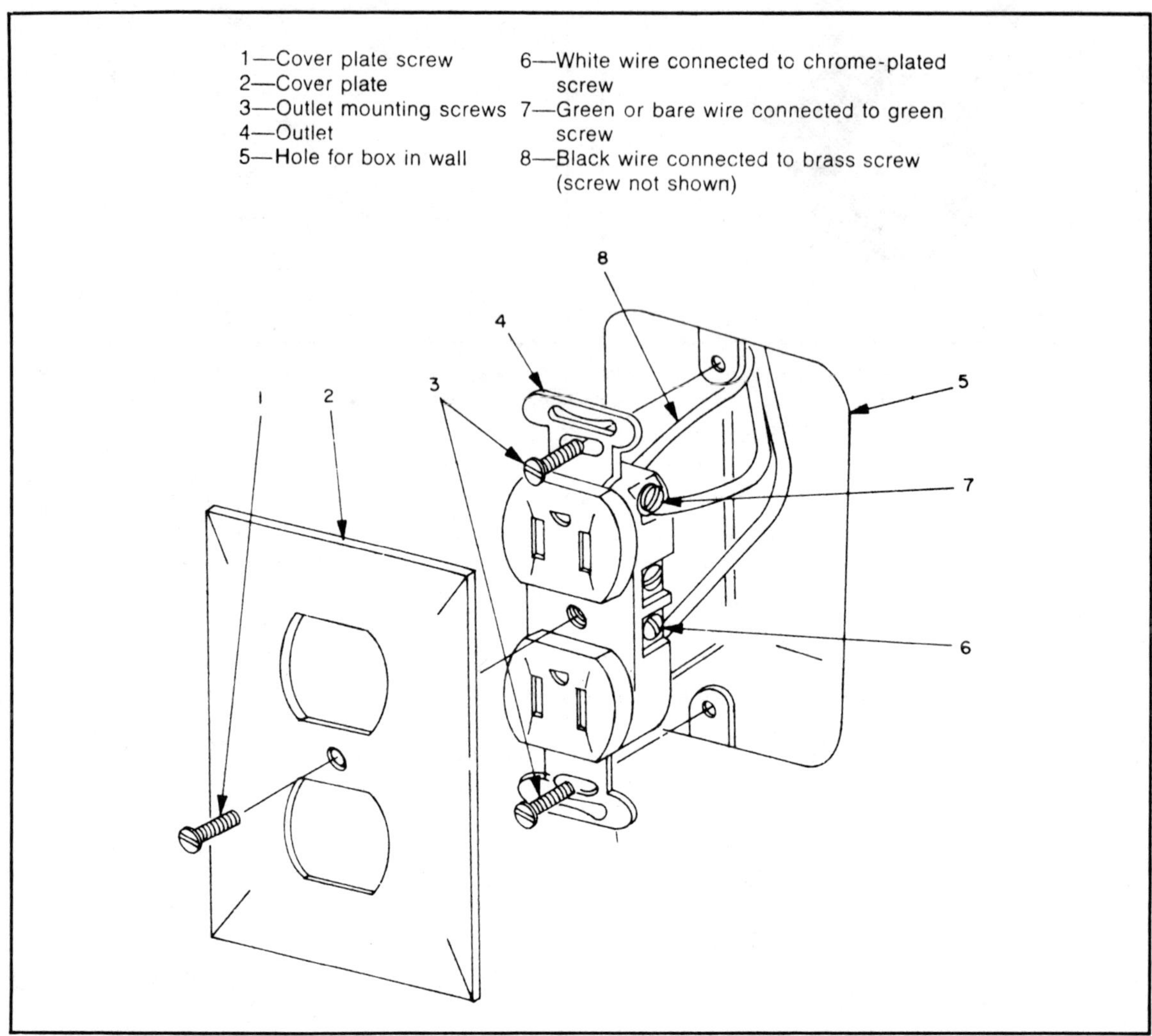

Fig. 2-4. This is what the inside of your wall outlet box should look like. Use your head and stay alive—do not open up the wall outlet until you disconnect the power to the outlet. Check wiring carefully. It will be easier in new homes. Be sure screws are tightly secured on the wire leads and the cover plate is installed.

tions *reversed, open-neutral* connection, *neutral* connection *hot* while the *hot* connection is *open,* and *hot open* or no-power. Each possible fault has its own light pattern indication which is given in the table. When you make your own Test-Out unit, copy the table and paste it on the case for rapid trouble and correction information.

Assembly of Test-Out. Building Test-Out is simple. The black plastic case with aluminum cover measures approximately 5 × 2½ × 1⅝-inches and has three rectangular neon lamp sets with external limiting resistors mounted on the aluminum cover, with the green lens on the left, red in middle, and orange (amber) at right. A hole is drilled

Table 2-1. Ac Outlet Fault Table.

WHAT IT MEANS	GREEN	ORANGE	RED
WIRING OKAY	●	●	◎
HOT & NEUTRAL REVERSED	●	◎	●
OPEN GROUND	●	◎	◎
HOT & GROUND REVERSED	◎	●	●
OPEN NEUTRAL	◎	●	◎
NEUTRAL IS HOT HOT IS OPEN	◎	◎	●
HOT OPEN OR NO POWER	◎	◎	◎

BULB ON ● BULB OFF ◎

in the box for the heavy-duty line cord to pass. The line plug is also heavy-duty type with built-in wire clamp. Overbuilding here is important because the line cord and plug will take considerable pulls and strain in the normal course of using Test-Out. Don't get cheap material here! A three-terminal strip will make wiring easier.

Paint the aluminum cover any light color and screw cover to box when wiring is complete. Check unit by applying power first to hot prong and neutral prong on the plug. The green light should go on. Now switch the neutral connection to the ground prong. The orange light should come on. Lastly, the power leads should be connected across the neutral and ground leads—the red light should come on. If all is well, Test-Out is ready for work after the handy-reference table is copied and cemented on Test-Out's aluminum panel.

Put Test-Out to work at once. You will be surprised how many outlets are improperly wired. Be sure to throw off the correct circuit breaker before rewiring an outlet.

LOW-COST RESISTANCE DECADE

Refer to Figs. 2-5 and 2-6 and Table 2-2.

Every home electronics experimenter should have a set of decade resistance boxes, with values from one ohm to one megohm. Looking through the electronic parts catalogs, you will find decade boxes in many sizes, colors, and degrees of accuracy. They all have switches from four to 10 resistors for each decade. Some use as few as four resistors (by use of a special switch or complex circuit they add the resistors together), while other decade boxes use separate resistors for each switch position.

Here is the simplest, cheapest, and most easily made resistance decade box imaginable. You will be able to get resistance values from 1 to 10 units (1 ohm, 2 ohms, 3 ohms, etc.; 10 ohms, 20 ohms, 30 ohms, etc.) with five resistors per decade. If you have to buy everything you can build each decade for very little money.

There are no switches used here; the different values are determined by which of the six banana jacks you insert a banana plug. The application of Ohm's law for resistors in series (simple addition) will

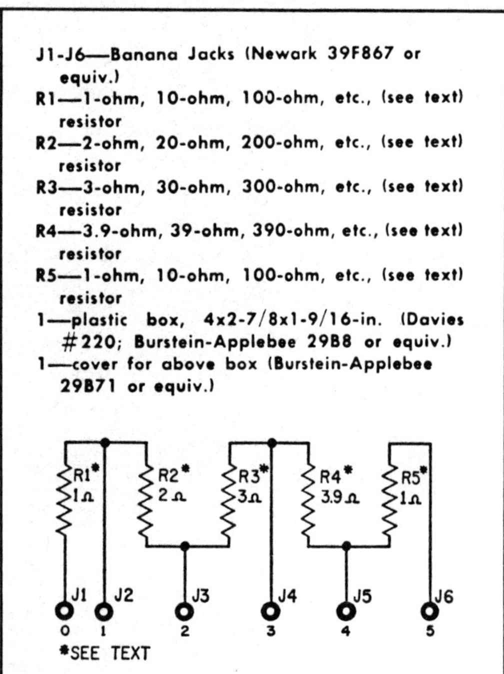
J1-J6—Banana Jacks (Newark 39F867 or equiv.)
R1—1-ohm, 10-ohm, 100-ohm, etc., (see text) resistor
R2—2-ohm, 20-ohm, 200-ohm, etc., (see text) resistor
R3—3-ohm, 30-ohm, 300-ohm, etc., (see text) resistor
R4—3.9-ohm, 39-ohm, 390-ohm, etc., (see text) resistor
R5—1-ohm, 10-ohm, 100-ohm, etc., (see text) resistor
1—plastic box, 4x2-7/8x1-9/16-in. (Davies #220; Burstein-Applebee 29B8 or equiv.)
1—cover for above box (Burstein-Applebee 29B71 or equiv.)

Fig. 2-5. Low-cost Resistance Decade schematic and parts list.

Table 2-2. Jacks to Use for Specific Resistance.

Resistance	Jacks	Resistance	Jacks
1 ohm	0 and 1	6 ohms	0 and 3
2 ohms	1 and 2	7 ohms	2 and 4
3 ohms	2 and 3	8 ohms	2 and 5
4 ohms	3 and 4	9 ohms	1 and 4
5 ohms	3 and 5	10 ohms	0 and 4

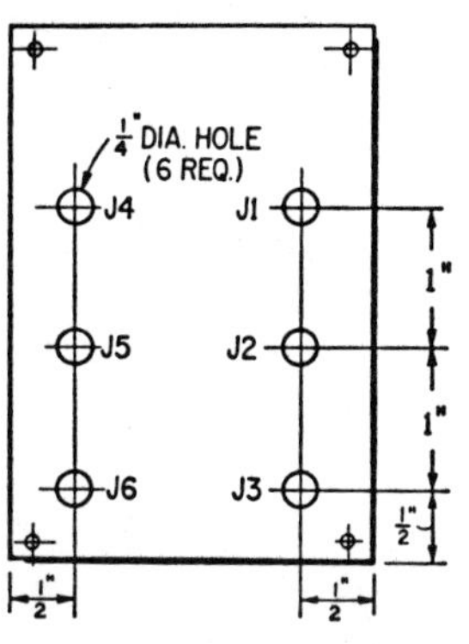

Fig. 2-6. Layout of phenolic cover is not at all critical. You can even change positions of jacks if it makes wiring any easier for you. Metal cover will need insulation shoulder washers.

show how we are able to get ten values using only four values of resistance.

The table indicates which jacks to use for each value. The decade box shown in the photographs uses resistors with a ±5% tolerance.

You can make the front panel according to the layout drawing shown. Uninsulated banana jacks can be used if you use the black plastic panel; otherwise, they must be insulated.

The table explains which banana jack to use for each resistance value. You can copy the table from this magazine and cement it to the side or back of the box for future reference. It can then be covered with transparent tape to protect it from dirt, dust, and handling.

After you have a set of decade boxes, you will find it much easier to breadboard experimental circuits or to find the value of unknown resistors needed to replace burned out resistors. You may use resistors higher than 1 watt if you desire and any resistance tolerance, depending on the ultimate need you may have for resistance decade boxes.

LO-CAP PROBE

"What you see is what you get" might be all right for a television comic, but it's not necessarily true when you use an oscilloscope. It is unfortunate, but true, that a scope's performance is specified from the input terminals to the scope itself, but does not include the test probe or connecting wires. For this reason, a service-grade scope rated out to 4 MHz, or 7 MHz, or even a laboratory scope rated out to 20, 50, or 100 MHz, might poop out on something as mundane as a 60-Hz square wave, delivering a CRT display with rounded leading edge while the real waveform is truly square. Worse than that, connecting your scope into an rf circuit may completely change the loading, or tuning of the circuit which is under test.

Refer to Fig. 2-7.

Here's Why. Forget for a moment that the scope has a frequency-compensated input. That has no bearing on your measurements, which is affected by the cable between the circuit being tested and the scope input. An ordinary shielded test lead approximately 3 feet long has a capacity of about 100-300 pF, depending on the type of shielding. If a "bare" test lead is connected into a circuit, it is effectively loading the circuit with 100-300 pF. Just imagine what this will do to an rf circuit, or any high-frequency circuit from about 10 kHz and upward. "What you see *isn't* what you get in this case."

Also, consider the average scope's 1-megohm "high-impedance" input. "High impedance" is a relative term: one equipment's "high impedance" is another's "low impedance." For example, imagine a transistor or integrated-circuit amplifier with a 500k or 1-megohm bias or feedback resistor. Connecting a scope's input across either value will completely change the operating parameters of the circuit. Or imagine what a 1-megohm "load" across a tuned rf input circuit will do: The Q might drop like a rock, not to forget the detuning effect of the test lead capacitance of the lead itself.

Follow the Labs. Commercial labs get around both the capacity loading and 1-megohm impedance by using a "10X low-capacity" test probe for the scope input. This device does two things: It makes the input capacity to the scope's test lead appear to be about 5-10 pF, and it raises the input impedance into the test cable—the impedance seen by the circuit being tested—to nominally 10 megohms (a value that won't affect any circuit being used or tested).

Easy to Build. A 10X Low-Capacity Test Probe circuit is shown here. Basically, it consists of two components: trimmer capacitor C1 and resistor R1. C1 is generally any small trimmer with a maximum capacity in the range of 25-50 pF. R1 should be 9 megohms for a precise 10:1 voltage division: ie: the scope will indicate 1 volt P-P if the input to the cable is 10 volts P-P. However, 9 megohms, or anything close, is usually unattainable by the hobbyist. If you substitute a 10 megohm 5 percent resistor for R1 the accuracy will be sufficient for almost all applications (nominal voltage readout error will be about 10 percent).

A Shielded Probe. The 10X probe must be assembled in a shielded test probe. If not shielded, hand capacity will induce "hum" into the signal and add capacity loading to the circuit.

Temporarily mount C1 to the perfboard and see if you can slide the shield over the assembly without having the shield short the trimmer capacitor. If it touches a metal part of C1, file the edges of the perfboard so it will sit lower in the sleeve and not short C1. When the shield can slide over the assembly, secure C1 to the board with flea clips. Install R1 across the C1 flea clips on the opposite side of the board (there isn't room for C1 and R1 on the same side).

Solder about 3 inches of solid No. 20 or No. 22 wire to the front flea clip, the one on the opposite end from the solder lug which is factory installed on the perfboard. This wire will eventually connect to the test probe tip.

Cut a piece of shielded wire to about 3 feet. You can use an ordinary audio patch cable with the phono plugs cut off the ends. Solder the center conductor to the rear flea clip. Solder the shield to the solder lug and bend the solder lug at right angle to the perfboard. Make certain

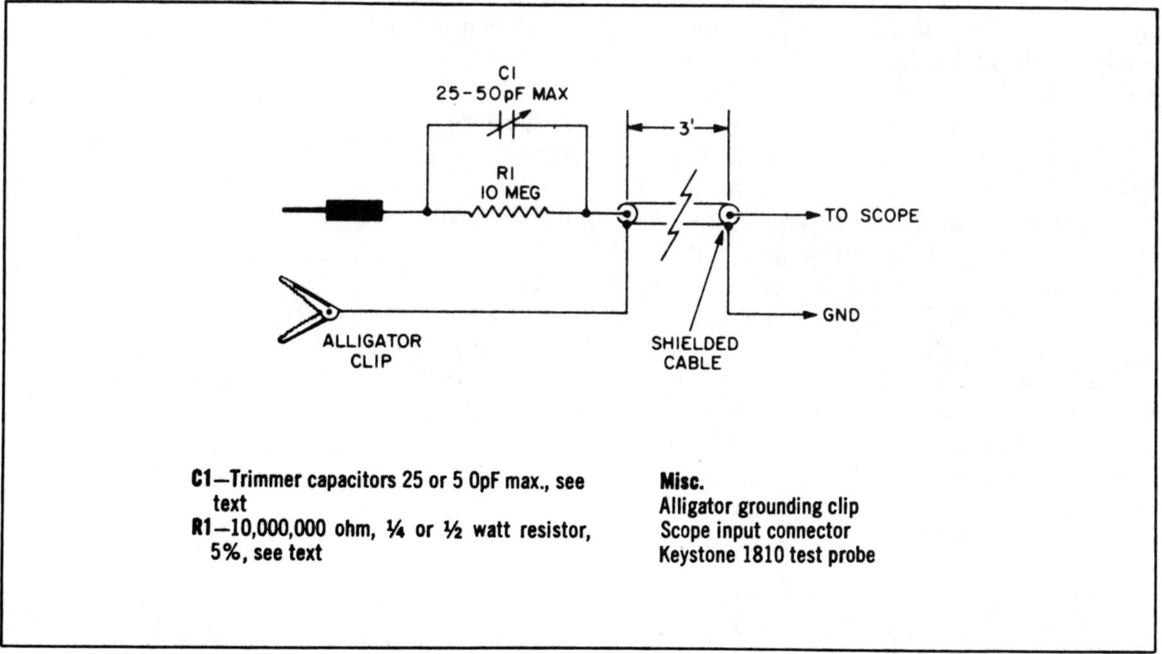

Fig. 2-7. Lo-Cap Probe schematic and parts list.

when you solder wires to the flea clips that C1 and R1 are also soldered to the clips.

Slide the probe shield over the perfboard from the front until it touches the solder lug. Carefully mark the sleeve directly over the trimmer capacitor's adjusting screw. Remove the sleeve and drill a ¼-hole at the mark (careful, the sleeve is very thin). Solder an insulated stranded wire approximately 8 inches long to the solder lug's grommet—where it's secured to the perfboard. (This wire will pass out the hole in the rear of the probe cover and will connect to an alligator ground clip) that you use.

Now slide the shield over the perfboard, press it against the solder lug, and tack solder the shield to the solder lug. Do not fold the lug over the shield as it might prevent the cover from being slipped into place. Screw the probe tip into the probe's front cap, and then thread the solid wire from the perfboard through the probe, pulling on the wire so the perfboard is tight against the cap. Secure the wire to the probe tip. Measure the distance from the cap to the hole in the shield and transfer this measurement to the probe cover. Drill a ¼-inch hole in the cover at the mark. This will be the access hole for the capacitor C1. Next, assemble the probe and install the required connector (to match your scope's input) at the free end of the shielded cable coming out the back.

Alignment. You must align the low-capacity probe using some form of square waveform in the range of 60-1000 Hz. This can come either from the calibration voltage built into your scope or the square wave output of a sine-square signal generator. You can even use a broad pulse from a pulse generator if you have such an instrument in your workshop.

Touch the low-capacity probe to the square waveform output, adjust your scope for a convenient CRT display, and then using an insulated alignment screwdriver, adjust C1 for a perfectly square *leading edge*. If you have too much capacity the leading edge will be rounded. If you have too little capacity the leading edge will peak. Perfect adjustment is a *perfectly square leading edge*. Once C1 is adjusted it need never be changed as long as the same scope is used.

Using the Probe. Remember to multiply the CRT voltage indication by 10 to obtain the correct voltage at the test probe. For example, if the scope is set for 1 volt per division, and the peak-to-peak waveform is 1.5 divisions, the actual voltage at the test probe is 1.5 volts p-p × 10, or 15 volts p-p.

COUNT CAPACITA

Maybe your junk box looks like a haunted mansion? Full of mystery and intrigue? Do you sometimes wonder just what values all those surplus or unlabeled capacitors really are? All the VOMs, frequency meters, power meters, FETVOMs and tachometers in the world aren't going to help you here. What you need is a visit from the Count—*Count Capacita*—our own toothsome capacitance meter.

Refer to Figs. 2-8 through 2-11.

You can use this capacitance meter to separate good capacitors from bad ones in your junk box. In addition, if you ever have to repair a television or radio, Count Capacita will quickly put the bite on a defective capacitor, thus saving you the expense of a repair bill in the process. Last, but certainly not least, the Count will enable you to purchase surplus capacitors, and this is where you can really save money.

Surplus capacitors are sold at discount rates, usually by mail-order dealers, for several reasons. First, suppose an audio manufacturer decides to completely phase out his old capacitively coupled amps in favor of direct-coupled designs. His inventory of new and perfectly good capacitors is now useless to him, so he disposes of the lot on the surplus market. Second, sometimes a capacitor manufacturer wants to get rid of old, mislabelled, or out-of-tolerance units. He can do this on the surplus market. You can take advantage of the savings—often more than 75 percent if you know the Count. With our meter, you can spot the mislabelled or out-of-tolerance units, identify unmarked devices, and eliminate the occasional defective unit. If you do much experimenting, your savings may soon pay for your capacitance meter.

Transylvanian Circuitry. Let's begin discussion of this particular circuit with the block diagram. The circuit is driven by a free-running oscillator that generates short-duration, negative-going pulses. These pulses are spaced by a time interval T_2. Now, T_2 is controlled by the capacitor under test. Specifically, T_2 is equal to k_1C, where k_1 is just a constant of proportionality. At the monostable's output, there is a rectangular waveform that is high for a time T_2, and low for a time equal to $(1_1\text{-}T_2)$. This waveform is then time-averaged to yield a meter current equal to $(k_2T_2)/T_1$, where k_2 is another constant of proportionality. Since T_2 is equal to k_1C, it follows that meter current It must also equal $(k_1k_2\text{C})/T_1$. Therefore, there is a direct relationship between meter deflection and capacitance; by choosing the right values for k_1, k_2, and T_1, you get a capacitance readout.

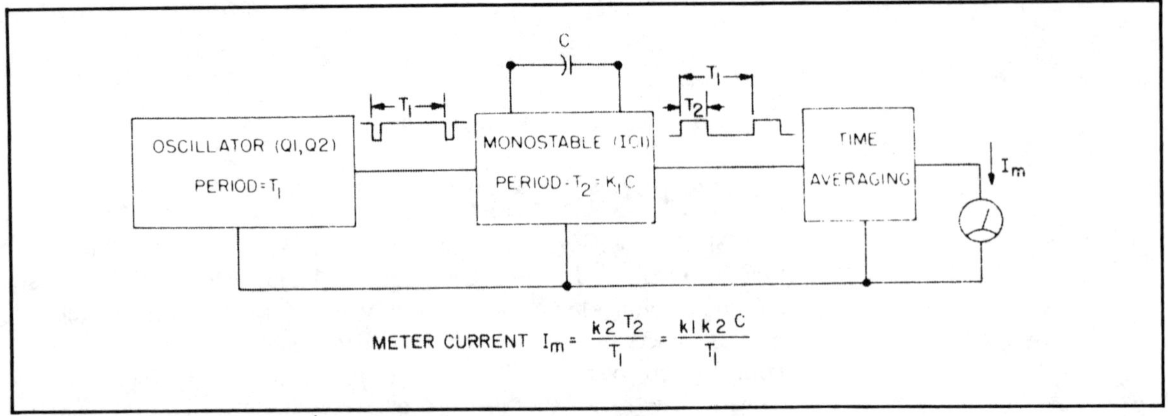

Fig. 2-8. Count Capacita's circuit is driven by a free-running oscillator that generates short-duration, negative-going pulses. The pulses are spaced by a time interval T2 controlled by the capacitor being tested. The waveform is averaged to yield a meter current that changes with varying capacitance moving the meter's pointer to the value.

The Count's various constants have been chosen to allow a useful measurement range that spans from less than 100 picofarads to 5000 nanofarads (5 microfarads). In case you are unfamiliar with the above nomenclature, one microfarad is one-millionth of a farad, the standard unit of capacitance. It takes a thousand picofarads to equal one nanofarad, and a thousand nanofarads to equal one microfarad. The scales on this meter measure capacitance in terms of picofarads and nanofarads; with this information, you should be able to easily convert between units when necessary.

Let's now consider the schematic diagram. Assume that switch S2 is in its *battery* position and that S3 is pressed down. Battery current will flow into meter M1 through resistor R2, and M1's deflection will indicate whether or not the batteries are good. Fresh batteries will provide a meter indication of about "45"; batteries should be changed when the indication drops below "33", or thereabouts. Now, flip S2 mentally back to its *capacitance* position, and let's proceed with the rest of the circuit.

Battery current flows through resistor R1 to yield a regulated 6.2-volt supply potential across zener diode D1. Capacitors C1 and C2 bypass the supply and stabilize the circuit. The free-running oscillator is composed of unijunction transistor Q1 plus associated components. Timing capacitor C5 is charged through R13 and R14, or R15 and R16, depending on the setting of *range* switch S1. When the voltage on C5 reaches a specific level, Q1's emitter breaks down to a low impedance, thus discharging C5 through resistor R11. When the capacitor has been discharged to a sufficiently low level, Q1 ceases to conduct, and C5 once again begins to charge. This charging and discharging of C5 proceeds alternately, causing a voltage spike to appear, across R11 each time C5 discharges. Transistor Q2 inverts and amplifies the pulse, which is applied to the inputs (pins 2 and 4) of monostable IC1.

The monostable's period is determined by the capacitor under test in conjunction with a resistor—either R5, R6, R7 or R8—selected by *range* switch S1. In operation, the capacitor being tested first gets connected across a pair of binding posts, and then S3 is pressed to take a reading. You will note that these binding posts are polarized, with BP1 being positive and BP2 (which connects to ground) being negative. This is an important consideration with polarized capacitors such as aluminum and tantalum electrolytics: The capacitor's positive terminal must connect to BP1. Reserve connection is harmful to such capacitors, so be careful. The standard non-polarized capacitors—mica, paper, mylar, polystyrene, ceramic and glass—may be connected across the binding posts in either direction.

Diode D2 functions to provide a quick discharge of the capacitor under test when S3 is released. Monostable IC1's output, pin 3, drives meter M1 through R3. Averaging of the pulses is accomplished by capacitor C3 across M1. Finally, diode D3 ensures that no current is emitted from IC1's output when it drops low (to about a tenth of a volt).

Since this is not a temperamental circuit, you should have few problems with its construction. One point that you should bear in mind, however, is that the binding posts must connect to the rest of the circuitry via short and direct wires spaced at least an inch apart. This

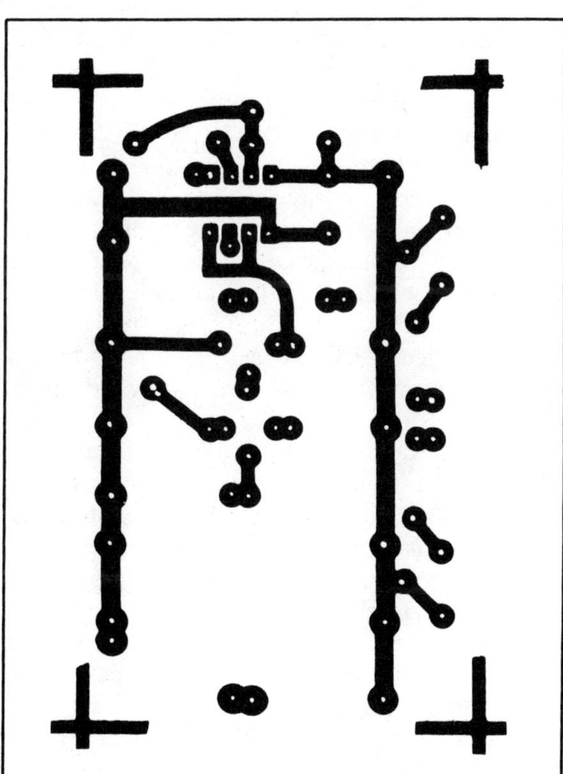

Fig. 2-9. PC board template for Count Capacitor. Use either photo-etch materials or just use a resist marking pen.

minimizes stray capacitance between the binding posts and maintains good accuracy on the lowest range (pf. X 10).

As specified in the parts list, resistors R2 and R3 must have 5 percent tolerances. Likewise, the tolerances of R5, R6, R7 and R8 must be at least 5 percent. If you desire, 1 percent precision resistors could be used for R5 through R8. This will improve accuracy somewhat on the four lowest ranges, but it will also be more expensive. You won't be needing hair-splitting precision, so 5 percent-tolerance resistors should be quite adequate here.

Although it might seem more difficult at first, printed-circuit construction is far and away the most convenient method of assembly. For your convenience, a PC foil pattern is provided, and it may be used in conjunction with a printed-circuit kit from any of the electronics retailers. An equally effective construction method involves the use of perfboard. Either technique is capable of turning out a small, neat circuit board.

When wiring the circuit, be careful to install all polarized devices in the correct orientation. This applies to all the semiconductors, meter M1, the batteries, and electrolytic capacitors C1 and C3. Basing diagrams for all the semiconductors are shown here. Lead identification for transistor Q1 applies *specifically* to a 2N2646. If you use a Radio Shack RS2029 for Q1, note that it uses a different lead orientation, which is clearly illustrated on the package in which it is sold. Though their lead orientations are different, these two transistors are electrically equivalent and interchangeable.

Although it is not absolutely necessary, the use of a socket is advisable for IC1, especially if you haven't had much experience soldering integrated circuits. The socket, as well as most of other components in the parts list, is available at Radio Shack. Two of the components, S1 and the case, may be purchased by mail from Circuit Specialists (see the parts list for their address). Circuit Specialists carry a tremendous assortment of electronic devices, and they cater to the experimenter by not imposing a large handling charge on small orders. You can obtain their catalog by writing to the address in the parts list.

Under the Lid. During construction, do not substitute for meter M1 unless the device you intend to use has a full-scale sensitivity of 50 microamps and an internal resistance of about 1500 ohms. As usual, you should make all connections with a 25-watt iron and resin-core solder. When wiring S1, make sure that the rotor of S1b engages R8 in the fully counterclockwise (CCW) position, and R5 in the clockwise (CW) position. Also, S1a's rotor must contact R16 when fully clockwise, and R14 in all other positions. You may then label S1 according to the diagrams provided here, with the lowest range in the extreme CCW position. Finally, be certain to label BP1 with a "+" and BP2 with a "–".

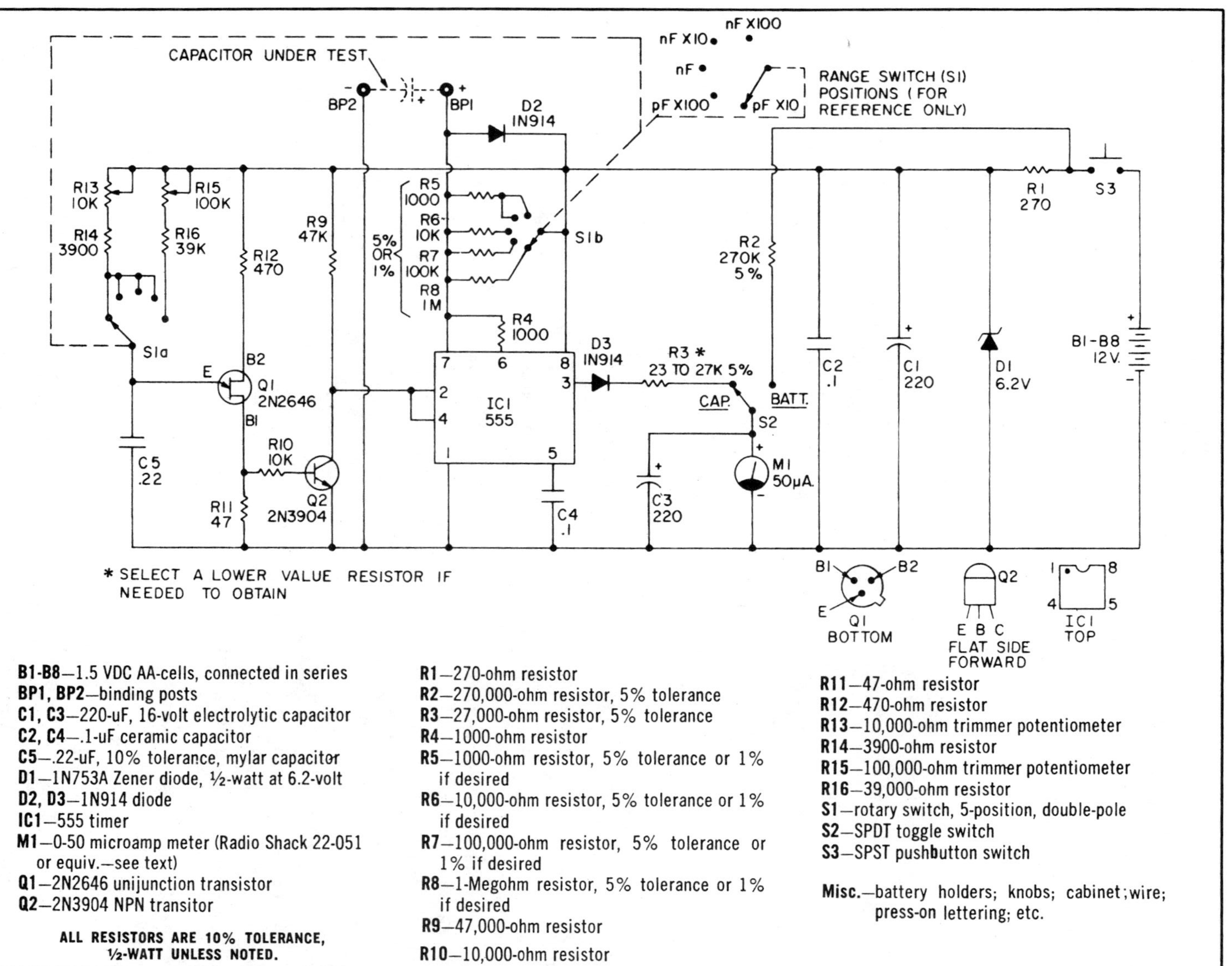

B1-B8—1.5 VDC AA-cells, connected in series
BP1, BP2—binding posts
C1, C3—220-uF, 16-volt electrolytic capacitor
C2, C4—.1-uF ceramic capacitor
C5—.22-uF, 10% tolerance, mylar capacitor
D1—1N753A Zener diode, ½-watt at 6.2-volt
D2, D3—1N914 diode
IC1—555 timer
M1—0-50 microamp meter (Radio Shack 22-051 or equiv.—see text)
Q1—2N2646 unijunction transistor
Q2—2N3904 NPN transitor

ALL RESISTORS ARE 10% TOLERANCE, ½-WATT UNLESS NOTED.

R1—270-ohm resistor
R2—270,000-ohm resistor, 5% tolerance
R3—27,000-ohm resistor, 5% tolerance
R4—1000-ohm resistor
R5—1000-ohm resistor, 5% tolerance or 1% if desired
R6—10,000-ohm resistor, 5% tolerance or 1% if desired
R7—100,000-ohm resistor, 5% tolerance or 1% if desired
R8—1-Megohm resistor, 5% tolerance or 1% if desired
R9—47,000-ohm resistor
R10—10,000-ohm resistor
R11—47-ohm resistor
R12—470-ohm resistor
R13—10,000-ohm trimmer potentiometer
R14—3900-ohm resistor
R15—100,000-ohm trimmer potentiometer
R16—39,000-ohm resistor
S1—rotary switch, 5-position, double-pole
S2—SPDT toggle switch
S3—SPST pushbutton switch

Misc.—battery holders; knobs; cabinet; wire; press-on lettering; etc.

Fig. 2-10. Count Capacita schematic and parts list.

When construction is complete, there are two calibration adjustments that must be made. In order to make these adjustments, you will need two accurate reference capacitors. The first, which will be used to calibrate the highest range, should have a value between 2 and 5 microfarads—the higher the better. Commonly available capacitors in this range are generally mylar or electrolytic. The mylar is your best choice; pick a unit with the highest tolerance you can find. In this capacitance range, that means about ±10 percent—sometimes better. If you must go with an electrolytic, choose a tantalum device and avoid the aluminum electrolytics, which tend to be leaky and have poor tolerances. Common tolerances for tantalums run about ±20 percent, so you can see why the mylar is the better choice.

For calibration of the lower four ranges you will need another reference capacitor; since calibration can take place on any of the four ranges, you have some leeway in your choice of a calibration capacitor for these lower ranges. One especially good choice is a 5000 picrofarad polystyrene capacitor, available from just about all of the large electronics retailers. This particular capacitor is cheap but precise (±5 percent tolerance). The steps that follow will use this capacitor, but remember that you can use any capacitor as long as it is accurate and its nominal capacitance falls at the high end of one of the scales.

Begin calibration of the lower ranges by connecting the 5000 picofarad polystyrene capacitor to BP1 and BP2. Set trimmer R13 to the midpoint of its range of adjustment. Make sure that S2 is in its *capacitance* position, and that *range* switch S1 is set to PF. X 100. Press S3 and adjust trimmer R13 for a full-scale indication of "50" on M1. This completes calibration of all four lower ranges.

Calibration of the top range is similar to the above. Hook up your capacitor, and set R15 to its midpoint. Make sure that S2 is set to *capacitance*, and that S1 is fully clockwise. Press S3 and adjust trimmer R15 until your meter indication corresponds to your capacitor's marking. This finishes the calibration.

Use of Count Capacita is fairly obvious; nevertheless, here are a few odds and ends that you might find helpful. The maximum voltage appearing across any capacitor under test is about 4.2 volts, which is well below the rated working voltage of almost any capacitor that you are likely to encounter. Because battery current drain is intermittent and moderate, the cells will last a long time, possibly for years. However, it might be a good idea to replace batteries once a year, even if they indicate more than "33," in order to prevent the possibility of a battery leak inside your meter.

Whenever you make a measurement, start on a range high enough to accommodate the capacitor being tested. If you have no idea of the capacitor's approximate value, always start on the highest range. Should a capacitor be opened up internally, it will provide a reading of zero on all scales.

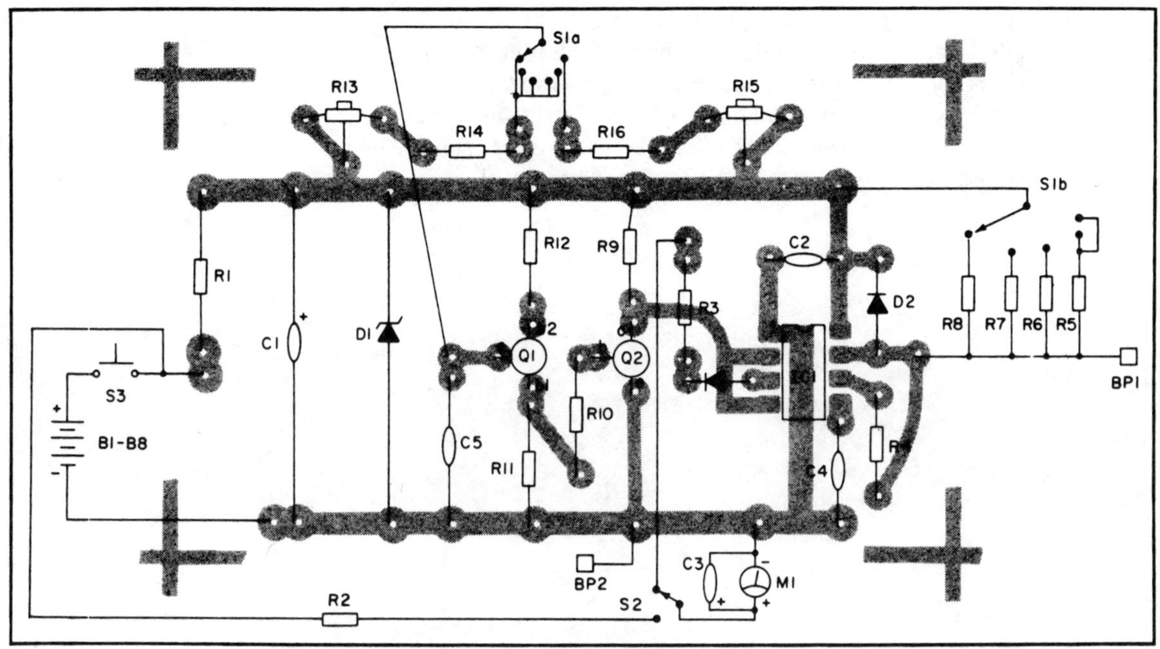

Fig. 2-11. Component layout. Count Capacita will bring back to life all of those once-useless, unmarked capacitors.

If the capacitance is leaky, its measured capacitance will be considerably larger than the value stamped on its case. This is because capacitor leakage is equivalent to having a resistor in parallel with the capacitor. This leakage resistance siphons off capacitor current, so the capacitor takes longer to charge, and monostable IC1's output stays high for a longer time. The result is an erroneously high capacitance reading. By the same token, you can expect an internally shorted capacitor to pin the meter's needle on all scales, since a short is, in effect, just a case of complete leakage.

Now, let's return to an important topic that was introduced earlier: stray capacitance between the binding posts. The construction details already presented should help to keep strays at a minimum; however, you can never completely eliminate stray capacitance or the errors it may cause. Fortunately, it is very simple to compensate for such errors.

After your meter is calibrated, turn to the most sensitive range: Picofarads × 10. This is where the effects of stray capacitance will show up. Without any external capacitor between the binding posts, press the pushbutton and note meter M1's indication. On the prototype, a reading of 30 picofarads was obtained. This represents the value of the stray capacitance in parallel with any capacitor under test. It also represents the amount by which any capacitance reading will be in error. To compensate, simply subtract the residual capacitance from any given meter reading. For example, a reading of 480 pF on the prototype meter would be corrected to 450 pF (480 pF minus 30 pF). Such corrections are significant and necessary only on the most sensitive scale. Finally, since stray capacitance can obviously affect accuracy on the most sensitive scale, it is preferable that you *not* calibrate there, but on one of the higher scales, as outlined previously.

BUILD A SIMPLE VOLTMETER AND SCOPE CALIBRATOR

Refer to Figs. 2-12 through 2-14 and Tables 2-3 and 2-4.

Precision voltage measurements require a calibrated source against which to compare the readings of the voltmeter of oscilloscope. In really high-class measurements, where absolute accuracy is needed, laboratories will use something like a Weston cell and a precision potentiometer. But to the hobbyist, such instruments are both too costly and, in most cases, more accurate than is necessary. In the past, the hobbyist had to be content with zener diode calibrators. Unfortunately, these diodes are not the best and tend to drift. But today, a new breed of regulator is available. Several manufacturers are now offering regulator/reference source ICs using *band gap* zener diodes, and internal amplifiers. These ICs give the hobbyist a low-cost method for building a reference voltage source.

Calculate Your Needs. The basic circuit shown is sufficient to operate as a hobbyist-grade voltage calibrator. Only a power supply (in this case a battery), a resistor, the regulator IC, and a means for turning it on and off are required.

The value of the series resistor depends upon the reference current selected and the power supply voltage. The reference current may be set at any point in the range of 2 to 120 milliamperes, provided that the overall power dissipation is kept to less than 300 milliwatts. In practice, however, one is advised to select a value in the 2 to 5 mA range. In the example of Fig. 2-13 we have selected 8.75 mA for a very special, high level, technical reason—we had a 4.2-volt battery and a 200-ohm resistor in the junkbox at the time.

The series resistor's value is computed as:

$$Rl = \frac{Eb - Eo}{Ir}$$

Where:
Eb is the battery voltage
Eo is the output voltage (1.26 or 2.45-volts)
Ir is the reference current
Rl is the resistance in series with the IC

Example:

In the basic circuit shown, we used a 4.2-volt mercury battery and

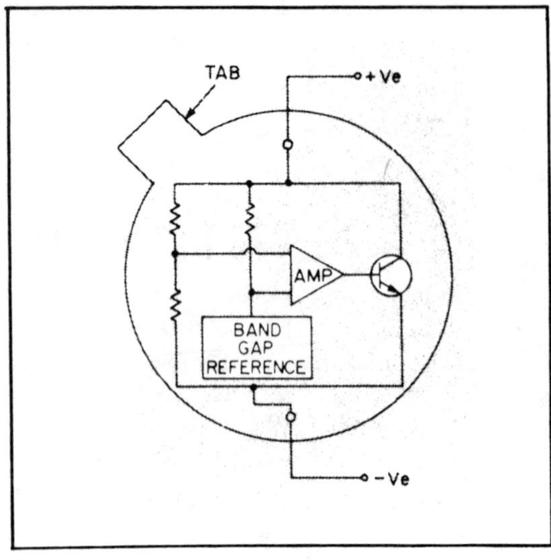

Fig. 2-12. The internal schematic of the band gap zener diode, which serves as the heart of the calibrator. Use the tab on the case as the reference point for making circuit connections. No heatsink is required here.

selected a reference current of 8.75 mA. Find the value of the resistor needed for *Rl.* A ZN458 (2.45 volts) is used.

$$Rl = \frac{(4.2-2.45)\ volts}{(0.00875)\ Amp}$$

$$Rl = \frac{(1.75)}{(0.00875)} = 200\ ohms$$

Construction. The largest part in the project is the battery, so a small LMB aluminum box was selected to house the calibrator. The electronic circuitry was built using the banana jacks as the points; no wire board is needed. The battery holder is ordinarily used with size "C" batteries, but the Mallory TR233 (4.2-volt mercury cell) fits nicely. The battery holder was fastened to bottom of the box using a small 4-40 machine screw. Small rubber feet can then be glued to the box to offset the "bump" created by the screw head. If you want to avoid this, however, it should be easy to superglue the battery holder flush to the aluminum.

The ZN458 has a 100 *parts per million* (ppm) drift specification, the ZN458A is a 50 ppm device, while the ZN458B is a 30 ppm device. The voltage output is nominally 2.45-volts dc. (measured at 2 mA reference current), but may have an absolute value between 2.42 to 2.49-volts. With no additional circuitry, then, these devices will produce an accuracy of ±40 milli-volts, or better. This voltage cannot easily be adjusted without external circuitry, but you can use any of the standard IC operational amplifier voltage regulator circuits to set the

Table 2-3. Zener Diode Selection.

Type	Voltage	Drift
ZN423	1.26	—
ZN458	2.45	100 ppm
ZN458A	2.45	50 ppm
ZN458B	2.45	30 ppm

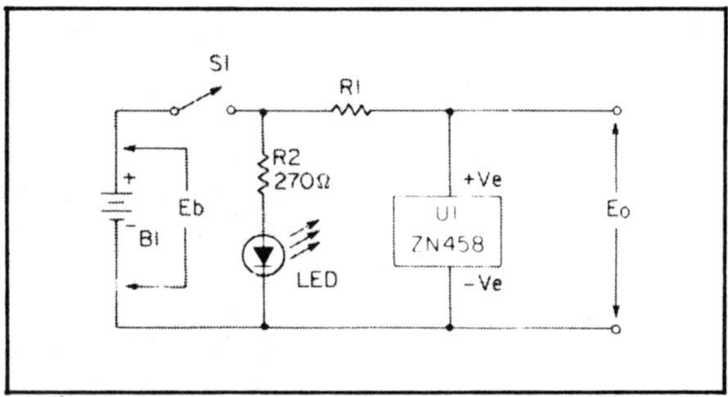

Fig. 2-13. Basic schematic used to demonstrate calculations necessary to determine the value of the associated components used in the regulator circuit. Refer to the text for a full explanation.

output voltage to a standard level. A circuit that is usable for this purpose is shown here. The ZN458 is used to set the voltage at the noninverting input of the op amp. The output voltage can then be trimmed to the desired value by potentiometer *R3*. This circuit is an ordinary op amp noninverting follower, so the desired output voltage can be derived in the following equation:

$$Eo = Eb\left(\frac{R3 + R2}{R1} + 1\right)$$

The table shows values for *R2/R3* needed for output voltages of 5 and 10 volts. Note that the resistors used in this circuit must be low temperature coefficient precision (1 percent) resistors, or drift will result. It is even more important in this circuit, than in the basic

Table 2-4. R2 and R3 Selection.

Output Voltage	R2	R3
5	1000-ohms	100-ohms
10	2600-ohms	500-ohms

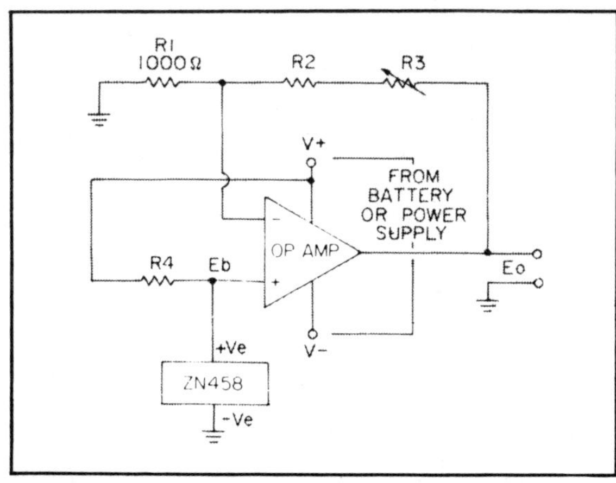

Fig. 2-14. This schematic depicts a variable regulated power supply, with the source being either a battery or a line-powered dc source.

circuit. The trimmer potentiometer should be a ten-turn, precision type, so that very tight control over the adjustment of the output voltage is possible.

There is, however, a hitch in this variable output circuit. It is not inherently "calibrated" as is the case of the basic curcuit. Although this circuit is capable of better accuracy, initially, it must be adjusted. You will have to find a very accurate voltmeter, or precision reference potentiometer to make the initial adjustment. After this adjustment, however, it should remain in calibration for a long time.

THE GO/NO GO TESTER

Most electronic devices can be divided into either analog (linear) or digital circuitry. *Digital circuits* operate largely by means of switches, which are most of the time in one of two quite different states (high or low, plus or minus, or one of two substantially different voltage levels). *Analog, or linear, circuits* generally function by amplification (or reproduction) of signals in as nearly an exact manner as practical; that is, signals within the circuit(s) are changed, amplified, or transformed according to an infinitely small series of changes of one or more input signals. This is the kind of circuit which we most often run into outside of digital devices, and it is in these circuits that the operational amplifier is most frequently used.

Refer to Figs. 2-15 through 2-17.

Operational amplifiers were developed into practical devices in the 1950s and first found their applications in scientific and industrial instruments, where their ability to amplify, add, subtract, multiply, or divide by precise quantities was valuable. Those op amps used vacuum tubes, of course, and required the usual accompanying capacitors and resistors. With the coming of integrated circuitry in the 1960s, op amps began to appear in chip form. Today, op amps are used in circuits more often than individual transistors are used.

How Op Amps Work. Briefly, an op amp has two inputs, one output, and plus and minus power supplies (most often +15 and −15 volts). The inputs are called the plus and the minus input, or the inverting input and the non-inverting input. If the op amp is to be used as a simple amplifier the non-inverting input (usually) is grounded. In addition, the op amp has feedback from the output to the inverting input, and the amount of feedback is precisely controlled, generally by careful selection of the value of the feedback resistor. The amount of feedback determines the amount of gain of the op amp in any particular circuit.

In recent years, then, op amps have become one of the most important building blocks in many circuits. They are produced in IC form at low cost, and are readily available on the surplus market, often for less than 10 cents a piece, in quantities, and *untested—as is.*

When you buy a bunch of op amps at a good price from surplus houses, most of them will turn out to be perfect, and you can use them as well as if they'd been bought from the factory, individually sealed and guaranteed (at much higher prices, of course). It's therefore necessary to have an easy way of testing op amps before they're put into working devices.

One way to test an op amp is to plug it into a typical circuit, feed it a signal and observe both the input and output signals on an oscilloscope. Since this takes a fair amount of time (and a generator as well as a scope), a simple plug-it-in, go/no-go tester has been developed by this writer, and you can make one just like it at low cost, in just a few hours. When you've finished it you'll know a fair amount about how op amps work and how to test them. And you'll be able to make each test in three to five seconds apiece, using this handy instrument.

How It Works. The circuit of the op amp tester may be broken down into five sections, as follows: power supply, voltage divider, test circuit, error amplifier and display circuit.

The operation of these sections can best be understood if the following paragraphs are studied in conjunction with the schematic diagram.

Power Supply. This is a conventional supply, consisting of a small power transformer (T1), which is the only part of the instrument not actually mounted on the printed circuit board. The low-voltage secondary of T1 feeds the bridge rectifier D1-D4. The bridge is shown here arranged differently from the customary one. But comparing its actual connections with the usual arrangement you will see that they are the same. This drawing has the advantage that it's easier to trace out the current flow to see the complete path for current flow from *both* sides of the center-taped secondary, at *all* times.

Voltage Divider. By tapping off the power transformer secondary winding we can obtain a 60-Hz sine wave signal for use as a test signal. The effective voltage is 18 volts, which must be dropped to 0.06 volts. This is accomplished by using a simple voltage divider, R3 and R4.

Test Circuit. The *Device Under Test,* DUT, is placed in the circuit as an inverting amplifier. In an ideal inverting op amp, the output is described by the simple formula:

Output voltage = –input voltage times

R8 divided by R7.

Put another way.

$$V_{out} = -V_{in} \times R8 \div R7.$$

With the values in this circuit, this would be –(0.06) times 1 megohm divided by 10,000 ohms, or –6 volts.

Notice that the output is the negative (inversion) of the input. This inversion is important in the operation of the tester, as is explained farther on.

Error Amplifier. The error amplifier in this circuit is a *summing* amplifier—it sums (adds together) the two signals applied to the junction of R9, R10, which come from output of the DUT and from R6, respectively. In other words these two signals are mixed together, where they add (or in this case, cancel, if the DUT is working properly). The summing amplifier is made up of IC1, R5 (or J1), R6, R9, and R10.

Display Circuit. The display circuit consists of diodes D3 through D6, and indicator LED 1. Diodes D3-D6 make up a full wave bridge rectifier (just like the bridge rectifier circuit in the power supply) which converts positive or negative signals, which are present at the display circuit input, into positive signals (just the way the bridge in the power supply converts ac into positive dc) which will turn on indicator LED 1. If bridge D3-D6 receives a positive input, current flows through D5, LED 1, and D8, and diodes 6 and 7 are reverse-biased. If the display circuit receives a negative signal, current flows through D7, LED 1, and D6, while diodes D5 and D8 are reverse-biased. The current through LED 1 is limited by the components inside IC1, the op amp, which is part of the test circuit.

How the Circuit Works. The input signal which comes from the voltage divider (R3 and R4) is applied to the DUT through resistor R7. The same signal is also applied to the error amplifier through resistor R6. The DUT inverts this signal and amplifies it. The amount of amplification is equal to −R8/R7, which is 100, since R8 is one megohm (a million ohms) and R7 is 10,000.

This is stated, for op amps, as V/V, a way of expressing gain. It means volts of output for each volt of input. If a circuit has 100 volts of output for each volt of input, it is said to have a gain of 100V/V. The error amplifier sums the voltage divider signal with the signal from the output of the DUT. The signal at R6 is amplified by a gain of −100,

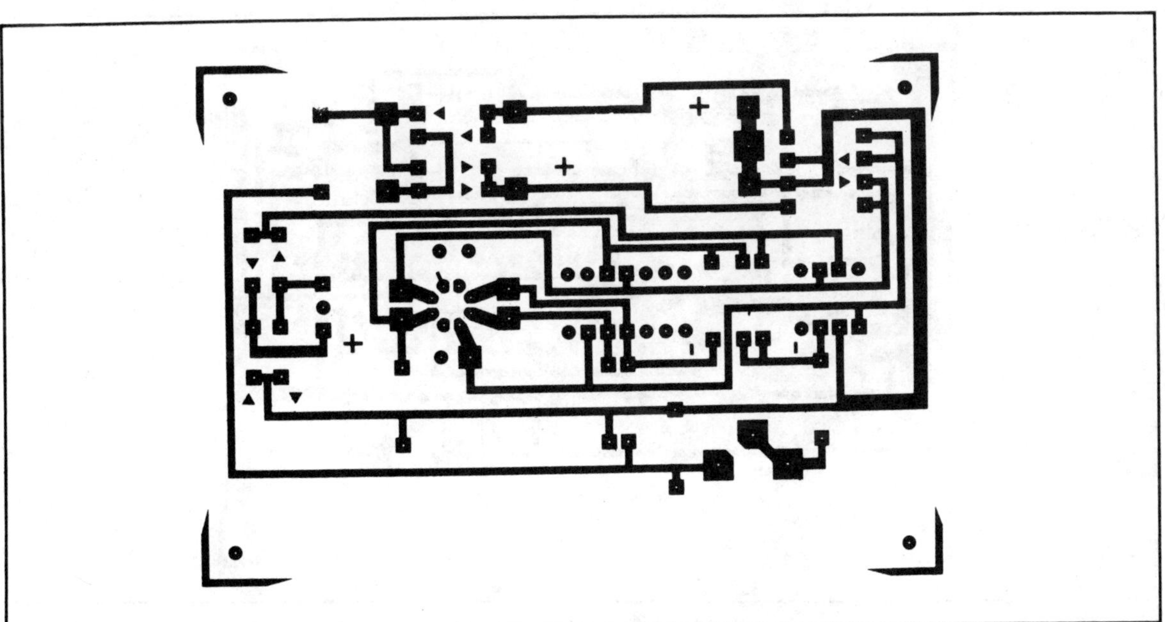

Fig. 2-15. Full-size pattern for making your own printed circuit board for the Op Amp Tester.

since R6 is 10,000 ohms, and R10 is one megohm. The signal at R9 is amplified by a gain of −1, the minus sign indicates a signal inversion.

If the DUT is working properly there will be no error (the two signals will exactly cancel out) at the output of the error amplifier, IC1. This is because the input to the error amplifier at R9 will exactly cancel the input to the error amplifier at R6.

Output Current. The DUT must be able to supply output current equal to at least Vp/Rt, which means, where Vp is the peak output voltage of the DUT, and Rt is the load on the DUT output (R8, R9, and R12 in parallel), we get 6V divided by 1,790 ohms, which is 0.0034 amperes, or 3.4 milliamps. If the DUT cannot supply at least 3.4 mA, an error signal will be developed, and the LED indicator will go ON, telling us that the DUT is faulty.

Input Voltage Offset. This is the input voltage required to provide zero output voltage in the absence of intentional signal input on the + or − inputs of the op amp under test. It will appear as an error signal at the output of the DUT, amplified by the gain of the test circuit, plus one; 1 + R8/R7. An error voltage of 4V will appear at the output of the DUT when the input voltage offset exceeds 40mV.

Input Bias Current. In an ideal op amp, no bias current flows between the two input terminals. In actual amplifiers, however, there is always some current flow between them. This current is similar to the base current which flows when a bipolar transistor is used as the input of an amplifier, or the gate current when an FET is used at the input of an amp. In this circuit input bias current will flow through R8,

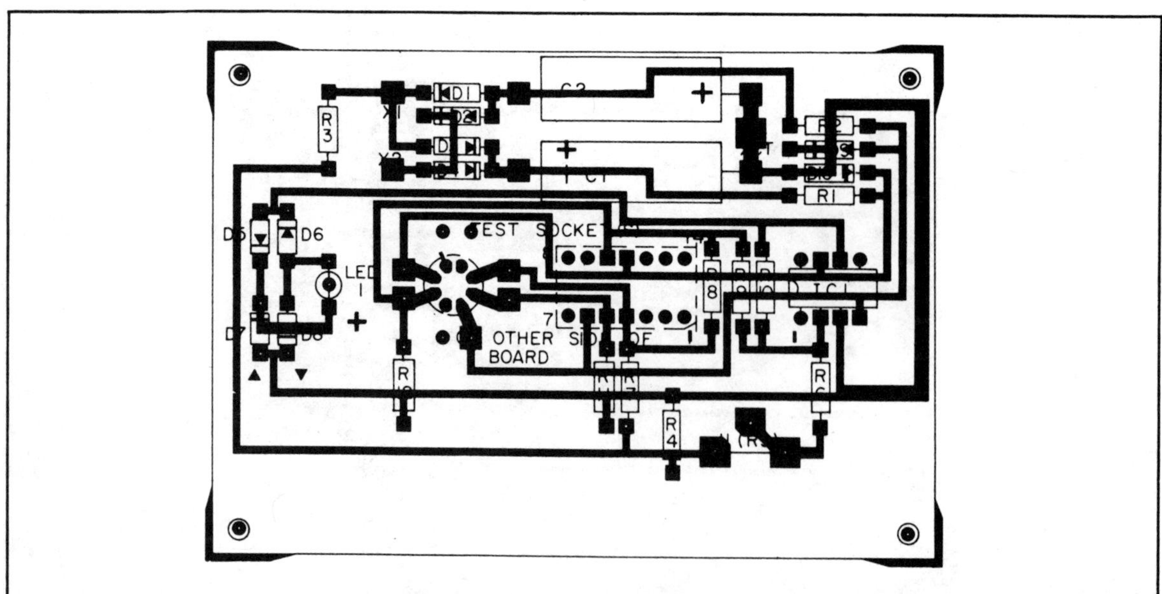

Fig. 2-16. Components placement go on printed circuit board. Drill holes halfway through board because parts mount on the foil side.

and will show up at the output of the DUT as an error signal. Bias flow from the positive input will cause a voltage drop across R11 which will be amplified by an amount equal to the gain: 1+R8/R7 which is 1,000,000 divided by 10,000 plus 1; or 101. This signal will appear at the output of the DUT as an error signal.

A bias current of 4 microamps will cause an error signal at the output of the DUT if R7 and R11 are equal, and R8 = 1 megohm. Notice also, if the bias currents at the input terminals of the DUT are equal, they tend to cancel, and no error signal is developed. Thus, only mismatches of input bias currents are treated as errors in this test circuit.

For the test circuit to work properly the actual values of resistors R6 through R10 must be pretty close—5 percent resistors at least, or resistors of greater tolerance which have been measured and found to be within 5 percent or better of the nominal value.

With an overall test gain of 100 or less, the 5 percent components specified will be adequate. In this case, jumper J1 is installed, and R5 will equal zero ohms. If higher test sensitivity is desired, then closer matching is required. This is accomplished by reducing R6 from its nominal value of 10K to, say, 8K and adding an adjustable resistor, R5 of about 3K ohms in series with it so that a closer match can be achieved by adjusting R5. When R5 is used, it can be adjusted by placing an op amp known to be good into the test socket and adjusting R5 to the center of the range in which failure indicator LED 1 stays off.

Errors at the output of the DUT will be amplified by the error amplifier at a gain of one (R10/R9) and will drive the display circuit. When the error at the output of the error amplifier, IC1, exceeds the forward voltage of LED 1 plus the forward voltage of the bridge diodes, LED 1 will turn on, indicating a bad DUT. This occurs at about 4 volts. The following DUT difficiencies can contribute to the error signal.

Open Loop Gain. If the open loop gain of the DUT is not adequate to accurately amplify the input signal, an error signal will develop. A gain error approximately equal to 3.0 V will be developed if AOL of the DUT is equal to 200.

Output Voltage Amplitude. The output of the DUT must swing at least as much as the amplitude of the input signal multiplied by the gain of the test circuit. If it does not, an error signal will be developed. Since R7 equals 10,000 ohms, and R8 is one million ohms, we get 1,000,000 divided by 10,000, or 100 times 0.06. This equals a swing of ±6.0 volts.

Stability. An unstable amplifier will oscillate even if it has no external signal input. This oscillation will directly cause an error signal. Op amps tend to be more stable at higher gains, and tend to be least stable at unity gain. So, this is not really a very strenuous stability test circuit, and op amps which are stable in this test circuit may not be in lower gain circuits. On the other hand any op amp which is supposed

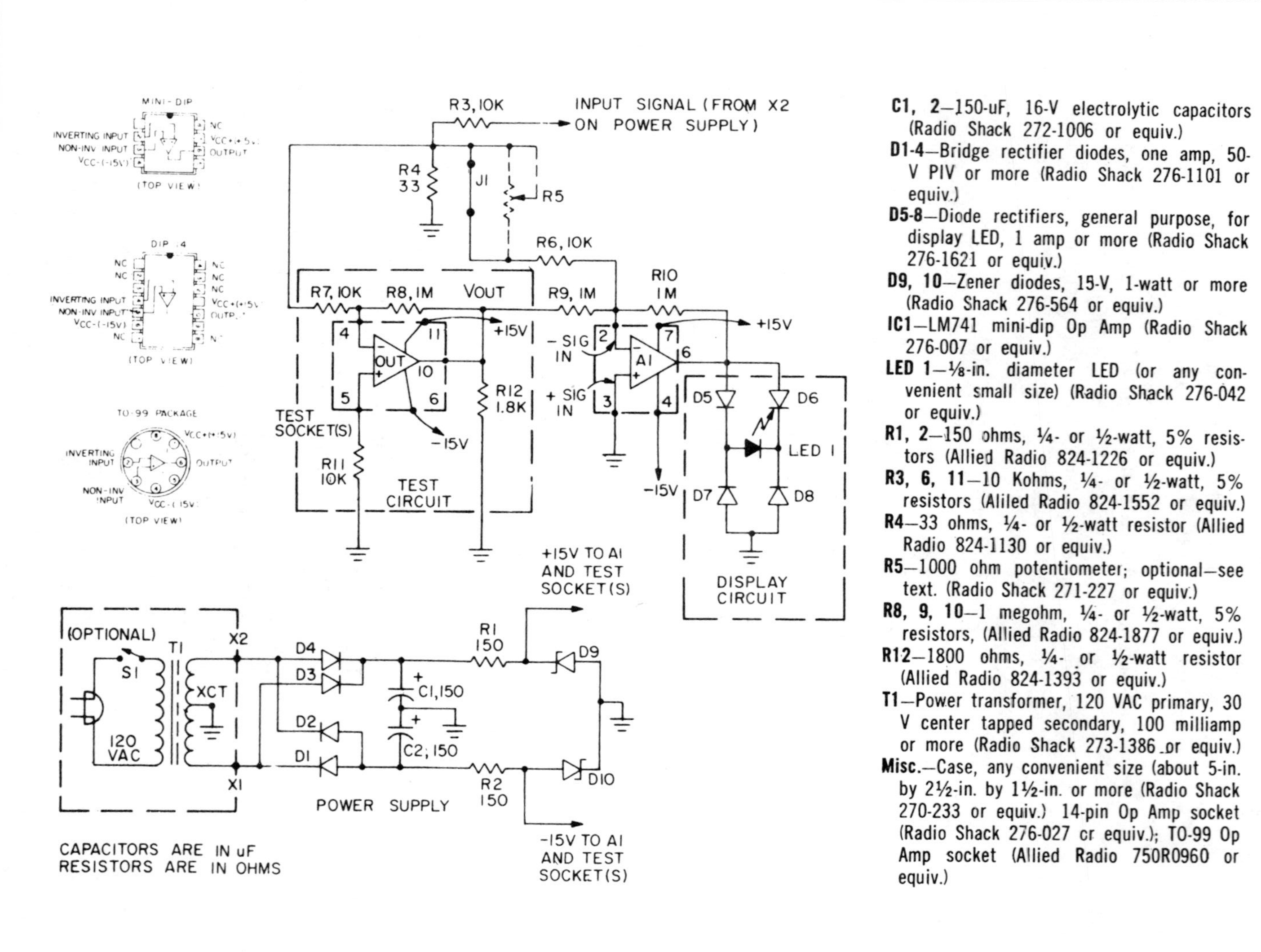

C1, 2—150-uF, 16-V electrolytic capacitors (Radio Shack 272-1006 or equiv.)

D1-4—Bridge rectifier diodes, one amp, 50-V PIV or more (Radio Shack 276-1101 or equiv.)

D5-8—Diode rectifiers, general purpose, for display LED, 1 amp or more (Radio Shack 276-1621 or equiv.)

D9, 10—Zener diodes, 15-V, 1-watt or more (Radio Shack 276-564 or equiv.)

IC1—LM741 mini-dip Op Amp (Radio Shack 276-007 or equiv.)

LED 1—⅛-in. diameter LED (or any convenient small size) (Radio Shack 276-042 or equiv.)

R1, 2—150 ohms, ¼- or ½-watt, 5% resistors (Allied Radio 824-1226 or equiv.)

R3, 6, 11—10 Kohms, ¼- or ½-watt, 5% resistors (Aliled Radio 824-1552 or equiv.)

R4—33 ohms, ¼- or ½-watt resistor (Allied Radio 824-1130 or equiv.)

R5—1000 ohm potentiometer; optional—see text. (Radio Shack 271-227 or equiv.)

R8, 9, 10—1 megohm, ¼- or ½-watt, 5% resistors, (Allied Radio 824-1877 or equiv.)

R12—1800 ohms, ¼- or ½-watt resistor (Allied Radio 824-1393 or equiv.)

T1—Power transformer, 120 VAC primary, 30 V center tapped secondary, 100 milliamp or more (Radio Shack 273-1386 or equiv.)

Misc.—Case, any convenient size (about 5-in. by 2½-in. by 1½-in. or more (Radio Shack 270-233 or equiv.) 14-pin Op Amp socket (Radio Shack 276-027 or equiv.); TO-99 Op Amp socket (Allied Radio 750R0960 or equiv.)

Fig. 2-17. Go/No Go Tester schematic and parts list.

to be unconditionally stable and is not so in this circuit, is certainly defective. Some experimenters may want to put a switch on the tester so that the DUT can be momentarily placed into a unity gain configuration for more exhaustive stability testing. This would be done by opening the inputs to R6 and R7. This would place the DUT in a non-inverting gain of one with a one meg-ohm resistor in the feedback, and the input tied to ground through R11 (10 Kohms). The error amp would be disconnected from the voltage divider and would just look for any voltage output at the DUT which exceeded 4V divided by R10/R9 (or just 4V if R9=R10).

With the circuit values shown, the DUT is tested for the following approximate conditions:

Open Loop Gain	200V/V (min.)
Output Voltage Swing	±6V
Output Current	±5mA
Input Voltage Offset	40mV (max)
Input Bias Current	±4μA (max)
Stability	Good at gain of 100.

While these test conditions are good for general purpose testing, an unlimited number of test parameters can be programmed by altering the component values.

Two test sockets are provided, wired in parallel, so that practically any op amp with standard pin configuration can be tested easily whether it is in a TO-99, 14 pin DIP, or mini-DIP. Op amps requiring external phase compensation can also be tested. In these cases, the phase-comp components are plugged into the socket not occupied by the DUP. Never plug in two op amps at the same time.

Construction. Neither parts layout nor lead dress is critical. When using the printed circuit board shown here, be certain that holes for the components are drilled at least half way, but not all of the way through the board. Bend the leads of the components and clip them flat so that the bodies of the components lie about 1/32 inch above the board, supported by the leads. Carefully solder the components in place. Take care when soldering the components not to get them too hot. Use a low-wattage soldering iron and rosin core solder.

SOLAR CELL TESTER

Refer to Figs. 2-18 through 2-21.

Photovoltaic solar cells may hold the promise for the future, but you can experiment with them today. Before utilizing the units in a project, it's necessary to know their capabilities.

All makers list a maximum output current rating, with some listing of power levels for a given light source (usually 100 mW per square centimeter). So what? You need to know what it will do for you—under your parameters, your light source, and your load. How? With a handy-dandy solar cell tester.

The Rating Game. Silicon cell output varies with the light level, and with the load as well. If a cell is too heavily unloaded, the power drops appreciably. Glance at the power graph. Notice it peaks when the voltage across the cell is 460 millivolts.

This is where the manufacturer tests his units, to determine

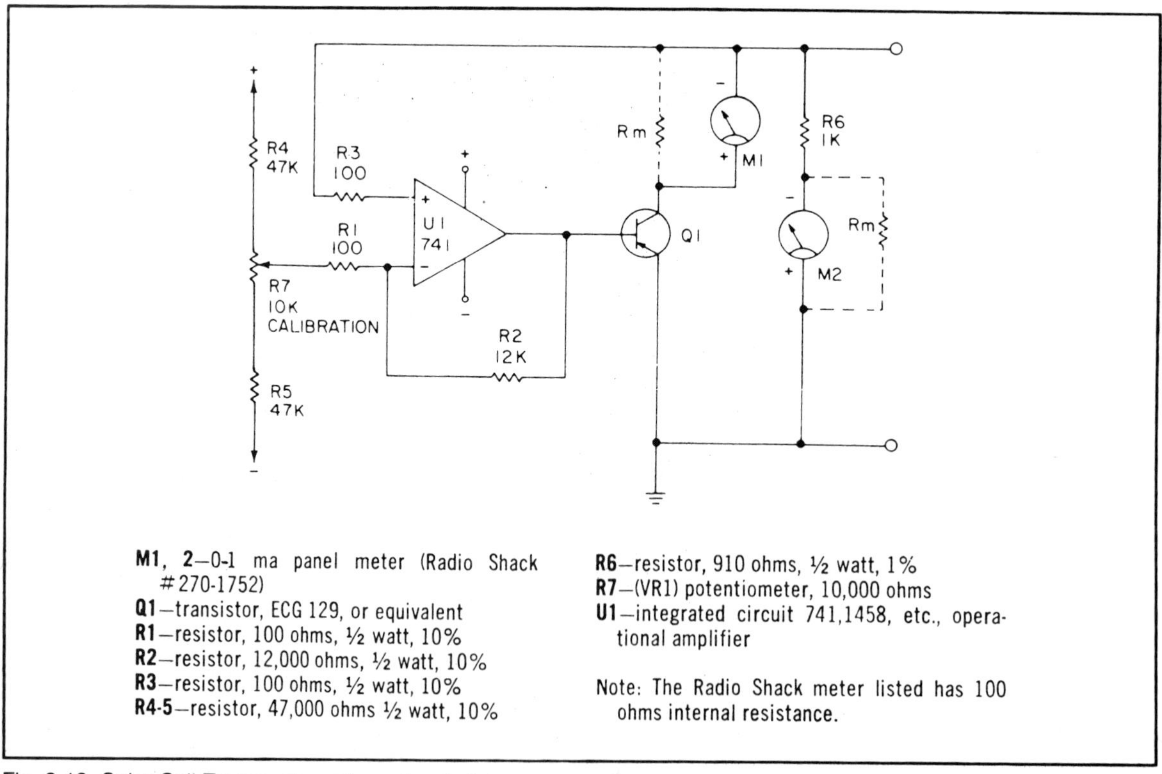

M1, 2—0-1 ma panel meter (Radio Shack #270-1752)
Q1—transistor, ECG 129, or equivalent
R1—resistor, 100 ohms, ½ watt, 10%
R2—resistor, 12,000 ohms, ½ watt, 10%
R3—resistor, 100 ohms, ½ watt, 10%
R4-5—resistor, 47,000 ohms ½ watt, 10%
R6—resistor, 910 ohms, ½ watt, 1%
R7—(VR1) potentiometer, 10,000 ohms
U1—integrated circuit 741,1458, etc., operational amplifier

Note: The Radio Shack meter listed has 100 ohms internal resistance.

Fig. 2-18. Solar Cell Tester schematic and parts list.

maximum performance. He uses variable load, placed across the cell. With no load the solar generator exhibits an open circuit voltage higher than its working voltage. As the load is increased (more current), the potential across the junction drops.

At one point, the current begins to dip along with the voltage—thus further reducing power. The maker sets the resistance so the voltage across the cell under test is optimum.

Figure 2-18 shows a simple circuit for performing just such a test. One meter monitors current . . . the other voltage. Adjusting the variable resistor to the peak power voltage (460 mV) will net you the device's current! But, the output current differs from cell to cell, necessitating a corresponding change in resistor value. That's fine if you're testing one or two units, but how do you efficiently check 20, or 50, or whatever? With a *dynamic variable load*, one that adjusts itself to the correct voltage.

About the Circuit. The easiest way to achieve a dynamic variable load (DVL) is using a transistor. In Fig. 2-19, the solar cell is connected across the emitter and collector. As current is metered through the base, the V_{CE} changes—loading the cell accordingly.

Now, add a feedback loop, an amplifier, a couple of meters, and we have a professional solar cell checker.

The feedback resistor, R2, determines the amplifier gain, while the noninverting input monitors the voltage across the cell and compares it to the reference voltage at the inverting input.

Let's Make One. The tester can easily be duplicated, using any method of construction available to you: perfboard, PC board, point to point, etc. You'll notice, a PNP transistor is used for the load, making

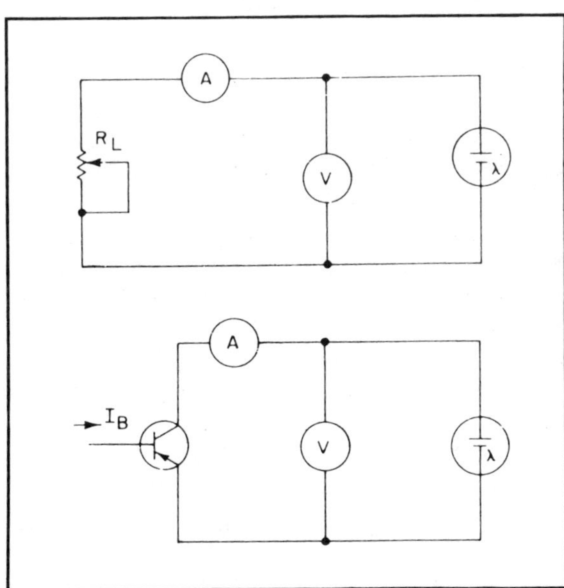

Fig. 2-19. The top diagram shows the basic circuit. The voltmeter shows cell voltage, while the ammeter(A) indicates the load, which can be adjusted by the potentiometer, called RL. In the lower diagram, the same output voltage can be obtained at different loads. This is accomplished by placing the calibration potentiometer in the base of the transistor.

WHAT'S YOUR Rm VALUE ?

XI	0-1	NONE
X10	0-10	5.00 10.00
X100	0-100	.500 1.00
X1000	0-1000	.050 .100
	RANGE (MA)	OHMS

TOP VALUE IS FOR METER WITH 50 OHMS, BOTTOM IS FOR 100 OHMS.

Fig. 2-20. R_m, or meter shunt value, is determined by the internal resistance of the meter. Using a shunt, you can multiply the face value of the meter by 10X, 100X, or 1000X. As you increase the multiplied range, the resistance value of the shunt (Rl_m) goes down.

the ground positive in respect to the cell's input voltage. That's because the silicon cell has a *positive* backing, with the front contacts *negative* polarity.

The IC amplifier is a 741, but any stable operational amplifier should suffice. Don't forget the external compensation, should your choice require it. The only requirement—output current. As the transistor reaches higher current levels, the H_{FE} (gain) decreases accordingly, requiring more base current through the transistor.

The sink transistor (Q1) may be any silicon PNP capable of passing 1 amp safely and able to dissipate about 1 watt. No heatsink is required.

My test instrument was designed to measure 1 ampere, but you can make it any range you desire by changing R_M. If you use a 0-1 ma meter with an internal resistance of 50 ohms, follow the chart for your selected value. If it is 100 ohms, double that figure—it'll be close enough to be accurate.

Connect R4 and R5 to the V_{CC} power supplies as shown. The power supplies should be tracking—or, at least regulated. Otherwise, the reference voltage at the inverting input will shift, throwing off your calibration setting. (Actually, I've even used two 9-volt batteries and had good results. If the cells are within reasonable tolerance, the shift is negligible. But, for precision, a well regulated power supply is a *must*).

Using It. Connect the cell under test to the input leads, observing polarity. Illuminate the surface with the light it will be subjected to (sunlight, desk lamp, etc.) and set the A1 control for the voltage—in most cases .46 volts. The current of the cell will be displayed on the other meter—don't forget your multiplication factor!

Fig. 2-21. This graph shows how the solar cell's power output in watts will vary according to voltages above and below rated voltage.

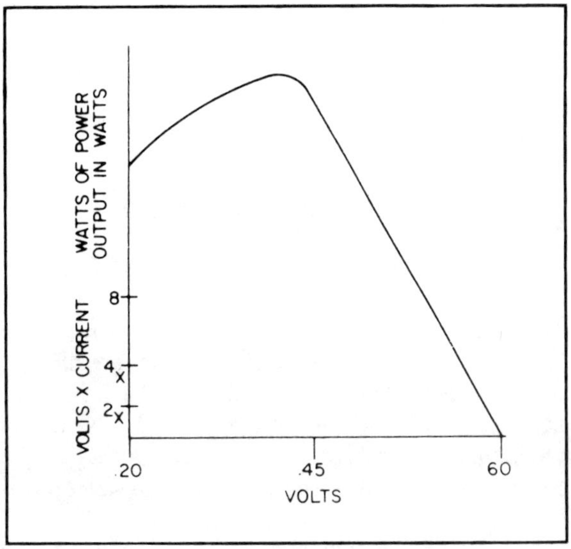

The tester will adjust to any cell automatically, regardless of the output current of the load. A nice feature about the instrument is you can change the calibration to give you the output voltage at a specific current. Twist the calibration knob to your current value; then read the voltage. Of course, it won't regulate at that current value as it does voltage, but you will know the output voltage under the actual operating conditions.

Obtain some solar cells you intend to use for your next solar project. Using your photovoltaic tester, you can now design your load to yield maximum power output.

DIGITAL DIODE TESTER

Just how many bullet diodes, miniature glass diodes, epoxy encapsulated diodes, unmarked diodes, unbanded diodes and stripped-from-equipment diodes have you run into? If you wanted to use any of these don't-know or not-sure-what-they-are diode types, you've had to drag out the ohmmeter for a front-to-back resistance check. There's nothing wrong with that, of course. But here's an easy-to-build and inexpensive digital IC gadget that blinks an *A* or *C* in a little window to tell you if the end of the diode you've selected is the *anode* or the *cathode*.

Refer to Figs. 2-22 through 2-25.

And if that's not enough for you, plug in an unknown transistor and the same window will come up with an *N* or a *P*—you guessed it—to tell you what type you have; NPN or PNP! If you try to fool this gadget with an open transistor, it pops up an *A* (for throw it away?) in the window for as long as you keep it there. (If an open diode is tested, the readout doesn't budge from its normal 8. A shorted diode blanks the display window for as long as the diode is connected.)

So, if you have ever wished for a simple gadget that would indicate the type of transistor, either PNP or NPN, if it has gain, and what lead of a diode is connected to the test terminal, wish no more, for this DDT will test almost every type of transistor made including germanium, silicon, low, medium, and high power devices. As we said, if the transistor under test is good, the readout will include a *P* for PNP transistor, and an *N* for NPN transistor. The readout will indicate an *A* for an open transistor, and will distinguish for a shortest transistor.

You can turn your white elephant collection of goodies into useable items by spending less than $20 and about four hours building an all-digital diode tester of your own. We believe that you will agree that this tester is the most valuable transistor and diode tester available for twice the money.

How the Circuit Operates. One-half of IC-1 operates as a ring oscillator. The output of the oscillator drives two sets of inverter stages.

A test terminal marked C is driven through two inverter stages, while the E terminal is driven with only one inverter. This makes the voltage present at terminal C always opposite to the voltage at the E test terminal. When a diode is connected (with the anode at the C test terminal) the output of the inverter, IC2 pin 4, is *low* pulling the six segments to battery negative. These six segments form the letter *A* for anode. The same is true when the cathode of the diode is connected to

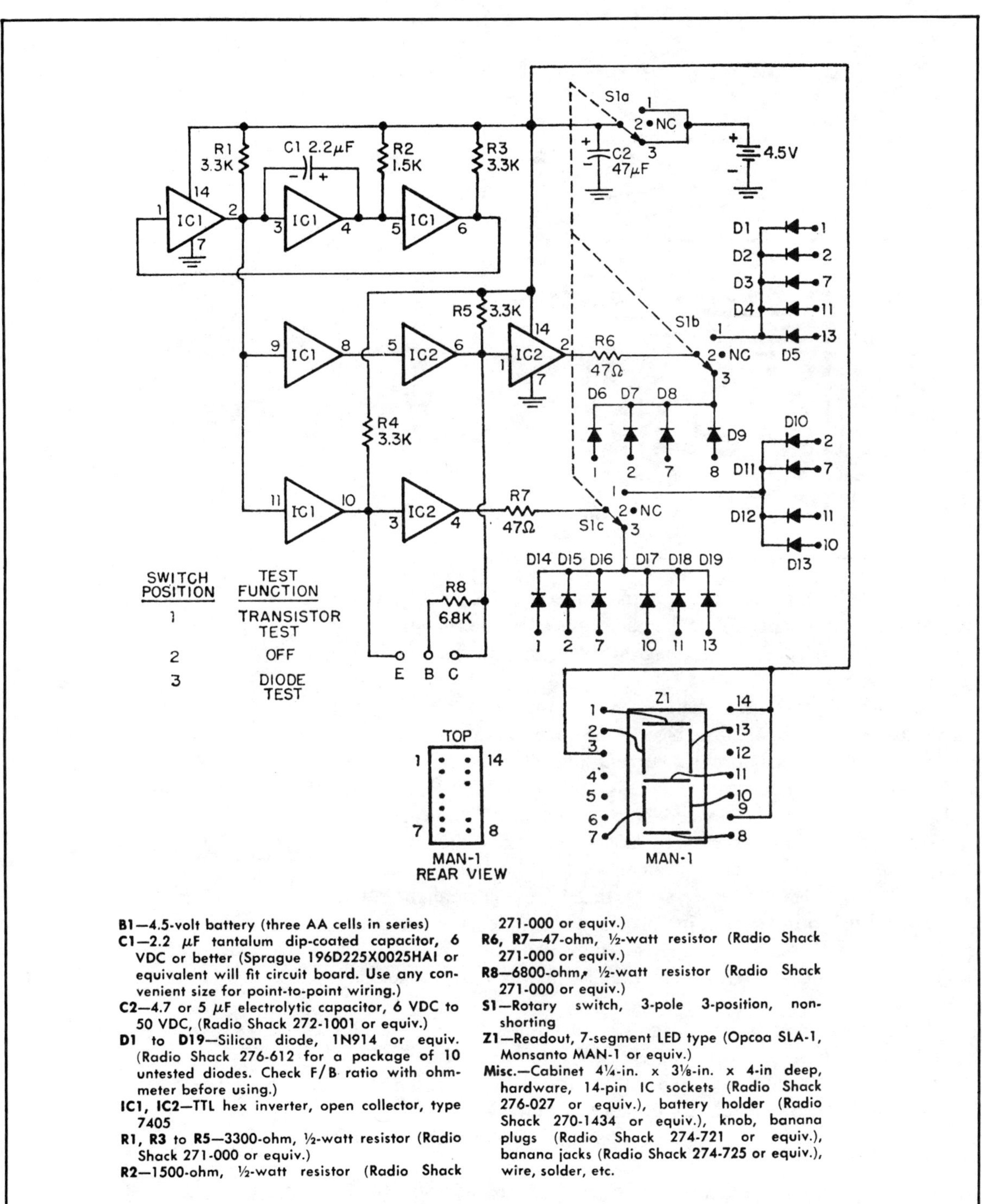

B1—4.5-volt battery (three AA cells in series)

C1—2.2 μF tantalum dip-coated capacitor, 6 VDC or better (Sprague 196D225X0025HAI or equivalent will fit circuit board. Use any convenient size for point-to-point wiring.)

C2—4.7 or 5 μF electrolytic capacitor, 6 VDC to 50 VDC, (Radio Shack 272-1001 or equiv.)

D1 to **D19**—Silicon diode, 1N914 or equiv. (Radio Shack 276-612 for a package of 10 untested diodes. Check F/B ratio with ohmmeter before using.)

IC1, IC2—TTL hex inverter, open collector, type 7405

R1, R3 to **R5**—3300-ohm, ½-watt resistor (Radio Shack 271-000 or equiv.)

R2—1500-ohm, ½-watt resistor (Radio Shack 271-000 or equiv.)

R6, R7—47-ohm, ½-watt resistor (Radio Shack 271-000 or equiv.)

R8—6800-ohm, ½-watt resistor (Radio Shack 271-000 or equiv.)

S1—Rotary switch, 3-pole 3-position, non-shorting

Z1—Readout, 7-segment LED type (Opcoa SLA-1, Monsanto MAN-1 or equiv.)

Misc.—Cabinet 4¼-in. x 3⅛-in. x 4-in deep, hardware, 14-pin IC sockets (Radio Shack 276-027 or equiv.), battery holder (Radio Shack 270-1434 or equiv.), knob, banana plugs (Radio Shack 274-721 or equiv.), banana jacks (Radio Shack 274-725 or equiv.), wire, solder, etc.

Fig. 2-22. Digital Diode Tester schematic and parts list.

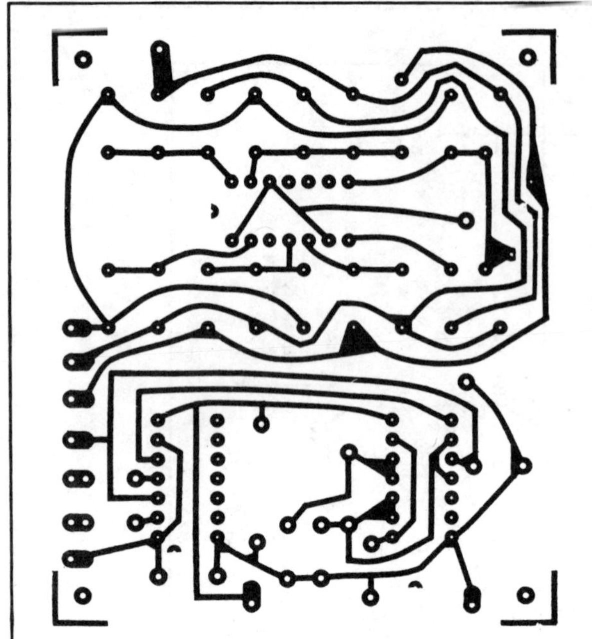

Fig. 2-23. Full-size printed circuit board.

test terminal C, but the output of another inverter, IC2 pin 2, goes *low* making the letter *C* appear. At the same time the other inverter that produced letter *A* goes *high* turning off the segments relating to that letter only.

Transistors are checked in a similar way when S1 is in the transistor test position, but the base is included by biasing it from the collector test terminal through current limiting resistor R8. This allows the NPN transistor to conduct only when its collector is positive, and a PNP transistor when the collector is negative. The inverters are connected to the proper diodes (through S1) to cause the letter *P* to light on the readout for a PNP transistor, and the letter *N* for NPN transistors. The tester is powered by three 1½-volt penlight cells.

Building Your Own. The circuit is a simple one and can be constructed on perfboard or printed circuit board; the choice is yours because the layout isn't critical and the circuit will work in most any configuration.

If a printed circuit board is used and the author's model copied, a metal or plastic cabinet about 4¼ × 3⅛ × 4-inch in size should do fine for an enclosure for the tester. If the kit of parts is used, just follow the layout of the author's model, and be very careful when soldering the semiconductors in place (if you are not using IC sockets) to avoid heat damage.

A printed circuit or perfboard is mounted to the front of the cabinet with a rectangle hole cut out on the LED readout (use a nibbler tool). This hole can be cut out and filed to a neat window for the read-

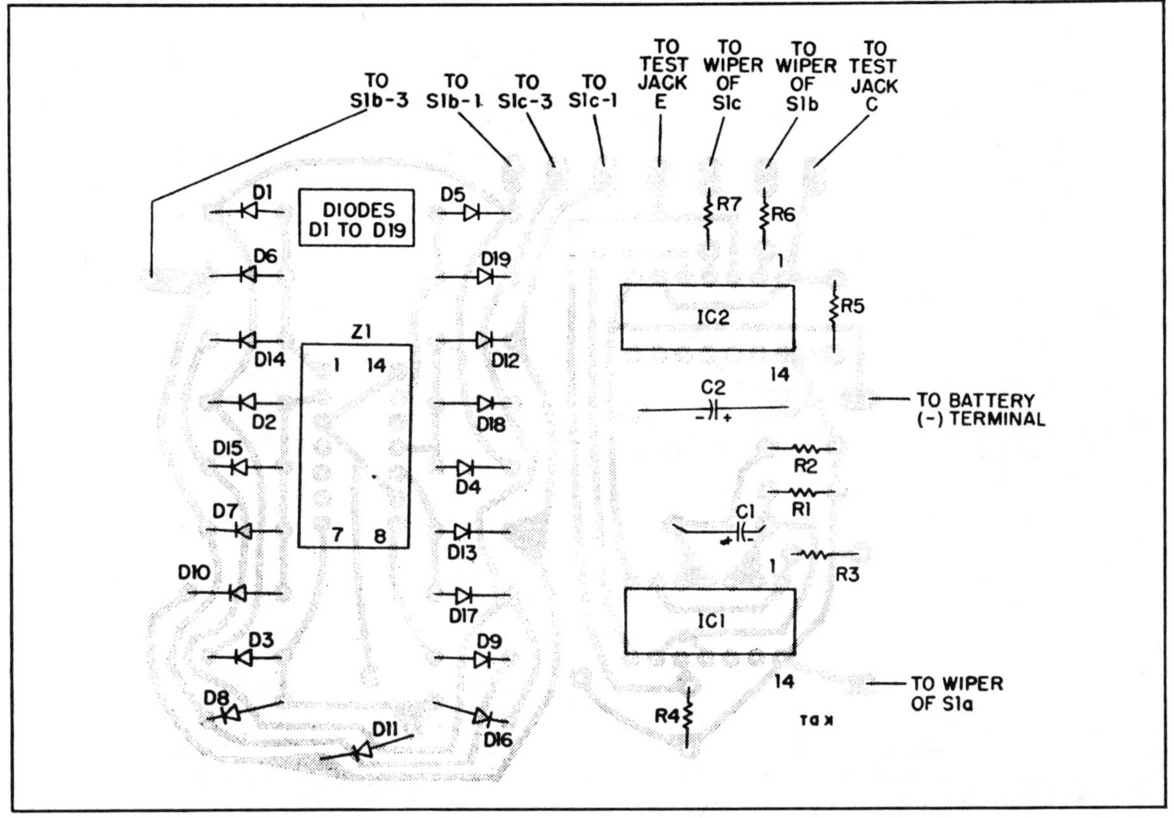

Fig. 2-24. This is the component (top) of the printed circuit board with an "X-ray view" of the copper pattern. In addition to components illustrated here, be sure to supply battery power to the LED readout by including a jumper wire between the two empty pads shown.

out. The three penlight batteries are located near the back of the bottom of the cabinet. The selector switch may be mounted in any convenient location.

Initial Checkout. With the batteries in place, switch the selector switch to the "D" position (diode test) and connect a good diode to the C and E test terminals.

The readout should present the letter *A* or *C* to correspond to the lead that is connected to the C test terminal. If the test leads are shorted together, the readout should go dark, and with the leads open the readout should display an eight.

The only precaution to take when testing transistors is to be certain that it is connected to the proper test leads or the test results will be misleading. A group of the most common transistor base diagrams is shown. If the transistor to be tested falls into one of the categories, no difficulty will be had in determining the type of transistor and its condition. Put those nameless functions to work for you now!

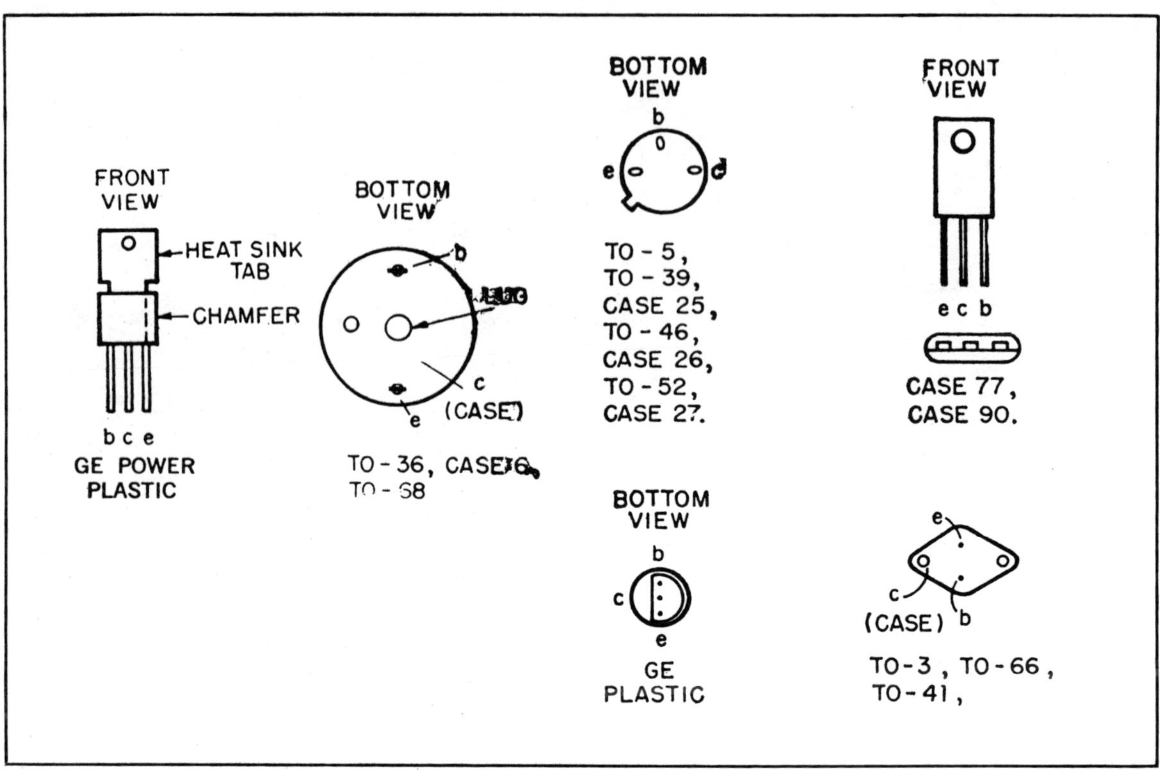

Fig. 2-25. To be sure you have the correct transistor leads connected to your tester, here are some of the more common transistor base pin configurations.

SIGNAL CHASER

One of the secrets of troubleshooting is to start at those circuit areas where there is no trouble, then to back your way through the circuit until you've reached the point where it isn't working. The same trick can work frontwards, letting you trace a signal through a circuit until you reach the point where it disappears. Here's a handy aid for troubleshooting in the frontward fashion, a signal tracer with a great deal of input sensitivity called Signal Chaser.

Refer to Figs. 2-26 and 2-27.

Built-in Demodulator. An ordinary amplifier could help you find signals in the AF (audio frequency) range, but the Signal Chaser can do more. D1, a 1N914 diode, acts as a demodulator, much like the diode in a simple crystal-set-style radio, to demodulate AM (amplitude modulated), rf and i-f signals directly to audio (or whatever the carrier is modulated with). On FM and PM (frequency modulated and phase modulated) signals, the diode acts as a slope detector, giving a suitable, if low-fidelity, audio output.

High Impedance Input. The one feature of this circuit that really makes it shine when compared to most signal tracers is its high impedance input. The input impedance of the Signal Chaser is close to 10 megohms. This is due to the use of a JFET (junction field-effect transistor) for Q1. Q1, a Siliconix 2N5458 or similar P-channel JFET, is configured as high-to-low impedance converter with an input impedance converter with an input impedance determined mostly by the value of R2, 10 megohms. Capacitor C1 blocks dc but passes af, rf, and i-f signals. Resistor R1 limits the input current to Q1.

A high-input impedance means that for a given signal voltage, very little current is drawn by the Signal Chaser. This means that under almost all conditions, the Signal Chaser cannot load down the current you are troubleshooting.

Speaker Size Output. The output of Q1 alone would be enough to drive a high impedance earphone, but keeping one in your ear while busy probing a suspect circuit can be, to say the least, inconvenient.

Instead, the output of Q1 (after demodulation) is coupled to the input of IC1, an LM38ON audio amplifier. IC1 provides enough drive to power even low-impedance speakers, around 8 ohms, to a good, healthy volume. Capacitor C5 provides dc decoupling between the speaker and the output of IC1.

Breadboard-Easy Construction. The entire circuit can be built up on a small solderless breadboard. I've used three tricks here I would especially like to share. For one, I used a pair of zig-zag mounting brackets (from Radio Shack) as battery hold-down clips. The mounting holes in the CSC EXP350 helped make this especially easy. At the far side of the breadboard, the mounting holes there happened to match exactly the holes on a small speaker I had on hand, and I was quick to take advantage of it. My third trick was to solder stiff wire (resistor leads I cut off some of the resistors in the circuit) to the breadboard end of the shielded probe cable. You may also want to use *headers* available from several sources and many parts stores for under a dollar a strip. The rest of the assembly is fairly straightforward.

Understanding Solderless Breadboards. In case you haven't tried solderless breadboards before, you may not know how easy they are to work with. The holes in the face of the breadboard are arranged on 0.1-inch centers (1/10th of an inch apart), which happens to be the lead spacing on standard DIP (dual inline package) integrated circuits and most other modern components.

The center channel (0.3 inch wide) is just right for IC's to straddle. On each side of the center channel are groups of five holes (columns, if you view the breadboards as widest on the horizontal, with the center channel running left to right). Behind each group of five holes in a spring clip with slits between the hole positions to allow a lead inserted into any one hole to be grasped firmly and independently,

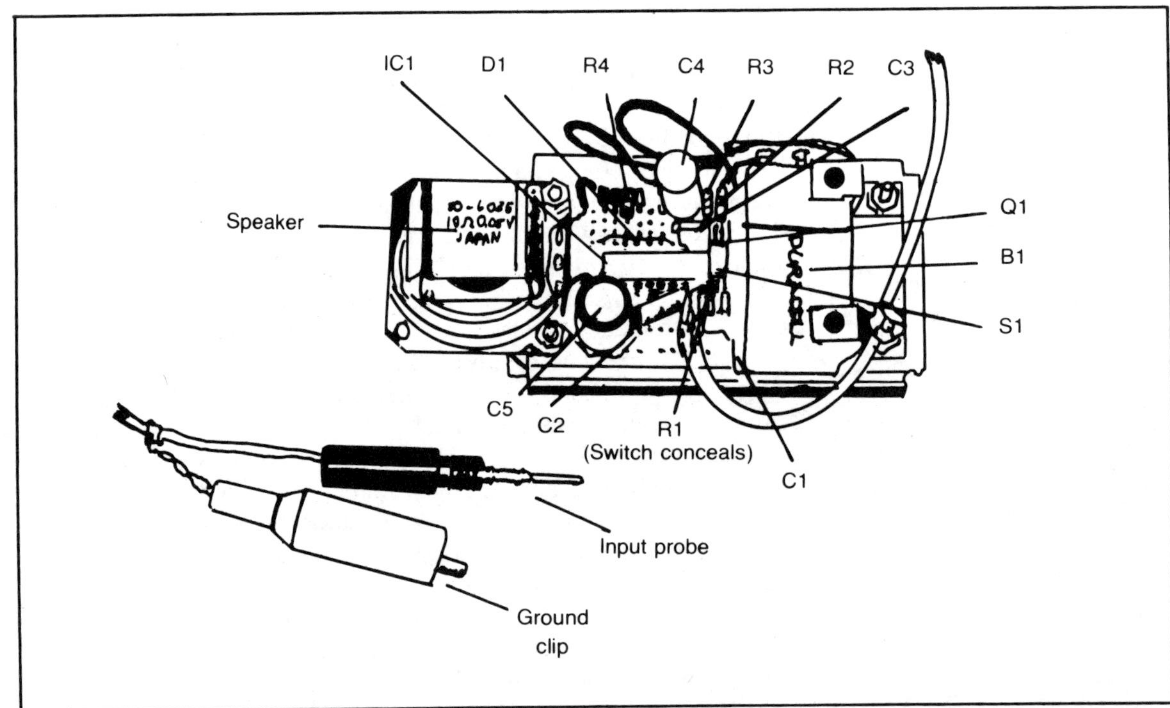

Fig. 2-26. Signal Chaser was built using a solderless breadboard and, as you can see, it made for a neat component arrangement. Be certain you don't forget about R1, which connects to the Gate of Q1 and to C1. The Signal Chaser should go together quite quickly, so if you start it after lunch you should be chasing your first signals before the dinner bell.

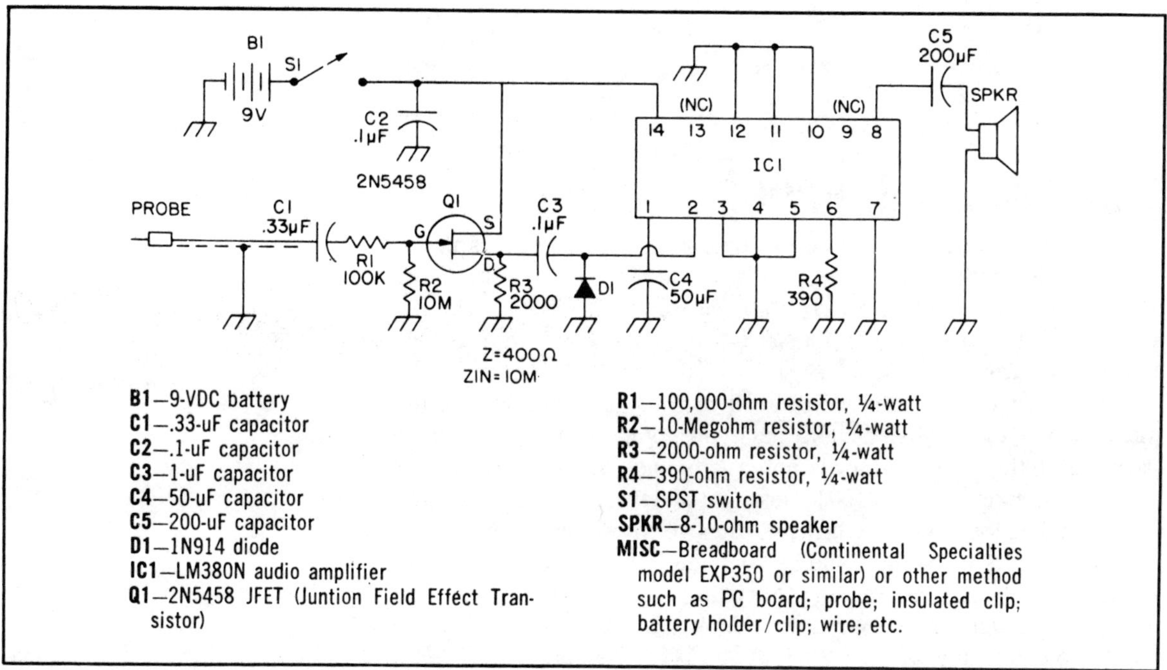

Fig. 2-27. Signal Chaser schematic and parts list.

and interconnected with anything grasped at any other position in the group.

Each five-position terminal can be interconnected with any other by simply using hookup-wire jumpers. The separate rows (at the top and bottom) are connected across their entire lengths, and can be used for power or signal busses. I use them to carry the battery plus and minus lines.

Using the Signal Chaser. For most run-of-the-mill signal tracing, clip the probe cable shield to a circuit ground near the area you're testing and touch the probe to each side of the signal path near each active or passive device in the signal path. Start at the front end and work your way to the output, if you like—but skipping a few stages on the chance they'll work can also help you localize a problem.

The high impedance of the Signal Chaser input means high sensitivity, which lends it to some useful applications. You can attach a coil of wire or a magnetic tape head to the input to inductively probe circuits and devices. You can "listen" to the magnetic stripe on the back of your credit cards, amplify a telephone conversation or pick off the signal on your transmitter's modulation transformer. Or attach a photocell to the input and listen to the sounds of light bulbs, LED readouts, the sun, street lights and then some.

Signal Chaser is not only a good introduction to solderless breadboarding, but once it's built you may also find it to be one of the handiest gadgets in your electronic bag of tricks-of-the-trade.

BARGAIN LOGIC PROBE

In the digital world, we do not find a variable signal, as in the analog world. It is either on or off, just as a switch would be either on or off. Another way of saying this is *high* or *low*, or 1 or 0. Each high or low bit is put together to make up a basic character on byte. Sometimes these bytes are called words.

Refer to Figs. 2-28 through 2-31.

If we have 1001, then we can call that a 4-bit byte. That is the smallest byte ever to be encountered in the computer world. It can be used where the data accuracy is not critical and the amount of data is small. To illustrate this, if 1001 were sent and interference generated a pulse at the moment of the third bit, then we have been left with false data of 1011. Its meaning would be completely different. To increase accuracy and handle more data, we could go to 8-bit bytes. Such as 10101010. A logic probe allows us to look at a particular point in the circuit to determine if a low (0) or high (1) is present.

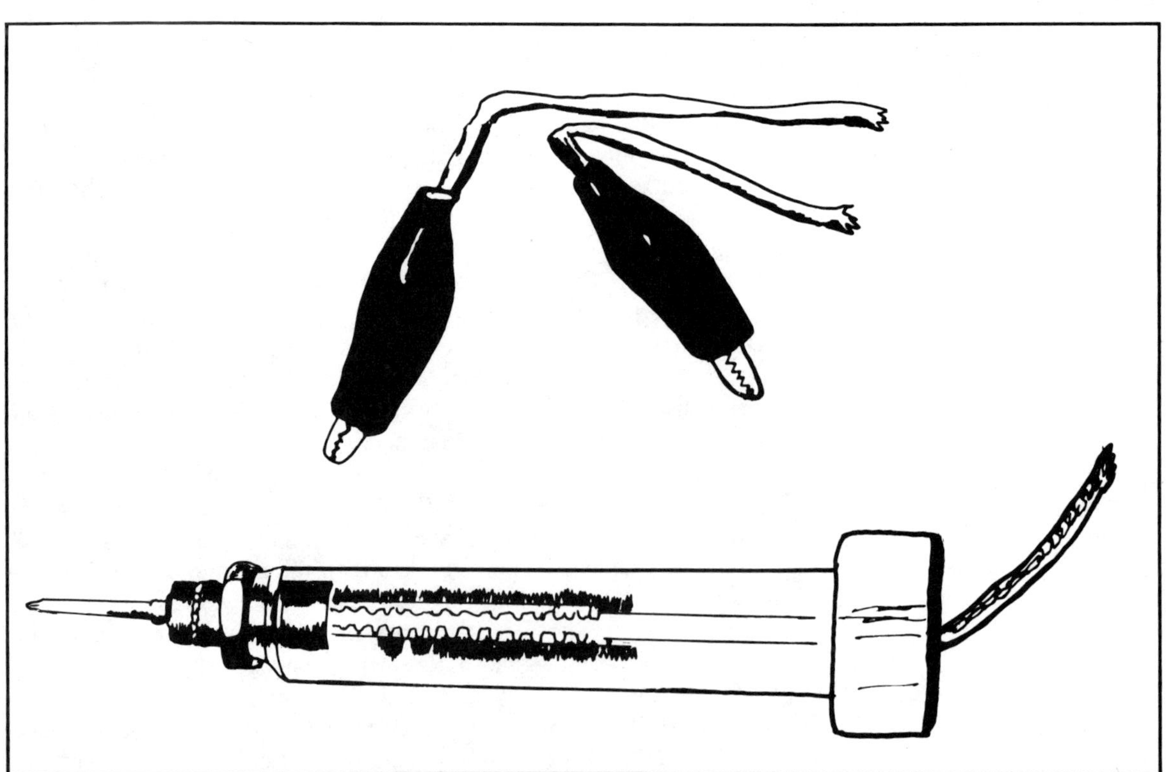

Fig. 2-28. Inexpensive logic probe duplicates its more costly counterparts.

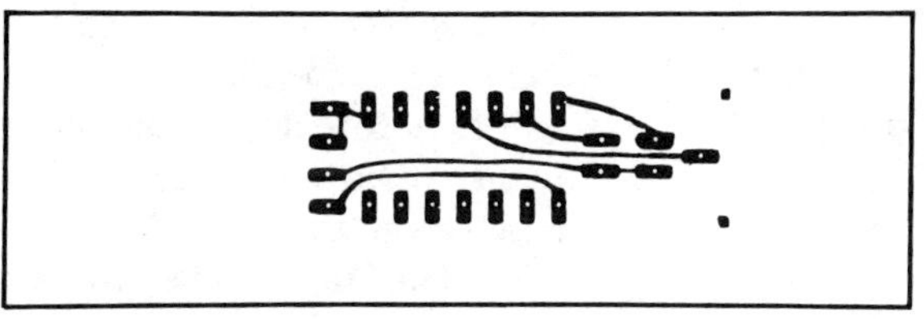

Fig. 2-29. Extra-small printed circuit board template for Bargain Logic Probe.

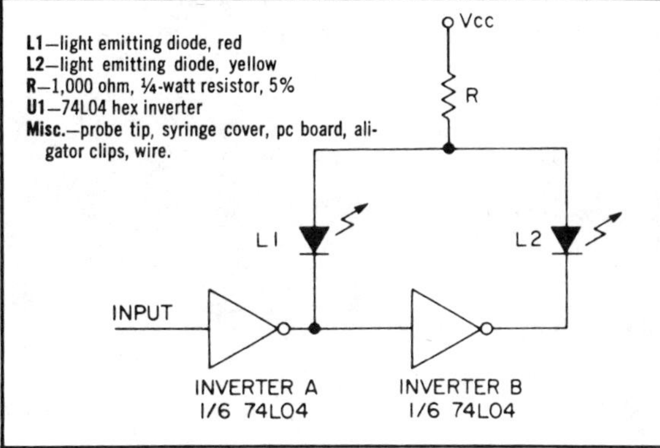

Fig. 2-30. Bargain Logic Probe schematic and parts lists.

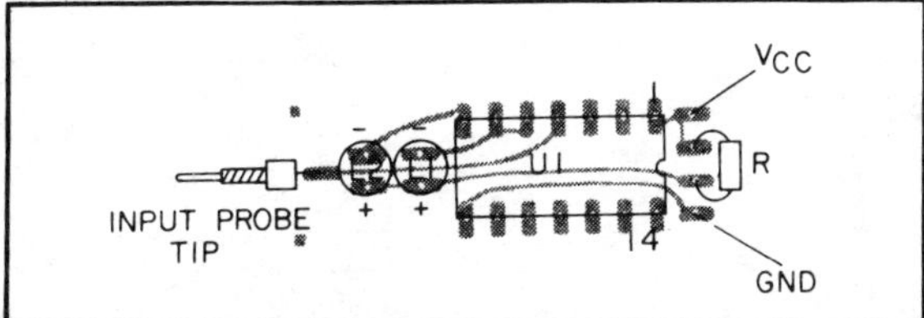

Fig. 2-31. Parts layout on the PC board. There are few parts.

For most of our electronic experiments, we don't need expensive logic probes costing upwards of $40. Here is a cheap unit which can signal high level (1), low level (0), and oscillation. No pulse detection feature was included thus keeping the size small and the price low, around $2. The probe is designed for TTL signal levels and can be used for 5-volt CMOS circuits although loading may occur.

How It Works. The Bargain Logic Probe uses only one IC, a 74L04 hex inverter shown in the schematic. The input to inverter A normally floats high, making its output low so as to light L1. The output of inverter B is high so L2 is off. If you now make the input of inverter A zero volts, L1 will turn off and L2 will illuminate. When oscillation is present at the input, both L1 and L2 will light at some intermediate brightness depending on the duty cycle of the signal being observed.

Using a 74L04 is important, the "L" series only requires the driving signal to sink 180 μA maximum, much below the 7400 series 1.6 mA maximum or even the 74LS00 series 400 μA requirement.

Construction. A full scale PC board layout is shown in addition to the parts layout on the component side. I slid the entire PC board inside a used syringe cover (available at hospitals for free), and attached a readily available test probe tip. Using different color LEDs to signal high or low will help to quickly distinguish the signal level. Power is supplied by the circuit under test, and runs around 10 mA. Note that voltage requirements for the "L" series are 5 ±.25 V nominal. So far, the Bargain Logic Probe works great. It fits in my pocket and gives me a quick handle on circuit performance. It can also be used to show oscillator output in low power transmitter stages, SW converters and receiver local oscillators.

UJT TESTER

The unijunction transistor, or UJT, cannot be used in linear circuits (such as amplifiers) in the same way that a conventional bipolar transistor can. Instead, you will find the UJT in timing and oscillating circuits for the most part. In order to test a UJT, therefore, it appears logical that a representative timing circuit should be used to do the job.

To operate the UJT tester presented here, begin by plugging your unijunction into SO1. This circuit is similar to the classical UFT relaxation oscillator, except that an LED and current-limiting resistor (R3) have been inserted in series with the emitter lead. Initially, capacitor C1 will charge up through R1, and the voltage on C1 will gradually rise.

Refer to Fig. 2-32.

Once the potential on C1 becomes large enough to force the UJT's emitter to break down, C1 gets discharged through LED1, R3 and the UJT's emitter. After discharge, the emitter terminal returns to a high-impedance state, and the capacitor charges once more.

Each time the capacitor discharges through the LED, a flash of light is produced. This serves as a simple GO/NO GO indication of the UJT's ability to oscillate. Resistor R2 is used to swamp the LED's high impedance in the OFF state, thus enabling the UJT to break down more readily.

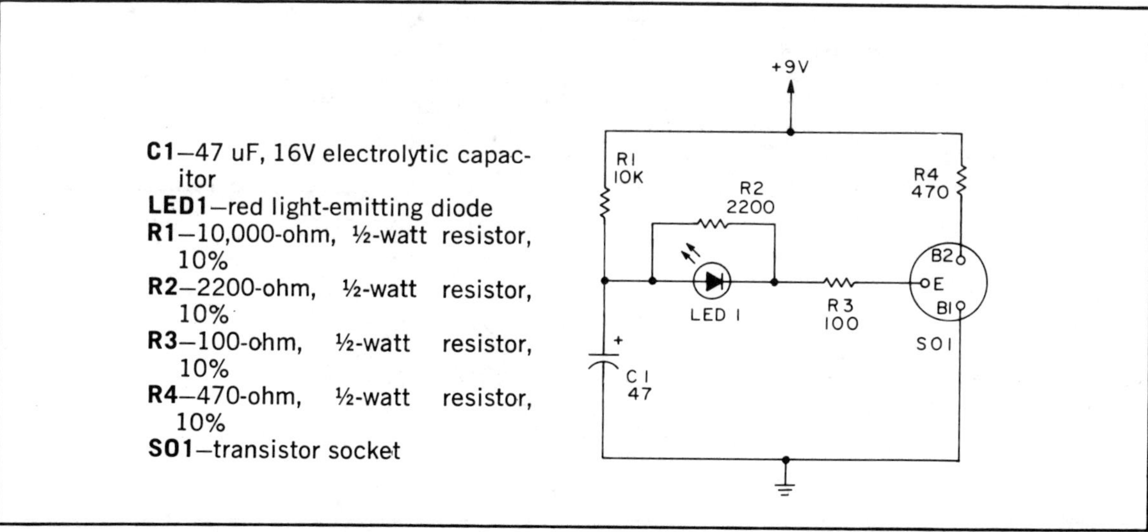

Fig. 2-32. UJT Tester schematic and parts list.

SQUARE-WAVE GENERATOR

Here is a versatile square-wave generator capable of surprising performance. It can deliver clock or switching pulses, act as a signal source, and more. And because the outputs take turns switching, it can be used as a simple sequence generator or as a multiple-phase clock.

Refer to Fig. 2-33.

The component values indicated will support a range of output frequencies from a few pulses per second up into the high audio range. And this square wave output is rich in harmonics. If you use a 5-volt power supply, this circuit can trigger TTL.

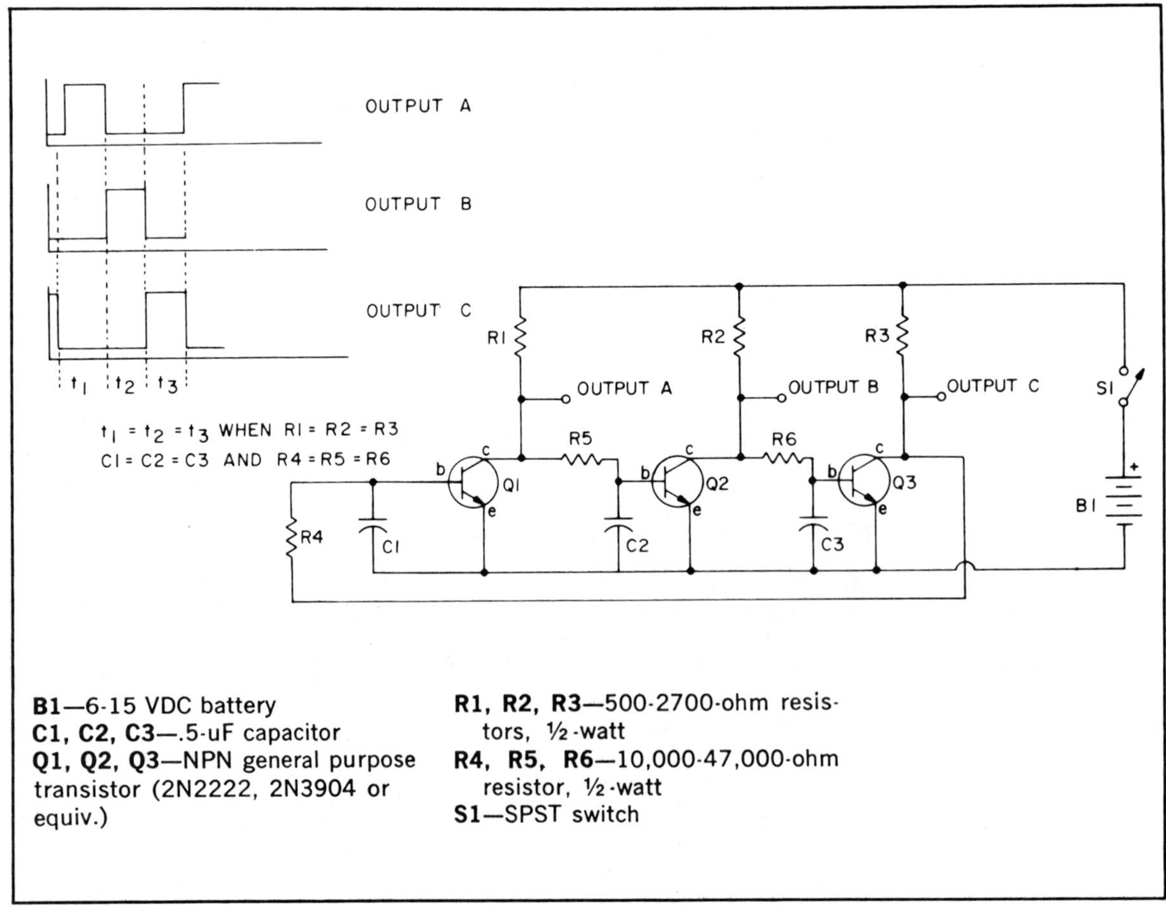

B1—6-15 VDC battery
C1, C2, C3—.5-uF capacitor
Q1, Q2, Q3—NPN general purpose transistor (2N2222, 2N3904 or equiv.)
R1, R2, R3—500-2700-ohm resistors, ½-watt
R4, R5, R6—10,000-47,000-ohm resistor, ½-watt
S1—SPST switch

Fig. 2-33. Square-wave Generator schematic and parts list.

WIRE TRACER

Problem! You've just snaked a multi-wire computer and/or intercom cable through two floors, five bends, and two "pull" boxes, and you have the creepy feeling that one of the wires broke in the process. Then, you discover upon trimming away the outer jacket, that all of the wires are the same color. What to do? Simple, just check them all with this simple wire tracer. Clip one end of the LED1/LED2 circuit to the same ground source and touch the other end to each wire. When you find the wire being tested, one of the two LEDs will light.

Refer to Fig. 2-34.

It doesn't matter which LED lights. We use two only to prevent confusion in the event a polarity gets reversed. This way, one LED is certain to light. The LEDs can be any general-purpose type available. Battery B1 is a 9-volt transistor battery.

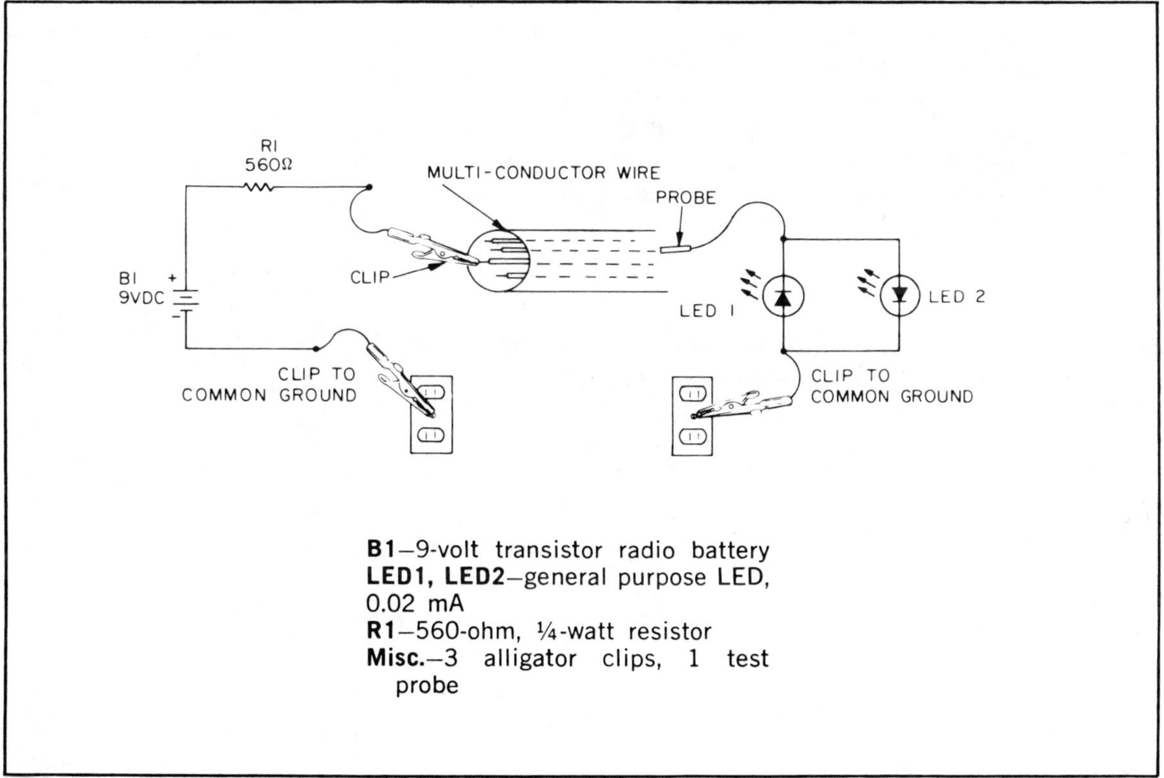

Fig. 2-34. Wire Tracer schematic and parts list.

SAWTOOTH GENERATOR

This simple sawtooth generator should be useful as a general-purpose source of audio test signals. Here we have a basic UJT oscillator, but instead of using a resistor to charge the timing capacitor, a transistor constant current source (Q1) is employed instead. This results in a sawtooth that rises linearly as a function of time, since the capacitor's charging rate is constant. When a simple resistor is used to charge a capacitor, the waveform produced is curved like a shark fin, since the charging current falls off as the voltage on the capacitor increases.

Refer to Fig. 2-35.

The charging current available from constant-current source Q1 is adjustable by means of R3. The higher the current, the quicker C1 gets charged, and the higher the frequency of the resultant sawtooth voltage developed across the capacitor. Therefore, decreasing R3 increases the frequency. With the values shown, the generator's output frequency can be varied from roughly 100 to 1000 Hz.

Since unijunction Q2 breaks down and discharges capacitor C1 very quickly, we get a near-perfect sawtooth shape: slow, linear ascent, and rapid, vertical decline. Emitter follower Q3 acts as a buffer between C1 and whatever load you connect. Maximum peak-to-peak amplitude is roughly 6 volts, which level control R6 allows to be cut down to any convenient voltage level needed.

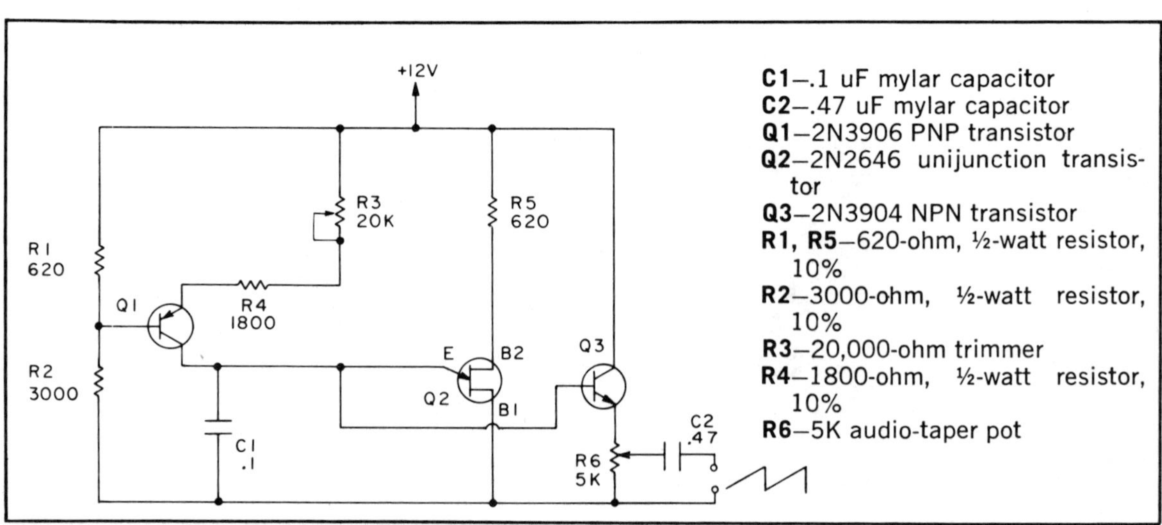

C1—.1 uF mylar capacitor
C2—.47 uF mylar capacitor
Q1—2N3906 PNP transistor
Q2—2N2646 unijunction transistor
Q3—2N3904 NPN transistor
R1, R5—620-ohm, ½-watt resistor, 10%
R2—3000-ohm, ½-watt resistor, 10%
R3—20,000-ohm trimmer
R4—1800-ohm, ½-watt resistor, 10%
R6—5K audio-taper pot

Fig. 2-35. Sawtooth Generator schematic and parts list.

Chapter 3
Radio Projects

MYSTERY RADIO

Build this mystery radio. It uses a radio vacuum tube that doesn't glow in the dark. See if your friends can figure out how it works before you reveal the trick. It's a fun project, educational as well as entertaining—a good choice as first receiver for the beginning constructor.

What everyone will see is a one-tube radio tuned by a common ferrite loopstick antenna coil. Nothing too unusual about that. But wait a minute—all there is to power the tube is a small penlight battery of 1.5 volts. There is no B battery for the plate of the tube, and there is no grid leak and capacitor going to the grid of the tube—yet you get a good signal in the phones from local broadcast stations! How can this be?

Refer to Figs. 3-1 and 3-2.

The secret is revealed when we examine the pictorial drawing, which shows a germanium crystal diode detector, with a transistor serving as an audio amplifier. These two units are hidden inside the base of the tube. The schematic diagram shows that there is a perfectly conventional crystal detector radio driving a one-transistor amplifier. The tube is a dummy, and can even be one that has burned out. It isn't in the circuit at all. If your friends can't guess the answer, there's the mystery in this "one-tube" radio receiver.

Construction. You can use most any 4-prong (or 4-pin) tube you can find which has a loose bulb. The larger the base the better as it gives you more room inside to wire in the diode and transistor.

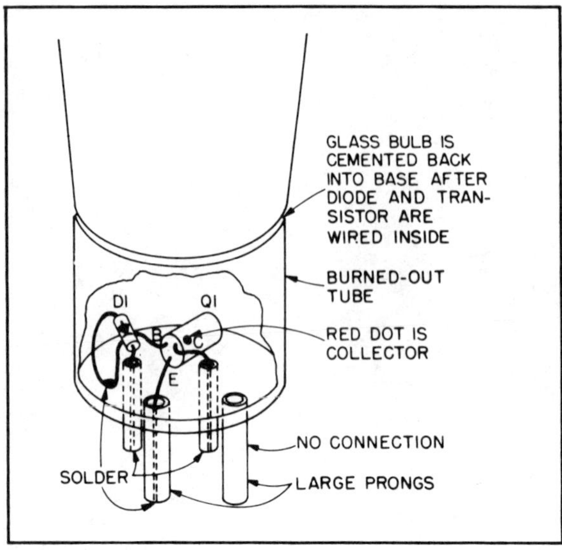

Fig. 3-1. The trick is in the base of the radio vacuum tube. Diode and transistor in its base replace the tube in receiving circuit.

To remove the glass bulb from the base, heat the ends of the pins with a soldering iron and shake out the solder, then carefully twist and pull the bulb out of the base. Caution: For safety, wear a pair of gloves when removing the glass bulb from the base of the tube. Use a tube that has a loose bulb to start with. Clip off the leads going up into the glass bulb. Now solder the diode (D1) and the transistor (Q1) into the pins as shown in the pictorial drawing. To do this put a drop of solder on the end of each pin, as shown. Note that the right-hand filament pin is not used at all.

Check to see that there are no shorts in the wiring. Then cement the glass bulb back on the base.

Use any 4-pin socket you can get. If the socket is a wafer type use metal collars as stand-offs, and mount the socket with round-head wood screws. If the ferrite loopstick antenna coil (L1) that you buy doesn't have a mounting bracket, simply bend an L from strap brass, drill the necessary mounting holes, and screw the L-bracket to the base.

To Listen. Use a pair of *magnetic* high-impedance (1000 to 2000 ohm) headphones. A single headphone is fine, too, only these aren't as easy to find as they were when crystal detector radios were all the rage. For your ground connection run a wire to the nearest cold water pipe. A length of wire 25 feet long (or more) will serve as the antenna for local stations.

This radio will pick up some local stations if you're near any strong ones without the long outdoor antenna which you'll need for weaker stations. Just hook any piece of wire to the antenna terminal. In some situations you can even get reception by connecting the antenna terminal to a water pipe and forget about the ground connection. For best

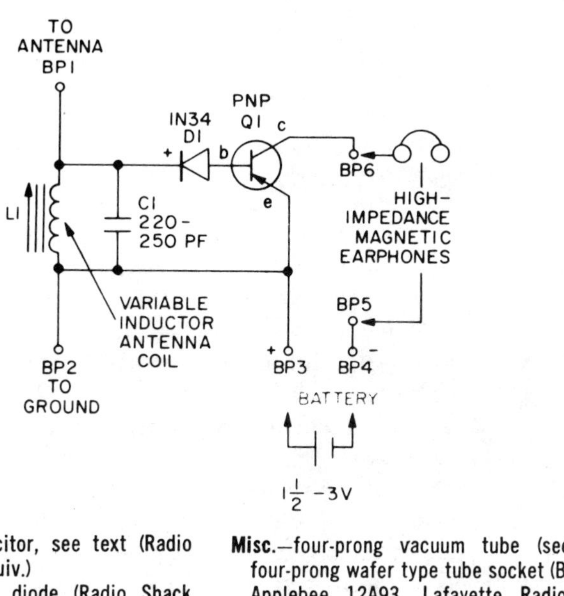

C1—250 pF. disc capacitor, see text (Radio Shack 270-1430 or equiv.)

D1—1N34A germanium diode (Radio Shack 276-821 or equiv.)

Q1—CK-722 or similar general purpose PNP transistor (Radio Shack 276-2004 or equiv.)

L1—Ferrite rod antenna coil for broadcast band, preferably with mounting bracket (Radio Shack 270-1430 or equiv.)

Phones—Any high impedance magnetic single or pair of phones, 1K to 2K or so. Cannon model CF (Lafayette Radio 40R81048-AN-151 or equiv.)

Misc.—four-prong vacuum tube (see text), four-prong wafer type tube socket (Burstein-Applebee 12A93, Lafayette Radio 32 E 20415 or equiv.), several feet of insulated hookup wire, 3-in. length of strap brass ⅜-in. or ½-in. wide for making battery holder (see text), nine ⅜-in. round-head wood screws, two metal collars for mounting tube socket (see text), 6¼-in. x 5-in. x ½-in. hardwood breadboard, knob to fit L1 adjustment screw. Binding posts (Fahnestock clips). Medium or large size (Radio Shack 270-393 or equiv.)

Fig. 3-2. Mystery Radio Schematic and parts list.

results, of course, use the longest antenna you can, as high as possible, and use a good water pipe for ground.

To turn the set on and off just remove one headphone plug from its terminal. This saves the cost of a switch to turn the penlight battery on and off.

SEE-THROUGH CRYSTAL RADIO RECEIVER

Have you ever wanted to recreate those old days of listening to AM radio on a crystal set and headphones? No tubes, no batteries, no hum, no nothing but pure clean sound drifting out of the ether into your headset? If you have the yen to get a crystal set which has the advantage of using a crystal diode instead of the old unreliable cats-whisker and galena crystal, this radio is the one for you to build. In addition to being about as good a power-supplyless AM receiver as you can make, it's also a pleasure to look at.

It closely resembles the beautiful glass-enclosed radio receivers that were custom-built by manufacturers for display in radio exhibitions in the 1920s manufacturers of radio receiver kits mounted and wired their kits in glass cabinets so the visitors could see the "works" from all angles instead of lifting the lid to look inside. Those glass cabinet radios are now rare collectors items.

Refer to Figs. 3-3 and 3-4.

This crystal radio also saves you the work and expense of making a wood cabinet, and it is low-loss for radio frequencies because the cabinet and coil form are made of styrene plastic which is a good dielectric material. The cabinet is simply a clear plastic 4-inch square photo display cube, and the coil form is a clear plastic pill container about 2-inch in diameter.

Circuit Description. The simple schematic diagram shows a time-tested hookup which is still widely used, but it's improved here by connecting one side of the crystal diode to a tap on the secondary, L_2. This greatly increases the selectivity of the receiver and helps you separate the powerful local stations.

Making the Coil. To make coils L1 and L2 the four ends of the coils are anchored in small holes drilled through the wall of the plastic container and spotted with Duco cement. You can also make small holes by pushing a hot sewing needle through the plastic. To make the tap on the coil, simply twist a small loop in the coil and spot it with glue. Scrape the enamel insulation off the loop, and solder to it.

Mounting The Parts. The coil form is mounted to the rear of variable capacitor C1 by means of a brass angle-bracket. Use lockwashers where needed to hold the screws, binding posts, and phone tip jacks securely to the plastic material. When assembling and wiring this receiver be careful not to scratch the plastic, and keep the soldering iron well away from the plastic. If you use rosin core solder, protect the plastic by covering it with pieces of paper taped in place to keep the rosin from splattering on the plastic.

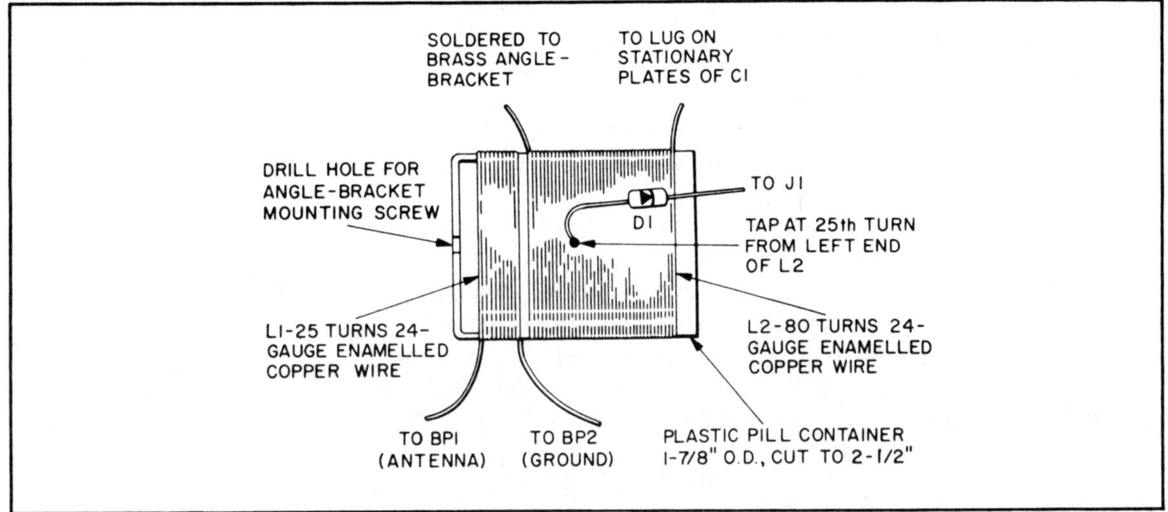

Fig. 3-3. Apart from getting hold of the parts, construction of the See-Through Crystal Set should take you less than two hours in all. You can use any kind of box you want, but the lucite box shown will make it a real showpiece.

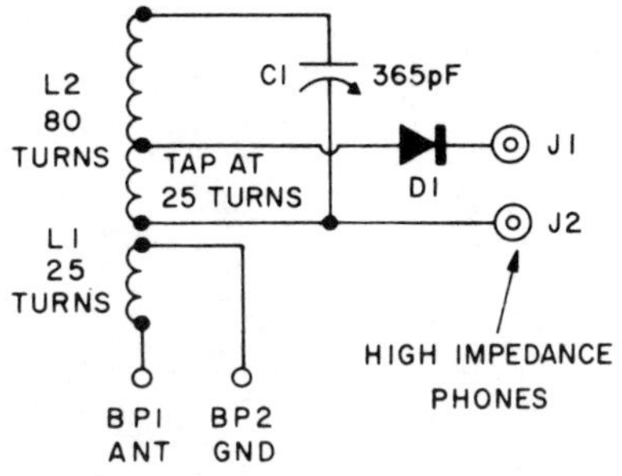

Fig. 3-4. See-Through Crystal Radio Receiver parts list.

BP1, 2—Binding posts for antenna and ground connections; may be any convenient type

C1—365-pF variable tuning capacitor, single-gang.

D1—Small-signal, general purpose diode, similar to 1N34.

J1, 2—Jacks for headphones (dependent on phone(s) selected.

Misc.—Headphone(s), high impedance. May be crystal or magnetic, or small earphone as supplied with transistor radios and portable tape machines; plastic photo display cube, approx. 4-in. each dimension; plastic pill container, 1 7/8-in. diameter, for use as coil form; 1/4-lb. 24-gauge enamelled copper magnet wire (Radio Shack 278-004 or equiv.); brass mounting strip, assorted screws, nuts and lockwashers.

The completed crystal radio is mounted on a fancy walnut base purchased from a woodworking shop. The plastic box is secured to the wood base by spotting the four corners of the bottom of the box with plastic cement. The dial for the pointer knob is simply a small disc of white double-weight paper held to the plastic box with a spot of plastic cement. Make pencil marks at the places where your local stations come in.

Use of a pair of sensitive high-impedance magnetic or crystal earphones, a good connection to a cold water pipe, and a long outdoor antenna (for best results, put up a long single-wire, random-length antenna). With a simple crystal set, it becomes particularly important that the antenna-ground system be the best possible. Remember, unlike its bigger cousin, the superheterodyne, the crystal set does not have rf amplifiers and other circuitry to help it pull in all those signals floating around out there in the ether.

OATMEAL BOX CRYSTAL

Ask just about any radio old-timer, including this writer, and he will probably tell you that his first radio was a home-brew slide tuning coil wound on an oatmeal box, a cat whisker and galena crystal detector, and a pair of earphones. We show you how to make such a radio.

Refer to Figs. 3-5 through 3-8.

First, make the coil. Remove the two end covers from an 18-ounce, round Quaker Oats box, and cut the tube to a length of about 6½-inch. Give the tube a coat of shellac inside and out to moisture-proof it.

The writer used No. 21 single-cotton-covered enameled copper magnet wire, and after the coil was wound the cotton was colored green by painting it with India ink to make it look like the old-time green silk-covered wire which is no longer being made. If you prefer, use No. 20 or No. 21 enameled or nylon-coated copper magnet wire, and one pound should easily do it.

Get Going. Punch two small holes through the tube at each end, about ½-inch from the ends, to anchor the ends of your coil. To do a

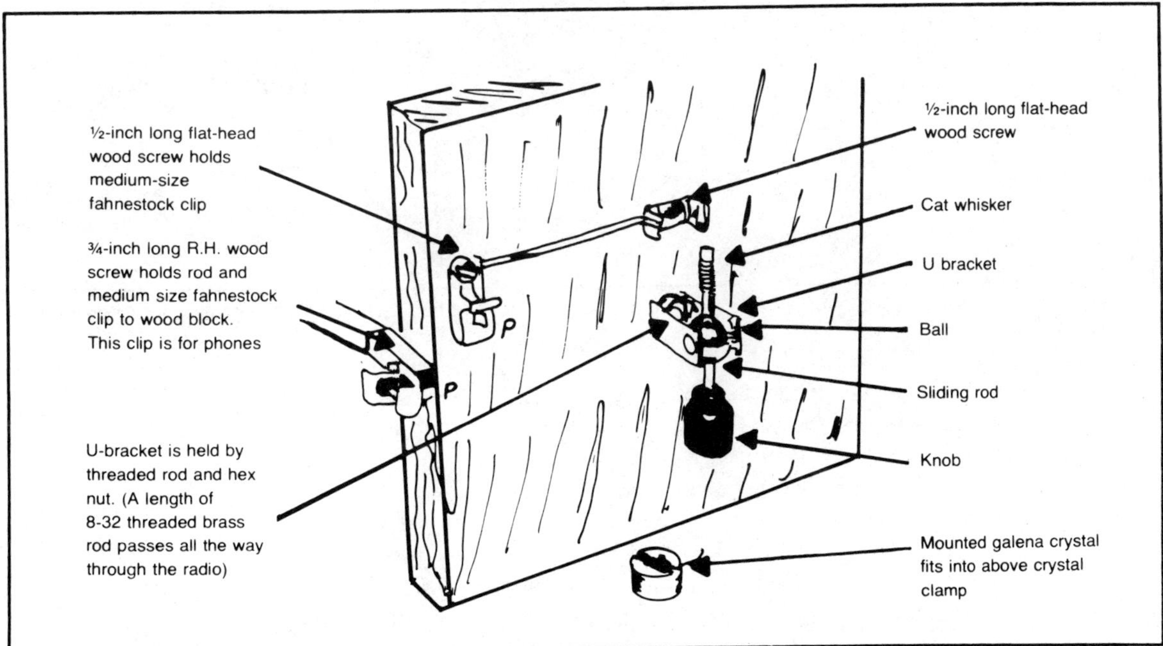

Fig. 3-5. Crystal detector end of the Quaker Oats radio.

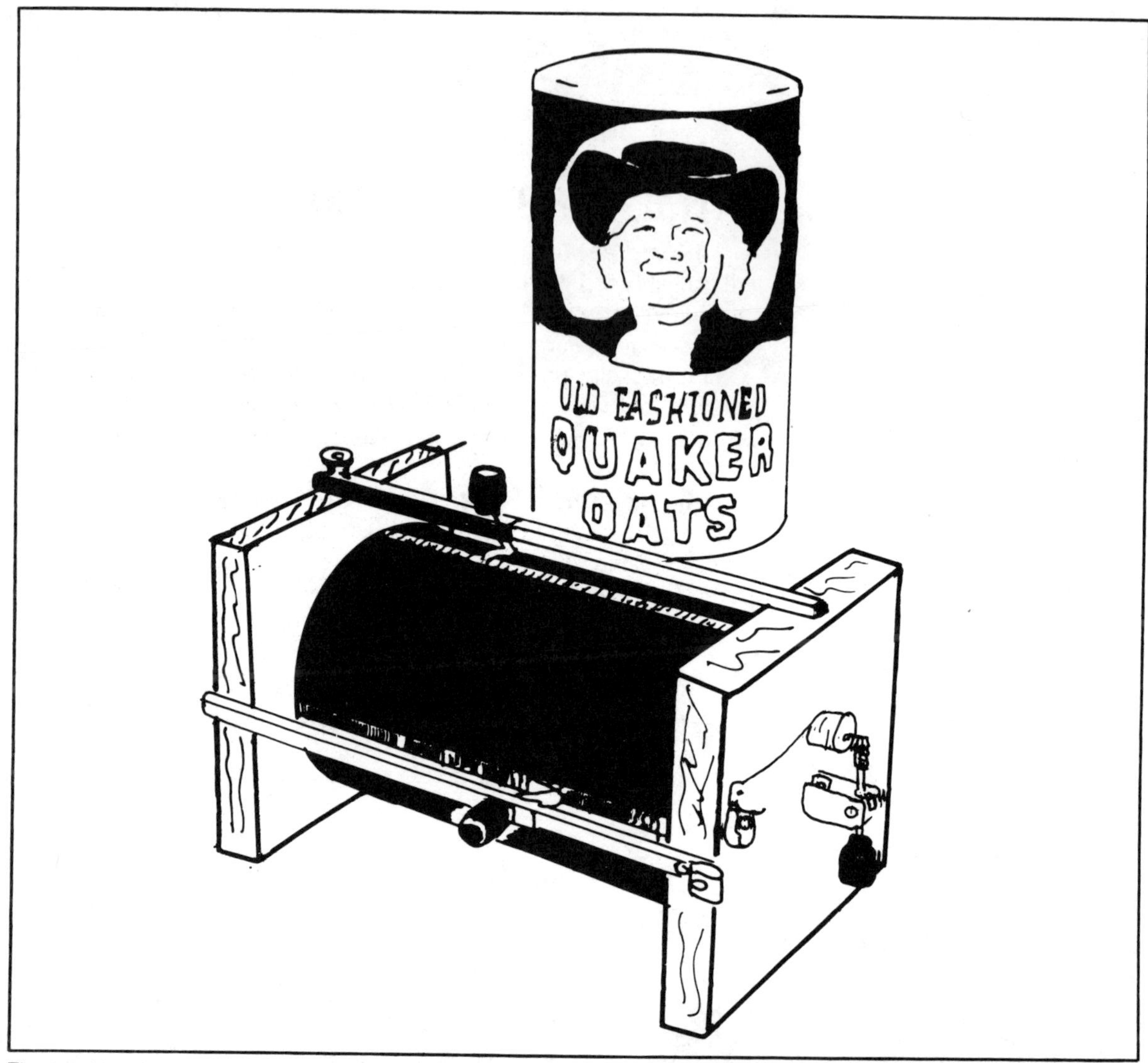

Fig. 3-6. Your basic materials may be the same, but the dollars required to buy them have certainly bounced upward from bygone days.

tight, smooth and neat job of winding the coil, tie the end of the wire to some object outdoors where there is plenty of room, and unwind a couple hundred feet of wire, and pull the wire tight to stretch out any bends in the wire. Cut off the wire and anchor the end in the two small holes near one end of the tube, and dab a bit of cement to hold it fast. Now wind the coil by turning the tube slowly while you walk towards the tied end of the wire, and when the tube is full of wire cut off the wire and anchor the end in the two holes at the other end of the tube and put on a dab of cement. This trick will give you a neat professional-looking coil.

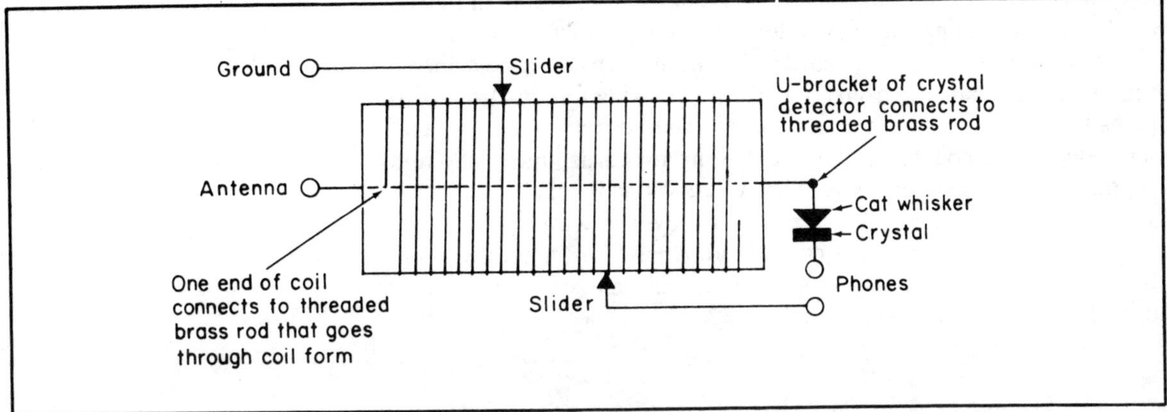

Fig. 3-7. Simple schematic for the crystal radio.

The two wood end blocks for the coil measure 5 × 5 × ¾-inch and are sanded smooth, stained, and varnished.

Bore a 3/16-inch hole through the exact center of each wood block; these are for the length of 8-32 threaded brass rod that passes through the coil and holds the wood end blocks. One end of the threaded rod holds the U-bracket of the crystal detector, and the other end of the rod serves as the antenna binding post.

Note that the end of the coil nearest to the antenna binding post passes through a small hole in the wood block and is clamped between the two washers of the antenna binding post; this automatically connects the coil end to the U-bracket of the crystal detector also.

The figures give details for mounting the slide rods, the earphone, Fahnestock clips, the grounding binding post, and the clamp that holds

1 round Quaker Oats box (18 oz.)
1 lb. #20 copper magnet wire, for winding coil (see text)
2 pieces 5-in. x 5-in. x ¾-in. oak, walnut, or mahogany (for coil end blocks)
1 foot of 8-32 threaded brass rod (to pass through coil form)
1 8-32 brass hex nut (holds crystal detector U-bracket to wood block)
2 12-in. lengths 7/32 OD square brass tubing or solid rod (for slider tracks)
3 ¾-in.-long round-head wood screws (hold brass rods to wood blocks)
1 8-32 x 1½-in. round-head brass machine screw, with hex nut and ornamental thumb nut to fit (for ground binding post)
3 inches of square brass tubing to fit snugly over slider rods (for making the two sliders)
2 6-32 x ¼-in. flat-head brass machine screws (to hold knobs to top of sliders)
3 inches ½-in.-wide brass band (for slider)
4 inches of ¼-in.-wide brass band (for making slider contact blades)
2 medium-size fahnestock clips (for phones binding posts)
1 ½-in. long flat-head wood screw (holds one fahnestock clip to wood block)
1 unmounted crystal detector stand (K/D Stand 9-14, Modern Radio Labs.)
1 mounted galena crystal for above detector stand (9-1 MRL Steel Galena, Modern Radio Labs., P.O. Box 1477, Garden Grove, CA 92642)
1 ½-in. long flat-head wood screw (holds crystal clamp to wood block)

Fig. 3-8. Oatmeal Box Crystal Radio parts list.

the galena crystal. The simple hook-up is shown. The first figure gives all details for making the two sliders that will contact the coil.

Contact. Perhaps the hardest job of all is to do a neat job of removing the insulation from the coil when making the two bare wire paths for the sliders. Use fine sandpaper and be careful not to sand off too much of the copper. When you are through brush away any fine copper dust between the turns of the wire. You will get a neater job if you use enameled wire instead of cotton-covered wire.

For best results with this crystal radio, use a long antenna, a cold water pipe ground, a sensitive galena crystal, and a sensitive high-impedance pair of magnetic earphones.

ANTENNA SYSTEMS FOR SWLS

If you're an armchair shortwave traveler, you've probably tried to hear Rabaul, Upper Volta, Yemen, Hanoi, or Vientiane. If you haven't been successful, be patient; that's what it takes—patience—*plus* a good shortwave antenna. We can't supply the patience, but here are some good antenna ideas which are sure to help.

No matter what shortwave receiver you're using, a good antenna is a must to bring in those distant stations. If it weren't, the manufacturer wouldn't have supplied antenna terminals! The problem is what kind of antenna—a hunk of wire? Or maybe something more scientific? We'll help you make the decision by telling you a little about shortwave (SW) antennas and how they work.

Refer to Figs. 3-9 through 3-13.

Shortwave antennas can be short and simple or they can be complicated and cover several acres. For shortwave listening most of us are limited to the short and simple ones—those that fit in a backyard and don't cost too much. But even a simple antenna, properly designed and installed, can work wonders.

The first figure shows several commonly used SW antennas with lengths shown for the SW broadcast bands. The antenna in Fig. 3-9A is known as an unbalanced end-fed longwire. It can be hung horizontally or vertically, or a combination of both. When hung horizontally it has some degree of directionality, while it tends to be omnidirectional when vertical. This antenna will work well on all frequencies if it is made long enough, or it can be cut to operate at only certain frequencies. It works best with an antenna tuner that can be located at the receiver, since the lead-in is part of the antenna's total length. The longwire has a high terminal impedance and always operates best with a tuner that matches the antenna to the receiver. With an antenna tuner, it is an ideal antenna to run around the eaves of the house, or across the attic.

Another Type. Figures 3-9B and 3-9C show the popular center-fed balanced dipole. This antenna can also be hung vertically or horizontally and uses two balanced leads to the receiver that can be any length. This antenna is always cut to a resonant length, though it will also work well at three times the resonant frequency. For instance, an antenna that works in the 90-meter SW band (3.2-3.4 MHz) will also work for the 31-meter band (9.5-9.7 MHz). If space is limited, its ends can be bent down (or even back) as much as 25 percent before reduction in performance becomes serious. Figure 3-9B shows a single-wire dipole fed with 75-ohm feedline which can be plastic appliance cord.

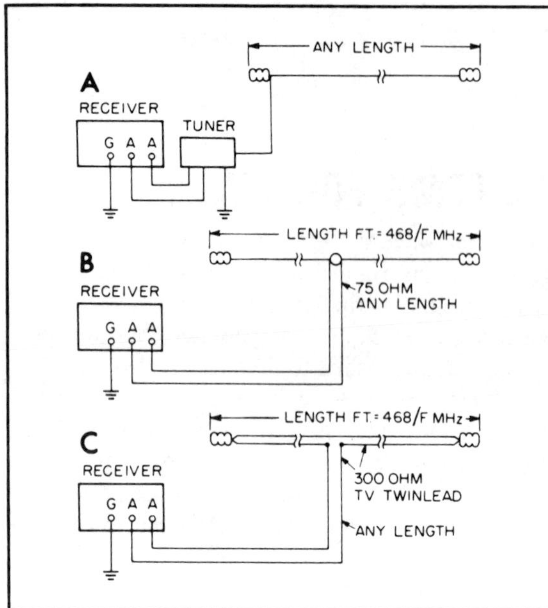

Fig. 3-9. Three porcelain antenna insulators are used for antenna A. One separates each section of the antenna at the point of feedline connection. The other two insulate the antenna ends from the supporting structure. The antenna at B uses 300-ohm twinlead for the transmission line and the antenna elements. Only the ends at the feedline connections are wound together and soldered. Outside ends are separated, with one partially removed as shown. At C the use of 75-ohm is called for.

Figure 3-9C shows the folded dipole version built from 300-ohm twinlead. The folded dipole, incidentally, will work well at half the resonant frequency.

As previously mentioned, a vertical antenna tends to be omnidirectional, while a horizontal antenna tends to have directional characteristics. For general around-the-globe listening the vertical antenna is probably best, though low frequency resonant antennas are difficult to orient in this position because of their length.

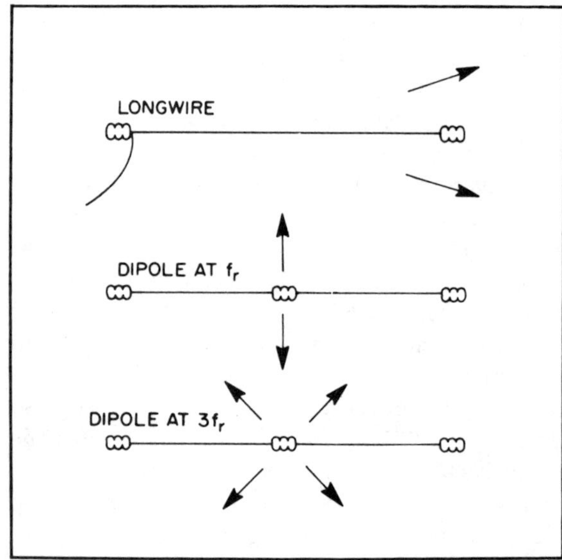

Fig. 3-10. Direction of maximum pickup for the horizontal longwire and dipole.

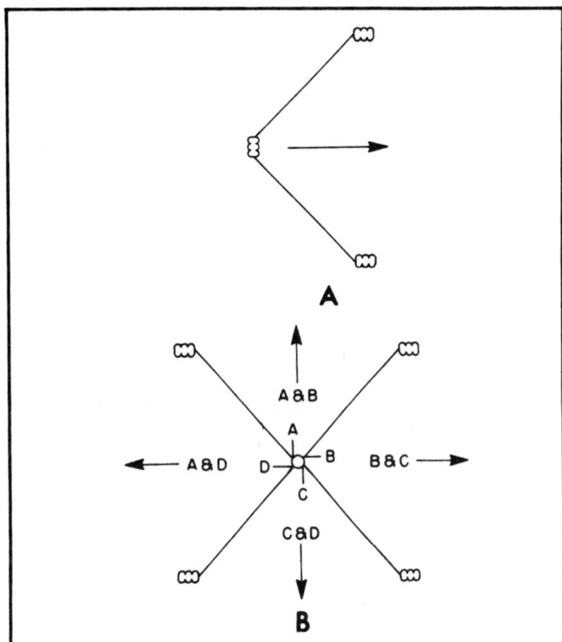

Fig. 3-11. Directionality can be further increased by bending the ends of the dipole inward.

Directionals. If you are interested in distance (DX) from a particular part of the globe, however, the directional characteristics of a horizontal antenna can work for you. Figure 3-1D shows the direction of maximum pickup for the horizontal longwire and dipole. By looking at a globe and determining the shortest path to the area you want to hear, you can position your antenna and use its directional characteristics to advantage.

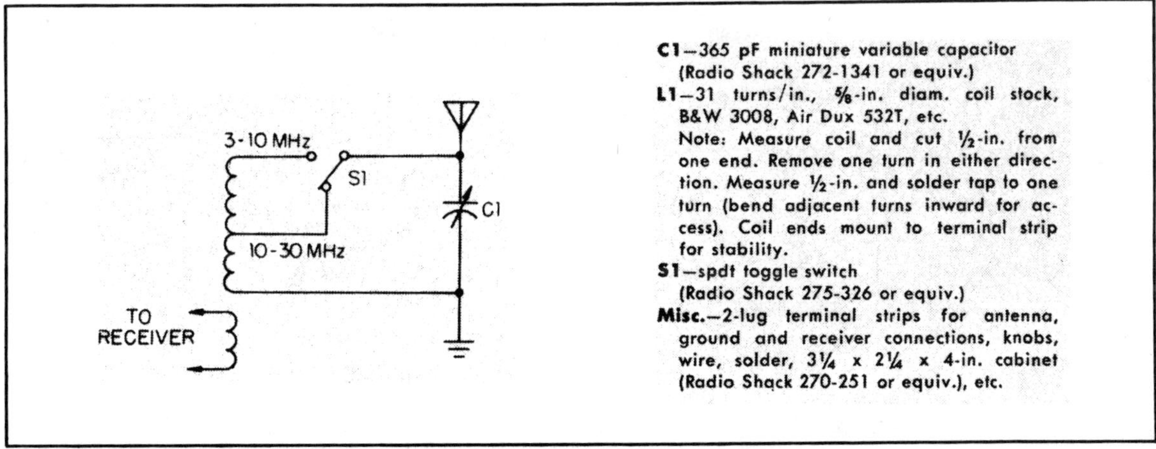

Fig. 3-12. Here's a simple antenna tuner any SWL can whip together in an evening and use for a lifetime. Neatness counts when monkeying with low level RF, so mimic the author's model for best results. Keep solder connections clean.

Figure 3-11 shows how directionality can be further increased by bending the ends of the dipole inward. This type of antenna can be easily built by using a center support, such as a TV mast, and bringing the ends in and down toward the ground. For best results, the ends of the antenna must be 10 feet or more above the earth.

A more elaborate antenna that will "look" in any one of four directions can be made by mounting two dipoles in this manner at right angles and connecting the lead-in to different elements to achieve the desired direction. This deluxe array has a disadvantage in that you must have easy access to the top, or center part, in order to change lead-in connections.

Fortunately, choice DX can be logged on any of the eleven international broadcast bands; but it is difficult (if not impossible) for the serious SW listener to come up with a good antenna for each of eleven bands. Few SWLs have the real estate or inclination to put up a single tuned antenna for each band, so a couple of multiband antennas running in different directions is often the answer. Figure 3-13 shows simple multi-band antennas that can be used; and, through compromises, they will give all around performance.

Still Around. The basic antenna shown in Fig. 3-13A was popular in the 1930s, and is known as the "windom" antenna. It can be fed with 300-ohm TV twinlead, and works well on even harmonics of the fundamental frequency.

Figure 3-13B shows how, by using 300-ohm twinlead, two antennas can be connected to the same lead-in to give satisfactory performance on the 60, 49, 41, 31, 25, 19, 16, 13, and 11 meter bands. This permits

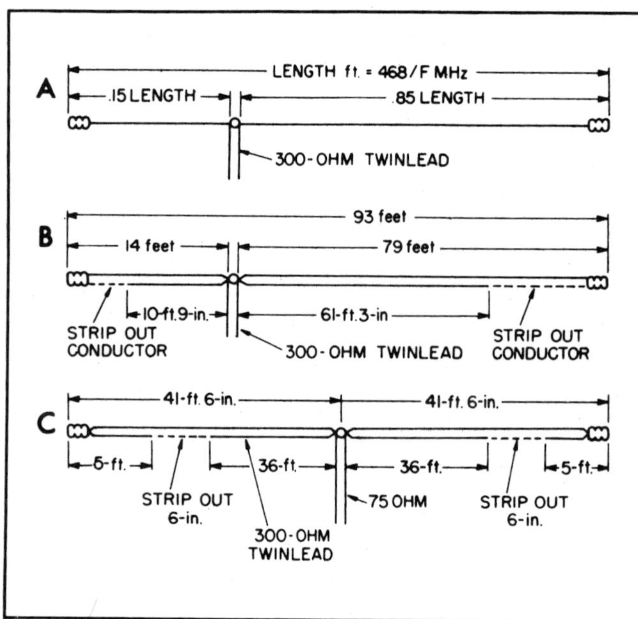

Fig. 3-13. Three porcelain antenna insulators are used for antenna A. One separates each section of the antenna at the point of feedline connection. The other two insulate the antenna ends from the supporting structure. The antenna at B uses 300-ohm twinlead for the transmission line and the antenna elements. Only the ends at the feedline connections are wound together and soldered. Outside ends are separated, with one partially removed as shown. At C, 75-ohm line is needed.

coverage of 9 of the 11 international broadcast bands with a single antenna. By tying the lead-ins together at the receiver and using an antenna tuner, this antenna becomes a longwire, making it probably the most versatile SW antenna available.

Another multiband antenna shown in Fig. 3-13C consists of two centerfed dipoles made from 300-ohm twinlead connected to the same feedline. This antenna has the advantage of being short (nice for small city lots, or apartment dwellers) and performs well on the 60, 49, 41, 25, 19, 16, 13, and 11 meter bands. Again, it can be connected as a longwire at the receiver and used with an antenna tuner.

The circuit of a simple SWL antenna tuner that you can easily build is shown in Fig. 3-12. Details of the tuner built in a small utility box is shown in the photographs. This SWL antenna tuner can be used to improve the performance of any longwire antenna. Select the proper range for C1 with S1 and peak C1 for best S-meter output on your receiver.

Summing Up. The best antenna for you depends on the type of DX hunting you want to do and the space available. A long-wire with the antenna tuner shown will work well for general listening. If you're interested in a particular part of the work and a particular band, a single frequency dipole pointed in the right direction will give excellent results. If you want one antenna that will do as much as possible, use a multi-band antenna. In any case, those difficult-to-log DX stations will come a lot quicker with any of these antennas, mounted as high as possible.

PQ MINDER

Ever been on the verge of a real DX catch or making a delicate adjustment on a project when all of a sudden a buddy drops in, "HIYA OM!! HOWZIT GOIN'?" Your DX catch is lost, or your adjustment is disrupted, and/or your train of thought is lost.

Take heart, friend! Our PQ Minder, mounted on the door of your operations center, will cause guests to think twice before barging in too quickly. The PQ Minder is a transistorized flasher with a bright red warning light that a fellow would have to be color blind to miss! To prevent prying fingers from playing around and turning if *off*, it incorporates a magnetic proximity switch whose location only you know.

Refer to Figs. 3-14 through 3-16.

Building It. Start with the code practice oscillator mode. Cut out the little disc capacitor that is connected between the base of Q1 and the collector of Q2. Replace it with a 25-μF electrolytic rated at 6 volts or better (C1). Be sure that the positive end goes to the junction of R2 and the base of Q1. This is the only modification required for the board. A No. 48 or 49 lamp in parallel with a 10-ohm, ½-watt resistor is connected in place of the speaker shown in the diagram that comes with the module. Power is supplied by three penlight cells mounted in the battery holder at the bottom of the case.

Mount the pilot lamp assembly in one end of the plastic box. Fasten the battery holder and the CP module inside the box to the 4 × 2½-inch plastic face.

The Mystery Switch. Mount the reed switch inside the box on the end opposite to the pilot lamp assembly. Cut its terminals short so that the switch will fit snugly into the end of the box. Fasten to this end, outside the box, two metal clips to provide a slip-in holder for the

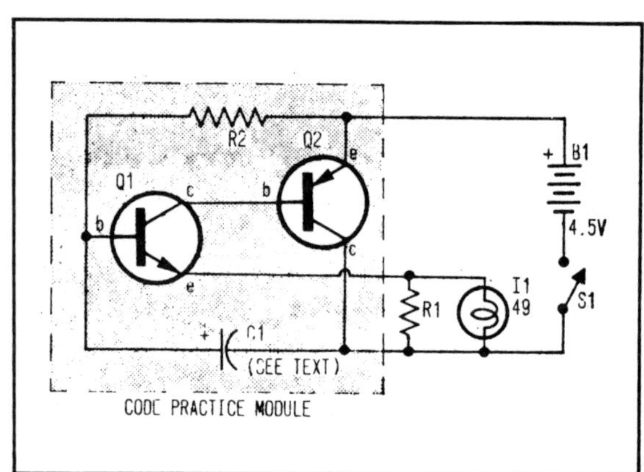

Fig. 3-14. Just four parts are all that are needed to make the PQ minder.

B1—Three penlight cells (Lafayette 99T6258 or equiv.)
C1—25-uF, 6-volt electrolytic capacitor (Lafayette 34T8429 or equiv.)
I1—Lamp assembly (holder, Lafayette 99T6339 or equiv.; bulb, Lafayette 32T6621 or equiv.)
Q1, Q2—Part of code practice module
R1—10-ohm, ½-watt resistor
R2—Part of code practice module
S1—Magnetic proximity switch (Lafayette 34T4401 or equiv.)
1—Code practice oscillator (Lafayette 19T1513 or equiv.)
1—4 x 2½ x 1⅝ in. plastic box (Lafayette 99T8078 or equiv.)
Misc.—Hardware, wire, aluminum for slide clips, solder, etc.

Fig. 3-15. PQ Minder parts list.

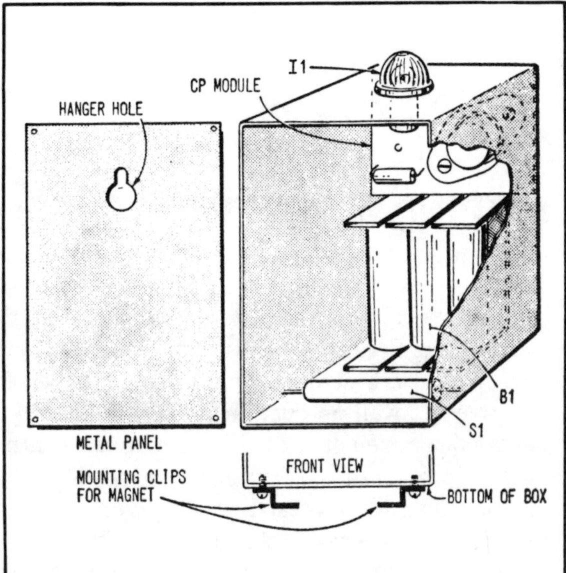

Fig. 3-16. PQ Minder parts placement and mounting methods.

magnet assembly. Paint the mounting clips and magnet assembly the same color as the plastic box so that magnet will not be noticeable. The idea here is to foil any attempt by a passerby to shut off the flashing lamp, since he will now be unlikely to see the magnetic switch.

Bore a hole in the metal panel which now becomes the back of the box, making it easy to hang the gadget on a nail or hook on the door. After mounting, wire all components in accordance with schematic diagram.

Now, remember that your PQ Minder is designed to be seen and not *overlooked*! It's a good idea to paint the box so it can be quite colorful and seen easily in a dark room. A bright yellow enamel will do the job. Now, if the PQ Minder does not stop that gate crasher, try wiring the door knob.

ROARING TWENTIES RECEIVER

During the roaring 1920s, the crystal set graduated from the experimenter's oatmeal box-breadboard, to wood-cabinet construction, as the quality of radio station broadcasting improved enough to interest the entire family. The crystal receiver was moved to the living room from the basement laboratory, and therefore had to appear attractive, as well as operate properly. The style of radio cabinet construction resembled the popular style of furniture of that era—heavy, ornate wood with a walnut veneer finish.

You can build a crystal set receiver similar to the ones in use during the 1920s, a wood cabinet model with a lift-up lid as shown in the photos. All of the circuit components are mounted in the back of the front panel like the old-time receivers.

The Receiver Circuit. The crystal set uses two high-*Q* spiderweb coils with variable coupling in a two-circuit tuner for maximum selectivity and sensitivity. Antenna signals at J1 are series-tuned by C1/L1 for maximum rf gain. L1 acts as the primary winding of a tuned rf transformer, with L2/C2 as the tuned secondary. Coupling is variable for best selectivity. D1 detects the signals, and the audio is fed to the headphones at J3.

Refer to Figs. 3-17 through 3-19.

Cabinet Construction. As you can see in the photos, the design of the crystal set cabinet is similar to the old radios of the 1920s. The top and bottom wood sections extend outward from the cabinet sides and have rounded edges. The top section is used as the cabinet lid, and is held on to the back of the box with two metal hinges. A section of black plastic is used for the front panel in place of the bakelite used by the early constructors. If black plastic is not available, you can substitute black-painted hardboard.

Begin construction of the cabinet by cutting the wood sections to the sizes indicated in the construction drawing. Then fasten the sections together with small finishing nails and glue. The front panel should not be fastened, but should be able to move free in or out of the cabinet. Position the corner moldings, before gluing them. Notch the top of the rear panel to allow clearance for the metal hinges.

After construction of the cabinet is completed, check for any rough edges and sand the wood surface. Then stain or paint the cabinet. Our model has a walnut stain with a clear plastic finish.

Spiderweb Coil Construction. As shown in the drawing of the spiderweb coil form, there are seventeen "vanes," ⅞-inch long, and each approximately ¼-inch wide, positioned around the perimeter of a

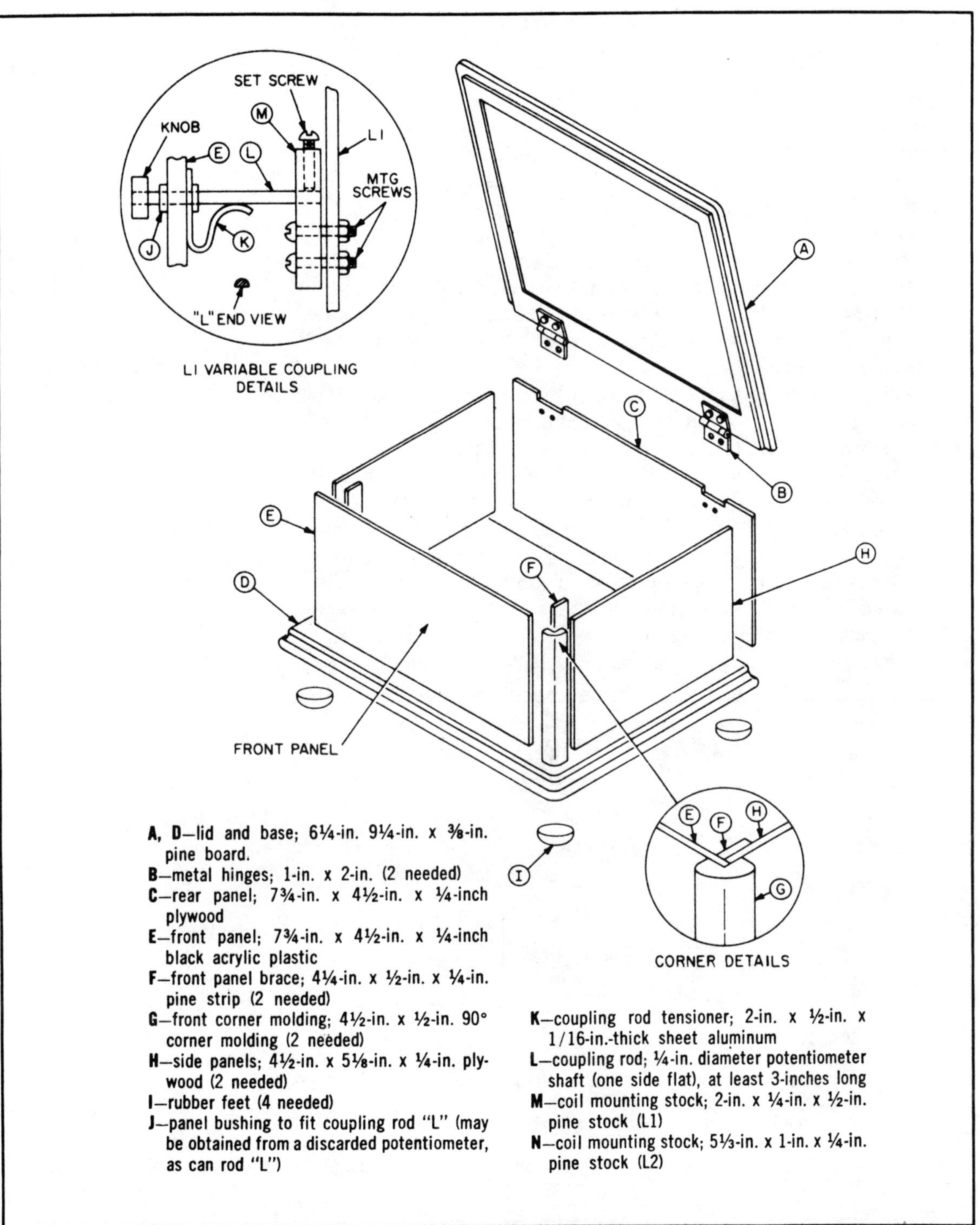

Fig. 3-17. Roaring Twenties Receiver parts list for the cabinet.

3⅝-inch disc. The coil form is made from the type of sheet plastic used for printed circuits (but without the copper coating) and is approximately 1/16-inch thick.

The easiest way to start construction of the coil forms, is to trace the outline of the spiderweb coil form drawing and then temporarily paste the tracing onto the plastic sheet. Fasten the plastic sheet firmly in a vise, and then cut out the vanes with a hacksaw. Remove the tracing paper from the plastic and round off any edges with a file.

Carefully drill two small holes at the center of the coil form, and mount two solder lugs. Wind as much No. 28 enameled magnet wire around the coil forms as possible (winding over one vane, then under the next, and so on) and solder the wire ends to the solder lugs. It is not necessary to count the turns of wire, as the coils can be pruned later after testing in the receiver.

Receiver Circuit Construction. All of the receiver components are mounted on the black plastic front panel. The layout of the parts is shown in the photos. Start construction by taping a section of graph paper to the panel and locating the mounting holes for the components. Install the two tuning capacitors (C1 and C2) on the panel after making allowances for the diameter of the tuning knobs.

Next, drill a hole for the shaft bushing of the L1 variable coupling rod (L). The one on our model came from a discarded volume control, with a three-inch flat (on one side) shaft. Bend a section (K) of sheet aluminum (after drilling a hole to fit the bushing) to fit against the flat side of the shaft, and install the metal tensioner and bushing on the

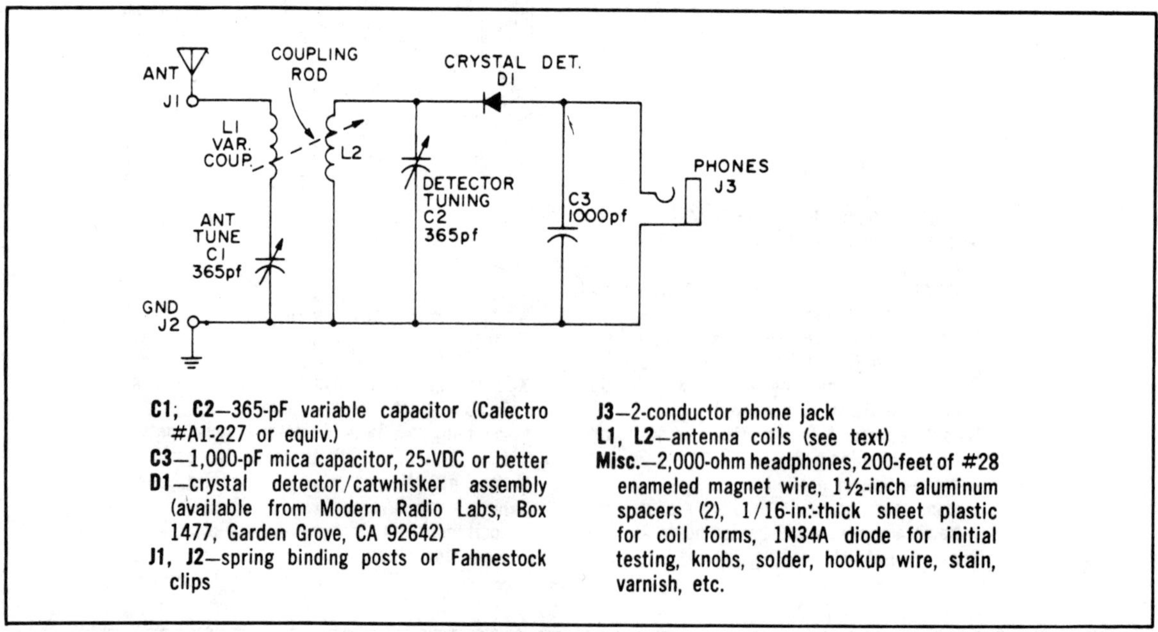

C1; C2—365-pF variable capacitor (Calectro #A1-227 or equiv.)
C3—1,000-pF mica capacitor, 25-VDC or better
D1—crystal detector/catwhisker assembly (available from Modern Radio Labs, Box 1477, Garden Grove, CA 92642)
J1, J2—spring binding posts or Fahnestock clips
J3—2-conductor phone jack
L1, L2—antenna coils (see text)
Misc.—2,000-ohm headphones, 200-feet of #28 enameled magnet wire, 1½-inch aluminum spacers (2), 1/16-in.-thick sheet plastic for coil forms, 1N34A diode for initial testing, knobs, solder, hookup wire, stain, varnish, etc.

Fig. 3-18. Roaring Twenties Receiver parts list for the radio.

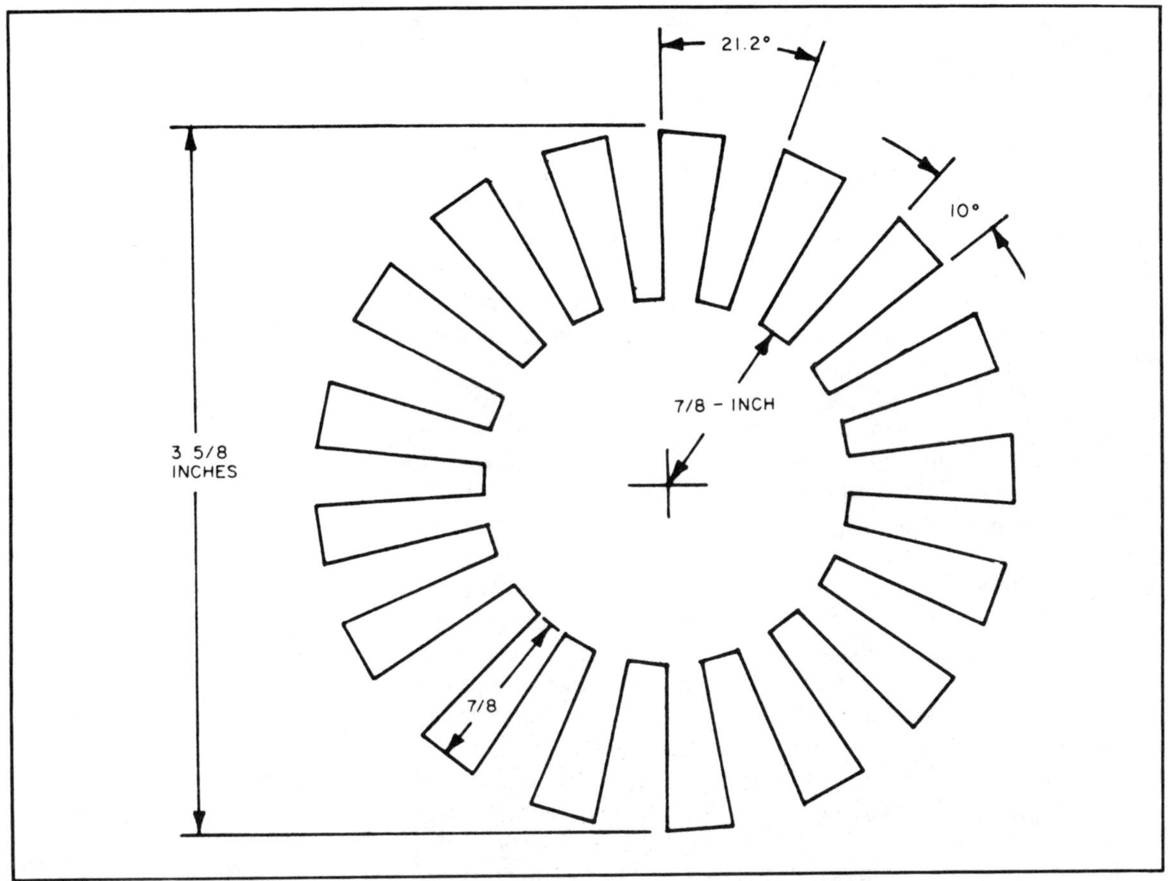

Fig. 3-19. Full-scale template for cutting the coil forms. Cut it off the page carefully, or photocopy it, temporarily fasten it to the plastic coil form stock, and make your cuts. Repeat this process again for the second coil form as well. After both forms are cut, compare them to make sure that there are no differences between the two—they should be made (and wound) as identically as possible to assure you of a proper match and optimum performance. Use No. 28 enameled wire for the coil windings and leave room at the beginning and end of the coils for connections to the solder lugs which you'll use later.

panel. The metal section is required to keep the coupling rod from rotating as it is pushed in and out. Drill a hole to fit the shaft near one end of a block of wood (Section M in the cabinet construction drawing detail) and install a set screw to hold the shaft in place. Install the block of wood with two screws and nuts on to the spiderweb coil L1 and insert the rod end into the panel bushing. Make sure that the coil can be pushed freely in and out to vary the coupling.

Install the other spiderweb coil (L2) on a wood section that is mounted on the rear panel with spacers and screws. The mounting of the wood section will depend on the size and shape of the variable capacitors in your model. The two spiderweb coils should be positioned so that they are approximately ¼-inch apart when the coupling rod is pushed all the way in. The wood section on our model was supported by two screws set in 1½-inch metal spacers mounted on tapped holes in the rear of the two tuning capacitors. If the two capacitors in your model can not be used in this way, increase the length of the wood section and the mounting screws as necessary and mount the wood section in the back panel.

Install the remaining parts onto the front panel as shown in the photos. If necessary, file down the head of the crystal cup screw to allow proper seating of the crystal. Wire the components as shown in the schematic. Use flexible stranded wire for the leads to L1, and position them so that there will be no interference with the tuning capacitors as L1 is moved in or out. Install solder lugs as required on all of the components, and use bare solid wire for the connections (for an antique wiring look; make square corners).

Next, install the knobs on the controls. Old-style knobs can be found at flea markets and hamfests. If none can be found, do what the old experimenters did—make your own. The big tuning knobs can be made with painted cardboard discs cemented onto the back of small plastic knobs. Fahnestock clips can be used in place of the terminals (J1, J2) on our model, and also in place of J3 for the headphone connections. Install the front panel on the cabinet and make sure that the spiderweb coils do not touch the top lid or bottom of the cabinet.

Operation. A good outside antenna and a good ground connection are required for best results with a crystal set receiver. There is no amplification as in vacuum tube or transistor radios, therefore the stronger the reception of the radio waves, the louder the signals will be. If you are located near a high-powered radio transmitter, an inside antenna will work. For more distant stations, an outside antenna, 25 feet or longer will be necessary. The ground connection can be made to a cold water pipe, or to a metal rod driven into the ground.

For ease of adjustment in the initial test of the receiver, connect a germanium diode (1N34A or equiv.) in place of the crystal detector, D1. Plug a set of 2000-ohm earphones into J3 (low-impedance, stereo-type earphones will *not* work), and connect the antenna to J1 and the ground to J2.

Tune C1 and C3 for a received station, and then adjust the L1 coupling for best selectivity. Retune C1 and C3 for best signal strength, and then readjust the L1 coupling. All of the controls interact, so it will require several tuning adjustments for optimum received signal audio volume. After a station is tuned in, remove the germanium diode without disturbing the tuning settings and then try your luck in finding a sensitive spot on the crystal with the catwhisker (like the old radio pioneers did). Compare the received volume with that of the germanium diode—you may be surprised at the reception you get. Now, if only we could find an antique radio *station* . . .

SIGNAL SNARE

An amplifier stage that is used at two widely separated frequencies (such as rf and audio) is called a *reflex amplifier*. This type of amplifier circuit was used in the early days of radio because of its economical use of the then expensive vacuum tubes. Later, between 1934 and 1937, reflex circuits were used in small home radios. Later, when transistors first became popular, two transistor radios were manufactured in Japan and sold here for very small prices. These transistor circuits employed reflex amplifiers, usually in a TRF type of receiver with a crystal diode detector. One transistor was employed as a reflex rf amplifier and first audio stage, and the second transistor was used as the audio power amplifier.

Refer to Figs. 3-20 and 3-21.

You can experiment with the reflex circuit by building our simple One-Transistor Reflex Receiver project. The circuit employs a j-FET as a tuned rf amplifier and also as a stage of audio (after the signal is detected by a germanium diode). The circuit is laid out breadboard style for easy construction.

The Reflex Action. The reflex circuit is a system in which an amplifying device (transistor or vacuum tube) is made to function at both rf (or if) and audio frequencies. As commonly used, the signal is amplified by the device, detected, and the resultant audio signal fed back into the same device for further amplification. Such a circuit has two inputs (one for each type of signal frequency) and two outputs, with filtering necessary to split the two sets of signals.

Look at the signal flow block diagram of the One-Transistor Reflex Receiver project. This is a diagram of a typical reflex circuit. Signals are amplified at rf frequencies and then fed through a signal splitter to a crystal diode detector. The detected output is filtered and coupled back through the tuned circuit to the amplifier, where the signal is now at audio frequencies. The audio frequencies are amplified and fed through the signal splitter to the low pass filter and to the headphones.

As shown in the block diagram and the schematic, radio signals are coupled through J1 and C1 to the tuned circuit of C3/L1/ and to the gate of the field-effect transistor (j-FET) Q2. C2 places the bottom end of L1 and the rotor of C3 at rf ground. R3 supplies the bias for Q2, and C8 is the rf/af bypass capacitor. R1 functions as the Q1 gate dc return and audio input load. The amplified rf signals from the drain of Q1 are coupled through C9 to R5 and detector D1. C7 is connected in shunt with the rf output of Q1 and is used to adjust the rf gain of the j-FET.

The detected audio signal is fed through the rc filter composed of R4, C5, and C6 to the volume control R2. The audio is then coupled through C4 to the junction of R1, C2, L1, and C3. C2 presents a low impedance to rf, but has a high impedance to audio. L1 has a high impedance to rf, but has a low impedance to audio. The two types of signals (rf and audio) are therefore applied directly to the gate of Q1, each being amplified therein.

The amplified audio signal at the drain of Q1 is coupled through L2 to the J2 headphone jack. L2 serves as both a component of the low-pass filter (L2/C10) and as an rf load for the signal splitting action of C9 to D1. L2 is chosen to have a high value of reactance over the broadcast band and serves to broadly tune the D1 detector circuit over the range of 550 kHz to 1500 kHz. The dc power for the circuit is supplied by an external 6-volt battery (or dc power supply), and C11 serves as an audio filter for the power input.

Construction. The receiver is built on a 6 × 3¾ × 2-inch plastic box with a perfboard section installed on top. Most of the components are mounted on the perfboard with push-in solder terminals. The input and output connectors J1 and J2, are mounted on the front and rear of the box. The 6-volt battery, or power supply, is connected via two leads fed through a hole in the rear of the box. The tuning capacitor C3 is mounted directly on the perfboard with machine screws, and the volume control R2 is mounted on a small bracket made from sheet aluminum that is also installed on the perfboard.

Begin construction by cutting a perfboard section to size to fit the top of the box. Locate corner holes to fit the threaded molded screw retaining extrusions located inside the top corners of the box. Mount C3 on the perfboard. The capacitor used in our model had threaded holes in the bottom for easy mounting with machine screws and

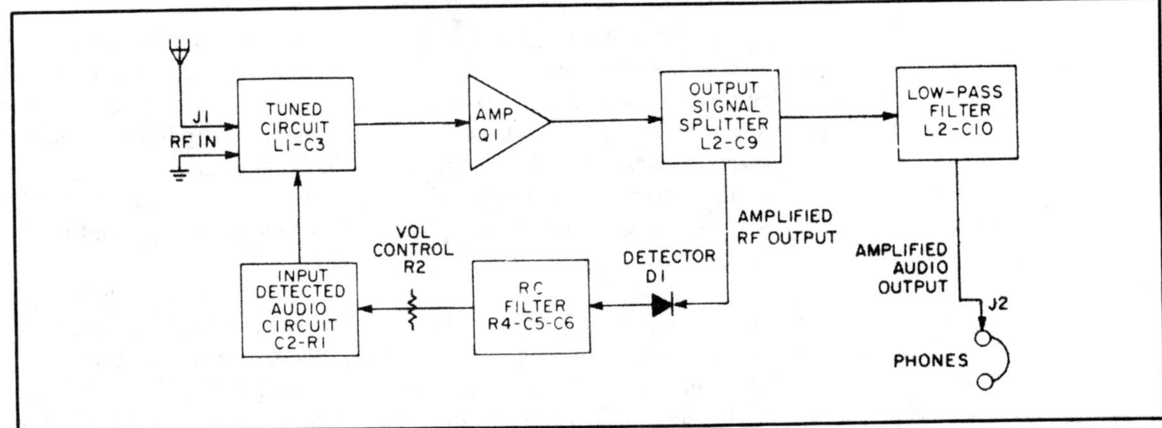

Fig. 3-20. This block diagram shows how the signal is amplified as af and rf, then split and fed back to the tuned LC circuit. The low-pass filter separates the amplified audio signal. The author used perfboard mounted on a plastic box as chassis for the signal snare. Push-in solder terminals make construction a snap. The input/output jacks are all mounted on box under the board.

washers to lift up the stator insulating panels from contact with the board surface. If your capacitor does not have these mounting holes, the capacitor can also be mounted by a small bracket cut from sheet aluminum to fit front mounting holes. Cut a bracket to fit R2 and mount it on the front of the perfboard in the general location shown in the photos. Use an internal toothed lock washer between the R2 mounting nut and the bracket surface to prevent accidental movement of the volume control.

Mount the remainder of the components on the perfboard with push-in solder terminals. The locations are critical, so follow the layout of our receiver model, as shown in the figure. Install a solder lug on the frame of the tuning capacitor C3 for connection to the rotor. Wire the board components as shown in the schematic and keep the connecting leads as short and direct as possible. L1 is mounted with two push-in terminals soldered to the coil connecting lugs.

Install the headphone jack J2 on the box front and the antenna/ground jack, J1, on the rear of the box as shown in the photos. Cut a hole in the rear of the box for the battery leads and complete the wiring of the perfboard with the interconnecting leads to the box components. Install lugs on the battery leads to fit your battery terminals. The leads on our model extended approximately 10 inches from the box, but the lead lengths are not critical and can be any convenient length to fit your particular installation. To minimize accidental breakage, the battery leads should be stranded wire (preferably color coded; red for positive and black for negative).

Testing Your Reflex. For best results, an outdoor antenna and a good ground should be connected to J1 (center connector to the antenna and the outer shell to ground). Because the Reflex Receiver project only uses a simple tuned circuit, you may have some problems with overloading on strong local stations. C1 value can be changed to adjust the antenna loading; small capacitance (3 to 20 pF) for light loading and better selectivity, and larger capacitance (25 pF to 47 pF) for heavier loading, higher sensitivity, but lesser selectivity.

Before connecting up the receiver project to the battery, check the wiring and then adjust the rf Gain control, C7, to maximum capacity (minimum rf Gain). Set the volume control R2 full counter-clockwise (for minimum audio gain). Check the lead polarity before connecting the 6-volt battery to the receiver, then plug in the headphones (2000-ohm type).

Tune the broadcast band with C3 while adjusting R2 for a comfortable audio volume. Adjust L1 for best band coverage for your particular location. Tune the slug in for more inductance. Adjust C7 for increased rf sensitivity over the band, and if oscillation occurs, adjust the maximum capacity (just short of the oscillation point).

Check the rf amplifier action in the reflex circuit by tuning in a weak station and then connecting a 500 pF capacitor (approximate

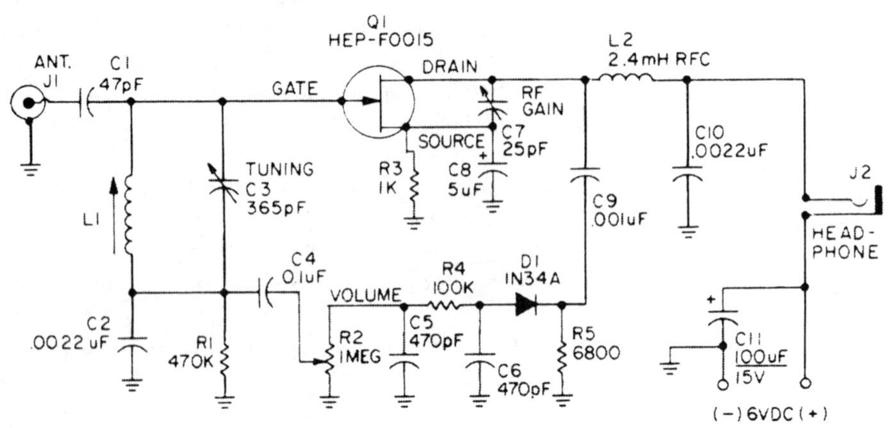

C1—47-pF ceramic disc capacitor, 15 volts
C2, C10—0.0022-uF ceramic disc capacitor, 15-volts
C3—365-pF variable capacitor
C4—0.1-uF ceramic disc capacitor, 15-volts
C5, C6—470-pF mica capacitor, 15-volts
C7—2 to 25-pF mica trimmer capacitor
C8—5-uF electrolytic capacitor, 15-volts
C9—0.001 uF ceramic disc capacitor, 15-volts
C11—100 uF electrolytic capacitor, 15-volts
D1—1N34A germanium diode (or equiv.)
J1—Phono jack (RCA type) for Ant. and Gnd. connector
J2—Phone jack (to fit your headphones)
L1—broadcast band loopstick antenna coil (Radio Shack #270-1430 or equiv.)
L2—24-millihenry RF choke
Q1—HEP F0015 NPN FET (Motorola) or equivalent
R1—470,000-ohm, ¼-watt resistor
R2—1-megohm potentiometer, audio taper
R3—1000-ohm, ¼-watt resistor
R4—100,000-ohm, ¼-watt resistor
R5—6800-ohm, ¼-watt resistor
Misc.—Plastic box, approx. 3¾ by 6 by 2-inches with perf-board top section, knobs, sheet aluminum bracket for R2, push-in clips for perf-board, hookup wire. Also: Hi-Z headphones (2000-ohms or higher) and a 6-volt battery or DC power supply.

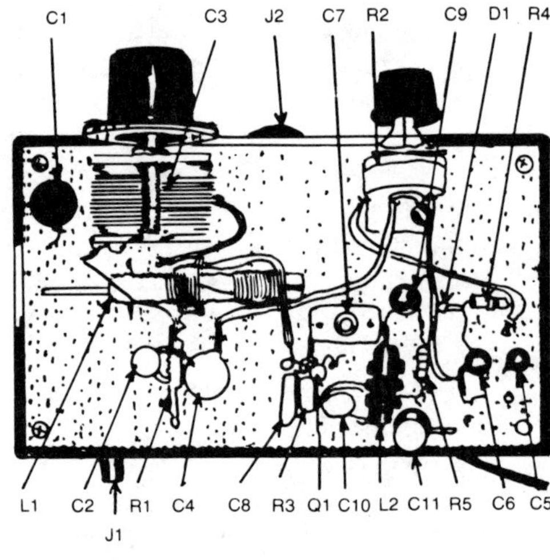

Fig. 3-21. Signal Snare parts layout, schematic, and parts list.

value) between the drain of Q1 and ground. The station signal should diminish. This indicates that the rf reflex portion of the circuit is operating correctly. The capacitor shunts the rf to ground, but does not affect the audio amplification as it has a high impedance to the audio frequencies.

Check the audio amplifier portion of the reflex circuit by tuning in a station and then removing the headphones from J2. Temporarily connect a 100-ohm, ½-watt resistor in place of the headphones to J2. Connect the headphones with a pair of clip leads across the output of D1 (across C6), and compare the signal level with the previous level at J2. The level at D1 should be lower than that of J2 to verify that the audio is being amplified by Q1.

SUPER BCB BOOSTER

Imagine your broadcast band receiver jammed from end to end with a solid wall of signals! Flea's-whisper stations, that normally can't be heard with headphones, booming into your shack at S9. This is the kind of reception you'll get with the Super BCB (broadcast band) Booster, a preamplifier specifically designed for BC DXers.

Whether you live in a concrete and steel tower, or out in the boondocks with enough space for a long-wire antenna, the Super BCB Booster will dig out stations you've never heard before because its average gain is almost 42 dB to 7 S units of extra sensitivity.

The booster can function as an electronic antenna with signals received only by loopstick antenna coil L1, or as a preamplifier, with a longwire antenna connected to binding post BP1.

Refer to Figs. 3-22 through 3-24.

How It Works. The signal voltage appearing across tuned circuit L1/C1 is fed to FET Q1, which provides approximately 20 dB gain on top of the L1/C1 resonant gain. Q1's output feeds transistor Q2, an emitter-follower that provides a 10-dB to 15-dB power gain, and also a low-impedance output for connection to the relatively low impedance output of a communications receiver.

Though intended for direct connection to a receiver's antenna input terminals, the Super BCB Booster can also be used with loop antenna radios by connecting the booster's output to a loopstick antenna (duplicate of L1) positioned near the radio. We'll show how both connections are used.

Powered by a type 9-volt transistor radio battery the current drain is less than 2 mA and a standard battery will last at least three-months, even under heavy service. An activator or heavy-duty battery can last a year or more. With such low power consumption there's no reason to build an external ac power supply for the Super BCB Booster.

Construction. Although the circuit appears simple, extreme care must be taken with the circuit board preparation since the high overall gain can cause instability if a single component, or printed circuit foil, is out of position. We suggest no attempt be made to use point-to-point wiring. Use a PC board that is an exact copy of the supplied template (any PC board material can be used). If you cannot make your own PC boards you can obtain a *plated board* (for easy soldering) from the source indicated in the parts list.

Avoid component substitutions: Q1 and Q2 should be the specified types. Though the booster might work with some general replacement transistors, it might not work with others. Worse yet, it might work only on very weak signals while distorting strong signals.

The specified components will provide distortion-free reception on signals as strong as 80,000 μV. It will deliver excellent performance with battery voltage falling as low as 6-volts.

The circuit board and a *very short* connection to output jack J1 are the only critical assemblies. You can make mechanical modifications as long as the general layout approximates the unit shown in the photographs.

We suggest that the unit be assembled in a plastic cabinet with an aluminum front panel, though a full plastic cabinet can be used because the PC board has a built-in hand-capacitance shield. Maximum stability, however, is attained through the use of a metal front panel because it reduces the possibility of feedback from booster's output to its input.

Drilling the PC Board. All of the component mounting holes except for tuning capacitor C1 can be made with a No. 58, No. 59, or No. 60 bit. Capacitor C1 requires a 5/16-inch hole. If you don't have a 5/16-inch bit use a ¼-inch drill and enlarge the hole very carefully with a miniature round file. The corner holes, which are used for the mounting screws, should clear No. 4 or No. 6 screws—whichever you prefer to use.

The PC board is best assembled in the following manner: Install capacitor C1 first, then all the remaining components except Q1. Then push Q1's leads through the board and solder. Note that Q1 is supplied with a shorting-clip around all the leads. *This clip must be left in position until assembly is completed and the booster is ready for operation.* If the clip is removed a high static voltage from the tip of the soldering iron,

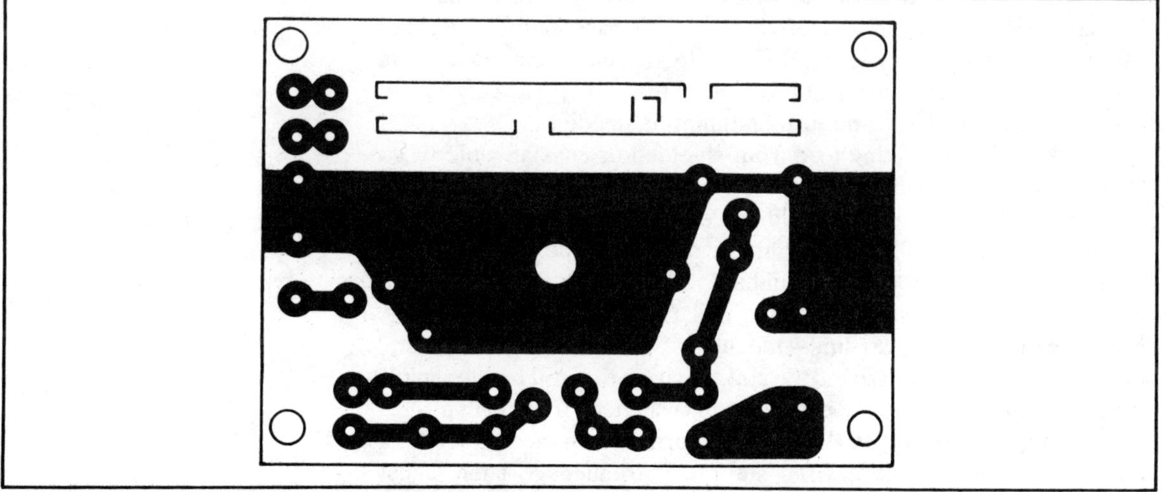

Fig. 3-22. Make a PC board that is an exact duplicate of this template. Do not try to use point-to-point wiring since the high gains could cause instability if any parts or strips of foil were out of position.

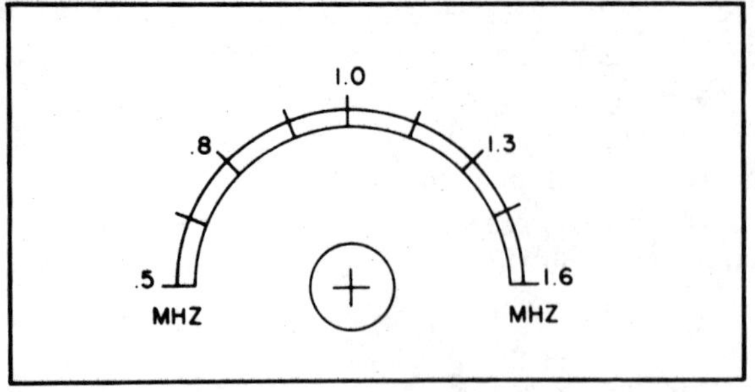

Fig. 3-23. The tuning dial faceplate can be used as is. Just cut it out and paste it down.

or a voltage generated through normal handling, might destroy Q1. Place a paper or tape tag on or near Q1 to remind you to pull off the shorting clip before applying power.

Double check that the tab sticking out from Q1's case faces the nearest edge of the PC board before soldering. The round side (opposite the flat) of Q2 should face the same edge of the PC board.

Note that L1's primary and secondary windings are independent though their *ground* connections are generally shorted together by a wire jumper on the PC board. If for some reason you prefer a separate antenna system ground, open the shorting wire and install a ground binding post on the panel. After L1 is wired to the PC board it can be secured with a few dabs of silicon rubber adhesive.

Since stand-offs space the PC board away from the panel to prevent shorts between the foil(s) and the metal panel, it will be impossible to add wiring after the assembly is installed on the panel. Install the wires for the connections to BP1, J1, SW1, and the battery connector before mounting the PC board. Insulated No. 20 or No. 22 solid wire is suggested. Mount the PC board to the panel using a ¼-inch spacer or stack of washers between the panel and PC board at each mounting screw. After all wiring to the panel components is completed adjust L1's slug so it protrudes between ¼- to ½-inch from the top of the coil form—no further tuning is needed.

Make up a connecting lead from shielded or coaxial cable to go from output jack J1 to the receiver's antenna terminals. For least signal attenuation the lead length should not exceed 15 inches.

If the booster will be used with a transistor-type radio having a built-in loop antenna and no terminals, connect the free end of the output cable to a loopstick antenna coil the exact duplicate of L1. Remove the primary winding—the heavy outer winding of plastic insulated wire wrapped around the coil. Position this coil on the radio's case opposite the built-in loopstick antenna and tape the coil in place.

Using The Super BCB Booster. Turn on both the receiver and booster and tune in the desired station or frequency. Then adjust turning capacitor C1 for maximum signal strength or highest S-meter reading. As a general rule the direct signal pickup by L1 will be sufficient. If greater sensitivity is needed connect 6 to 15 feet of wire to

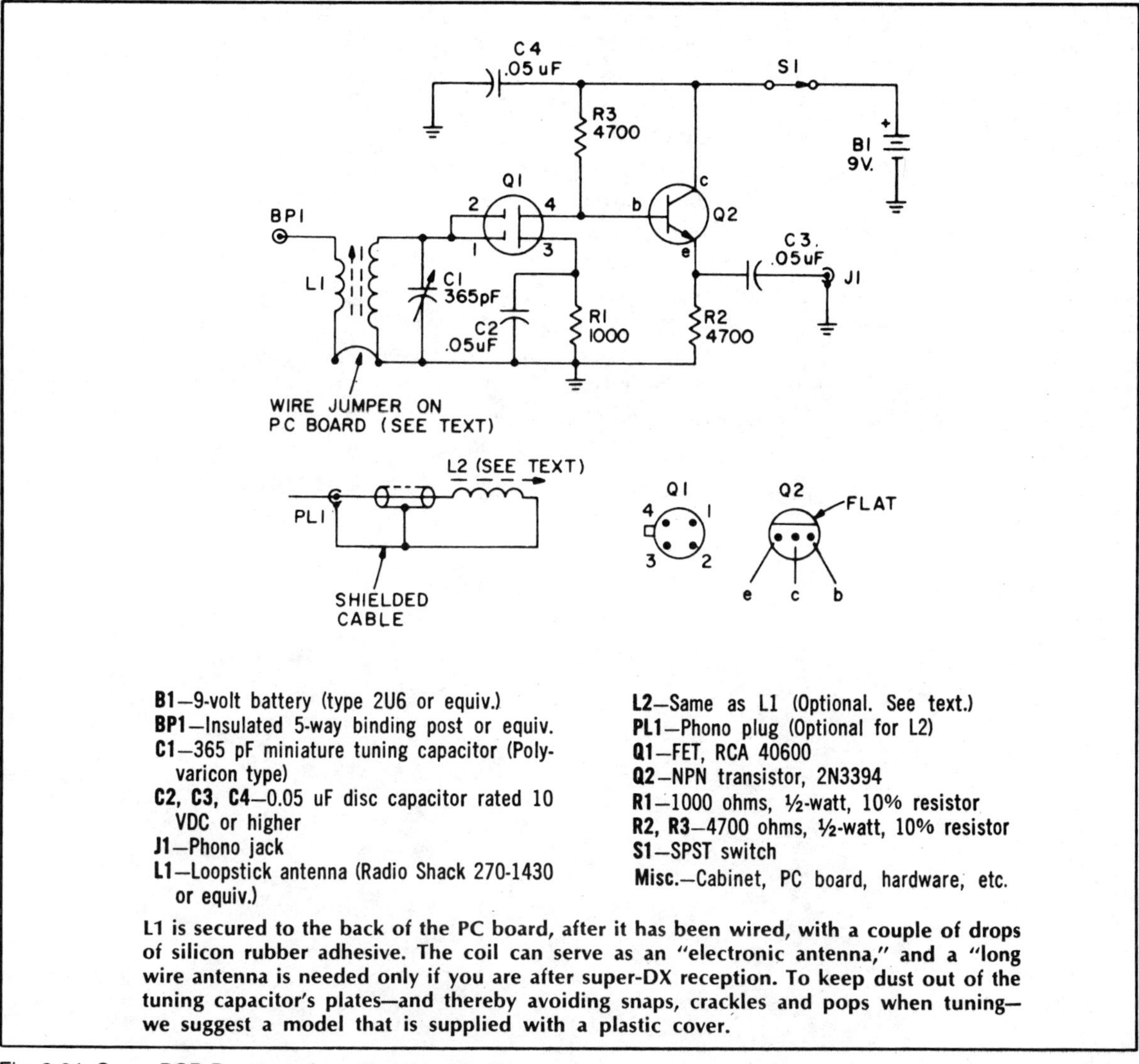

Fig. 3-24. Super BCB Booster schematic and parts list.

BP1. If you connect a longwire antenna take note; the signal level out of the booster can be so high as to overload the receiver.

If there is a strong local station in your area it's possible that the signal strength will be boosted to a level that swamps the receiver when listening to a weak signal on the other end of the dial. If this happens simply detune the booster away from the strong signal's frequency until the interference is reduced or eliminated. While this might sacrifice some gain on the desired station, the loss will probably be slight.

It is possible for the booster's output to radiate back to its input (particularly when using a loopstick coupling coil). You'll know when this happens—the booster breaking into self-oscillation—if the receiver gets "blocked," or off-frequency signals get tuned in and *lost* as C1 is adjusted. If this problem occurs, position the booster as far as possible from the receiver and keep any external antenna, if used, away from the receiver and booster's output cable.

Under certain conditions the Super BCB Booster will provide an additional benefit which should not be construed as improper operation. Some inexpensive receivers are highly prone to marine band image interference when signals at the high end of the broadcast band are received. The booster, by providing tuned preselection, suppresses or eliminates these interference-causing image signals while providing amplification to broadcast signals. Don't assume the loss of image signal reception means reduced sensitivity. Actually, the desired signal will be getting full boost while the image interference is squashed.

SHORTWAVE SUPERCHARGER

Some big changes are on the way for the shortwave listener (SWL), especially in the upper shortwave bands from the 25-meter band on up to 30 MHz and beyond. The Sun is now entering one of its periods of increased sunspot activity after a 20-year period of relative calm. This will make short range communications unreliable and long range DX an everyday affair. Signals from stations just down the road will be, literally, lost in outer space, and wishy washy signals from outer nowhere will come booming into your listening post like they were right next door.

Refer to Figs. 3-25 through 3-27.

Under these conditions many old and some not-so-old shortwave receivers will need a bit of help when they try to work the high bands. Their circuits tend to get a little frazzled. As a matter of fact, almost any SWL would appreciate a bit of signal boost now and then. It might just make the difference between a very good DX catch and a record breaking DX discovery.

If you decide you want a DX boost or you need to increase the versatility of your old set then you should build this Shortwave Supercharger.

This unit will boost selectively the rf signal by 20-30 dB and it will compensate for many deficiencies of your set. It will not improve the gain of the shortwave receiver but will also improve its selectivity and the image frequency rejection. Simple, single conversion superhet SW sets have the annoying tendency of receiving spurious signals separated by twice the if frequency from the desired signal. For example if you tune to 20 MHz you may also receive $20 + (2 \times 0.455) = 20.910$ MHz (image frequency) signal which will interfere with the 20 MHz signal. In addition you will be able to pull in many SW stations you didn't even know existed. With 10 to 15 feet of wire behind your sofa as an antenna you may receive stations as distant as Australia or China.

How Does It Work. The circuit is based on an inexpensive integrated circuit manufactured by Motorola. Its innards consist of three transistors, a diode and four resistors which together form an excellent automatic gain controlled (AGC) radio frequency amplifier. To build the circuit with separate discrete components would cost a bundle and the result would not be as good. The incoming rf signal is coupled with a few turns of wire to the coil L1. The tuned parallel-resonant circuit consisting of L1 and C1 selects the wanted signal by rejecting adjacent frequencies and feeds the signal to pin 1 of the integrated circuit. The amplified signal leaves the IC on pin 6. The

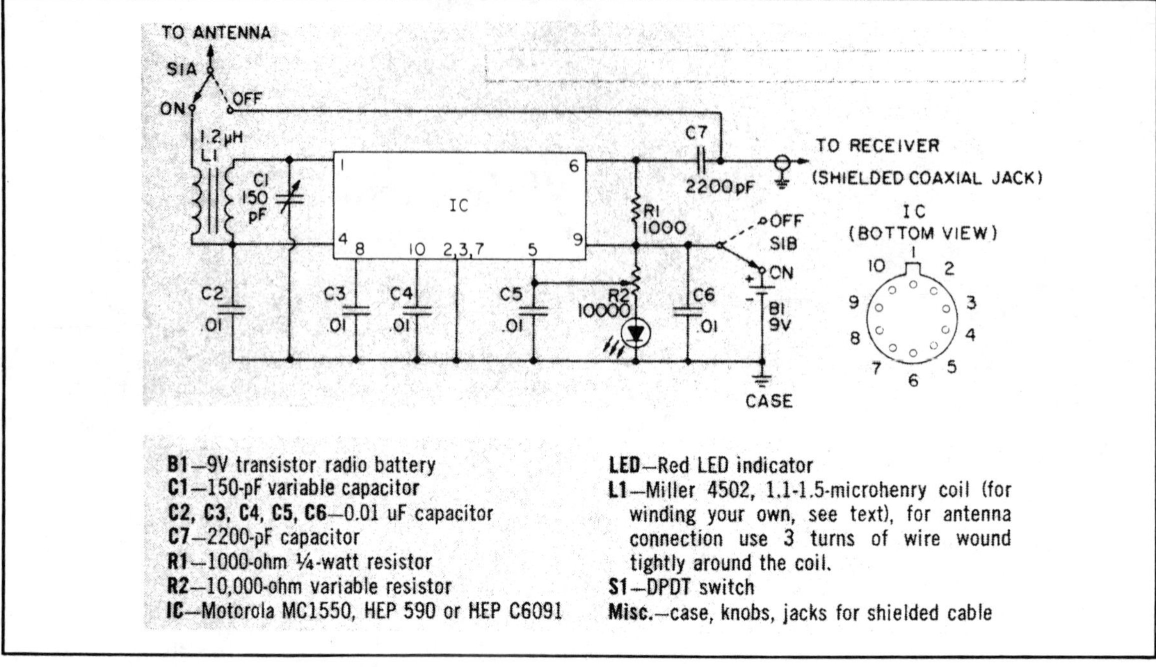

Fig. 3-25. Shortwave Supercharger schematic and parts list.

AGC input on pin 5 is used to control the gain of the amplifier when you turn potentiometer R2. The light emitting diode indicates that the circuit is on and that the battery is still alive. The DPDT switch S1 selects between straight-through connection, booster off and booster on.

Construction. This is a radio frequency project which requires a neat soldering job and short connections. However, if you do a half-decent job the supercharger should fulfill your expectations. The author used point-to-point wiring on a perfboard. If you have some experience with PC boards you might use the layout shown here. The Supercharger with the indicated component values will cover approximately 10-30 MHz. Using different values for L1 or C1 will

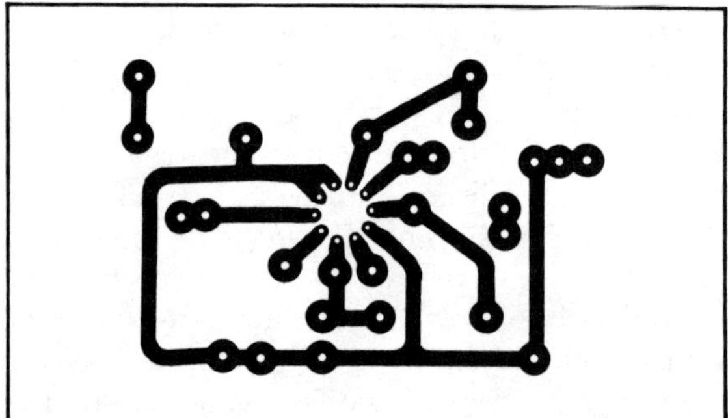

Fig. 3-26. Use this full-size circuit board template to build your Shortwave Supercharger. You can find etching materials at a radio or electronics shop.

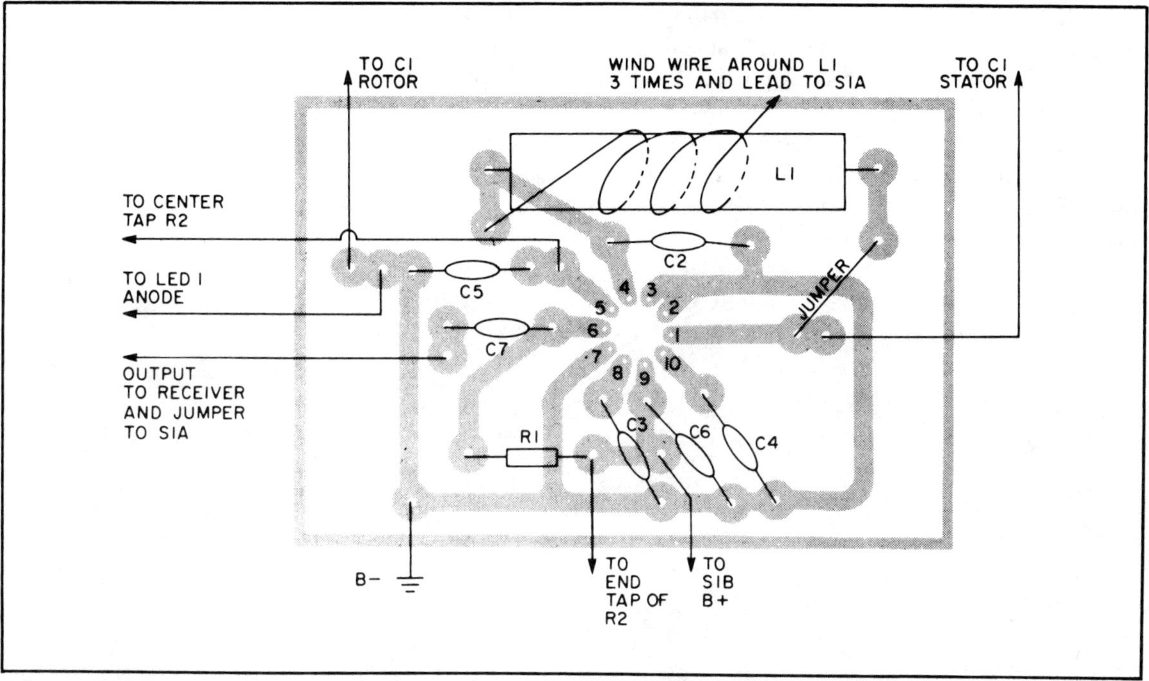

Fig. 3-27. The parts location overlay is twice the actual size in order to clarify the positioning. If you use a loop different than specified in the parts list, you may want to modify the appropriate spacing on the printed circuit board. Don't forget to wrap the L1-to-antenna wire around the loop stick three times. You might install an integrated circuit socket on the printed circuit board to simplify installation and repair.

change this range, though the ratio of minimum to maximum frequency will remain 1:3. Doubling the capacitor *or* inductance value lowers the frequency by 1.41 and lowering either value increases the frequency by the same factor. If you want to substitute some parts, or wind up your own coil or use a different capacitor, the circuit is quite flexible in this respect. For example you may want to replace the 150 pF capacitor C1 used by the author since this is often difficult to find. Use instead the oscillator half of the standard AM tuning capacitor from any pocket transistor radio. Instead of the coil mentioned in the parts list you might try to wind 15-20 turns of insulated copper wire on a pencil.

Mount the Shortwave Supercharger inside a metal case which you can find in most electronic supply stores. Use shielded cable between the supercharger and your receiver otherwise the connecting wire will behave like an antenna and some of the features of your supercharger will be lost. The final job is to make a dial. You can calibrate it with your shortwave receiver by tuning C1 to optimum reception.

If you find that the circuit "whistles" at certain frequencies (this may easily happen if you do not use a PC board or your connections are too long), the simplest cure is to thread a few small ferrite cores through pins 1 and 4 of the IC. Such cores can be purchased from many electronic surplus dealers.

Operation. Tune your receiver to the desired frequency and then tune C1 till you can hear maximum signal or noise, if no station is present. Returning your receiver with the fine tuning knob should require no readjustment of the supercharger. You can use R2 as your volume control or leave it in some intermediate position and use the volume control of the receiver. For strong signals, you may want to turn R2 back to prevent overloading the receiver with the corresponding increase in the background noise.

Once you get it working, start digging deeper into the higher shortwave frequencies. There is a lot going on out there and with the increased sunspot activity and a Shortwave Supercharger you can't go wrong.

BEAT THE ANTENNA SPIRAL

Beginners to the shortwave listening hobby have no difficulty in obtaining good receivers, either budget jobs or gold-plated specials, when starting their first listening shack. Putting up antennas is their downfall.

Antenna theory is beyond the grasp of most novices. It is very complex at the beginning and rapidly becomes incomprehensible as different antenna types are introduced. So, why not take a shortcut approach to your first antenna installation. Get your new receiver pulling weak signals as you pile up listening hours with exotic DXing. What about antenna theory? It'll come if you work at it by reading theory books, but in the meantime here are three recorded case histories to low-cost antenna shortcuts which may be profitable.

Refer to Figs. 3-28 through 3-30 and Tables 3-1 and 3-2.

Case No. 1—The Dangler. Harry is a youngster I met while giving a talk to the local high school student body during Science Fair Week. Harry was fascinated by the idea of English language newscasts from far-away places, so he bought a Realistic DX-160 receiver and set up a listening corner in his upstairs room in his folks' Colonial-style house. For an antenna, he dangled an odd length of wire out the window, letting it drop to the ground. BBC and Radio Moscow came in fine except on rainy nights. In fact, it was a rainy evening when he rang my doorbell for help.

Harry's long wire was long and that's all it had going for it. It was vertically polarized (wrong) by hanging down and shorted out to ground (not good either) on damp nights. What Harry needed was a length of wire extended from the window to a distant pole, outbuilding, garage, or tree. In Harry's case, some sturdy trees outlined the house's property line and he could run a 60-foot antenna with no difficulties. The antenna pointed due North-West and in his area of the U.S. was able to pull in Europe, North Africa, and the Near East with ease. Here's how we went about licking Harry's problem.

First, I told Harry that a good longwire antenna should be at least 30 to 100 feet long for good reception performance on 2 to 30 MHz. As mentioned earlier, a 60-foot run was possible. A sturdy tree was selected because it hardly swayed in strong winds at the 20-foot level where the antenna would be secured. Some slack (one foot of droop) was left in the antenna to compensate for tree sway and strong winds. Harry's antenna details can be seen in Fig. 3-28.

Antenna wire and antenna long wire kits are available everywhere. Harry actually used the Radio Shack shortwave antenna kit which consists of 75 feet of bare copper antenna wire, 50 feet of lead-in wire, four insulators, and instructions. Harry had no trouble at all getting the antenna up.

Harry was a little smarter than me. He remembered to protect against ligntning. Since shortwave lightning arrest kits are usually not available locally, Harry made do with lightning arrester parts made for TV. The parts available from Radio Shack include the arrester ground rod 40 feet of aluminum wire and other small parts. The TV arrester has two screw-tight terminals with star washers for the 300-ohm TV line, however Harry only used one for his antenna and the other was left unused. The whole lightning installation bit came to about $5.00. That's cheap. To bring the antenna lead-in into the house, Harry used a "Wall-Thru" tube available at Radio Shack stores.

Now I don't see much of Harry. Maybe once in a while he's at the pizza joint with a date, but you can be sure Harry's getting a lot of DXing and varies every week.

Case No. 2—The Specialist. I've known Mort for over 20 years. We knocked about through high school and somehow the paths of our lives are forever crossing. At one such juncture, Mort invited me to his home to see his new shortwave listening shack which sported a brand-new freshly assembled Heath GR-78 receiver. The GR-78 is a hot receiver. Unfortunately, I couldn't say the same for Mort's antenna. Mort was always inclined to specialize, and he had rigged up a dipole antenna with 300-ohm TV antenna lead-in wire. Mort wanted to

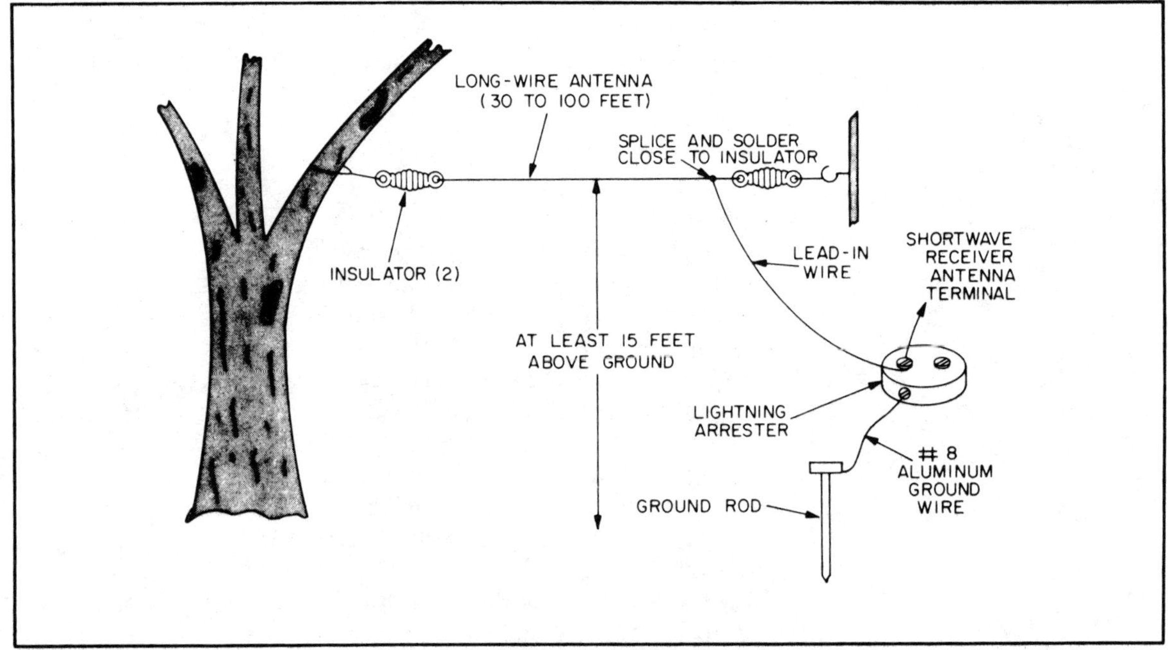

Fig. 3-28. A long-wire antenna with lightning protection provides an effective shortwave listening antenna.

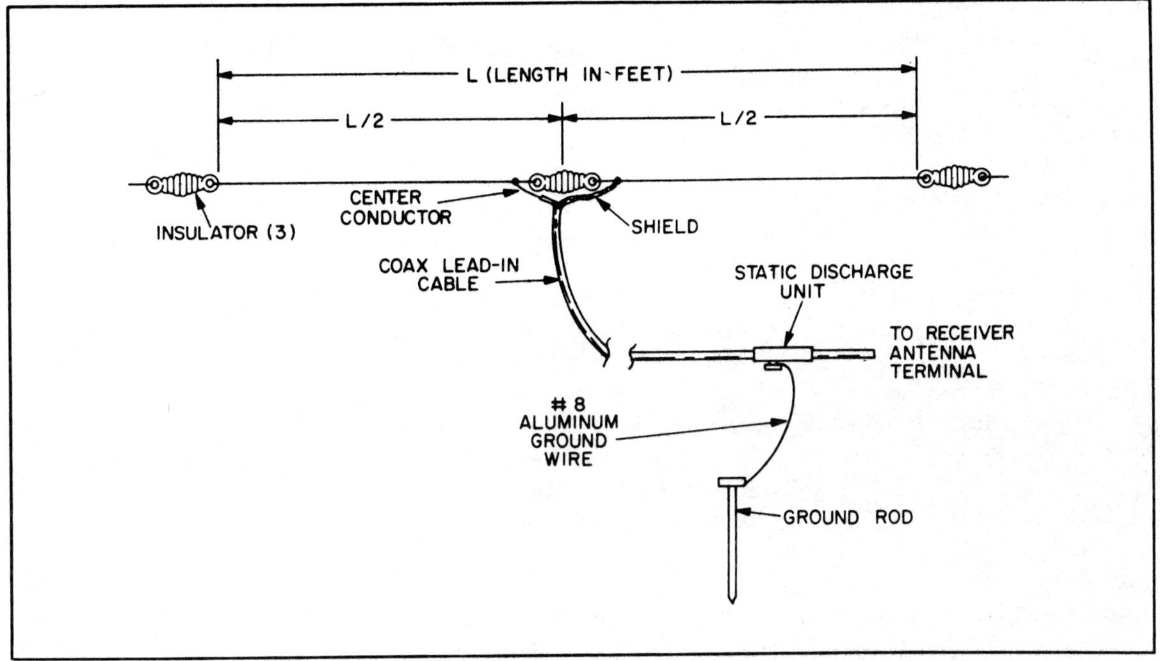

Fig. 3-29. If you want the kind of DX Mort is getting, put up a tuned dipole antenna.

log the 41-meter band and found it offered poor reception in his area. Besides, the noise was too high. He was asking, if not pleading, for advice.

First of all, I told Mort that dipole antennas are cut to exact dimensions for specific frequency bands as shown in Table 3-1. The dipole consists of a wire of a specific length which is cut in half. At the mid-point and both ends, each half of the wire is insulated from each other and insulated from ground. The lead-in cable from the antenna is actually two wires, and it's best to use a 73-ohm coaxial cable (coax) because it is inexpensive and commonly available. Without getting into theory, let me say that a 73-ohm coax lead-in cable "matches" a dipole antenna with less signal loss than does a 300-ohm TV twin-lead cable. On the design board, dipoles have a 75-ohm impedance and match pretty well into 73-ohm coaxes. The 300-ohm cable Mort was using was a bust.

The equation for determining the overall length for a dipole antenna at a given frequency is determined by dividing the given frequency in kiloHertz into the number 468,000. Or, as seen in the text books:

$$L = \frac{468{,}000}{f}$$

Where L is the overall length of the dipole in feet, and f is the desired reception frequency in kiloHertz (kHz). I computed the overall length for dipole antennas to receive the international shortwave broadcast bands using the mid-frequencies of each band and listed them in a table that appears on this page.

When buying materials for a dipole antenna, wire and insulators are the same type as required for the long wire antenna. The lead-in coaxial cable should be RG-59/U or RG-11/U, each of which exhibits 73-ohms impedance. Stay away from unknown coax types or those with different impedances (ohms). As a guide, a table given on this page lists commonly available coax cables and their impedances. Any coax exhibiting an impedance near 75 ohms is good for the purpose. Let price dictate your selection.

I did not forget the lightning arrester in Mort's antenna. At the window, out of the rain, I installed a Radio Shack coax static discharge unit. This gadget requires PL-259 connector on the coax lead-in cable. It's worth the trouble. A grounding screw on the connector attaches to the grounding wire and ground rod. An ounce of prevention can save your home.

A dipole has some bonuses. For example, a dipole works equally as well on frequencies three times the designed frequencies. Thus, a

Band	Frequencies (kHz)	Mid-Frequencies (kHz)	Length (feet—inches)
120	2300-2495	2397.5	195—2
90	3200-3400	3300	141—10
75	3800-4000	3900	120—0
60	4750-5060	4905	95—5
49	5950-6200	6075	77—0
41	7100-7300	7200	65—0
31	9500-9775	9637.5	48—7
25	11700-11975	11837.5	39—6
19	15100-15450	15275	30—8
16	17700-17900	17800	26—3
13	21450-21750	21600	21—8

Table 3-1. Dipole Lengths for Shoftwave Bands.

Cable Type	Typical Ohms
RG-11/U	75
RG-59/U	73
RG-59A/U	75
RG-59B/U	75
F-11/U	75
F-59/U	73

Table 3-2. Coaxial Cable Impedances.

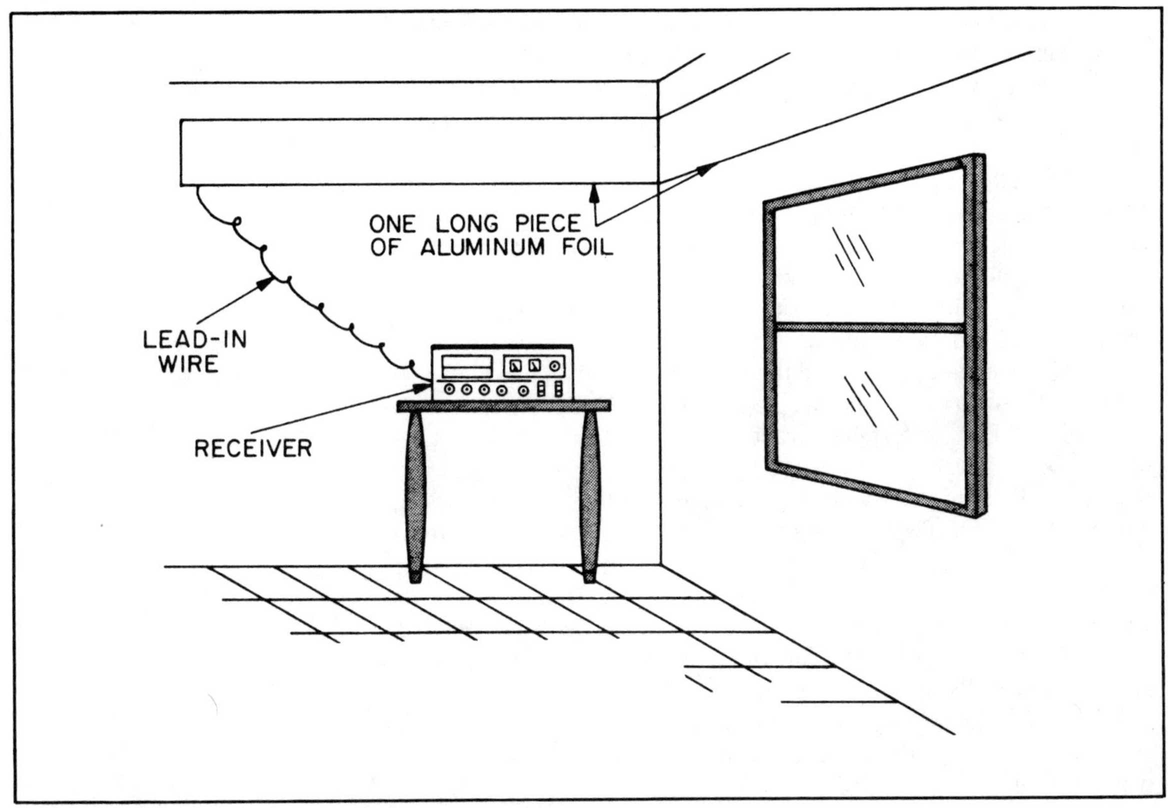

Fig. 3-30. How to install an aluminum foil antenna like the one that set Carl straight.

41-meter band dipole which will pull in 7100-7300 kHz signals will also receive 21300-21900 kHz which covers the 13-meter band. Or, if there is sufficient space to string a 195-ft. antenna for the 120-meter band (2300-2495 kHz), then you could pull in the 120, 41 and 13-meter bands. Of course, if you want all the shortwave bands, then your best bet is a commercial dipole antenna with built-in wave traps.

Don't see much of Mort anymore except at the supermarket. Seems he's a "stay-at-home" type lately. Happy DXing, Mort.

Case No. 3—The Cliff Dweller. Carl is a fun guy to know except when he's upset. For example, Carl drove over on Sunday afternoon to tell me a story he was barely capable of getting out. He had picked up a used Drake SPR-4 receiver at a fantastic price at a flea market and wanted to get involved with DXing in a hurry. It was important to Carl since he teaches French and German, and shortwave DXing would keep his foreign language skills sharp. Unfortunately, Carl lives on the 14th floor of a 24-story apartment house near the city center. His landlord, actually an agent representing the owner, refuses to let any tenant hang anything out of the windows, let alone permit Carl to install an antenna on his patio.

I heard his sad story and told him to have his lease available when I visited him the following weekend. When I came to visit, I could see that the lease was "ironclad," so much so that it made professional baseball's reserve clause seem wishy-washy. That was it—no outdoor antenna for Carl.

I did make him somewhat happy by showing him an old trick. I connected the antenna lead-in wire to the metal finger stop on his phone's dialing mechanism. Reception was good considering the construction of the building, which killed reception even for parts of the AM broadcast band. This was a temporary measure since Carl was soon to get pushbutton phones.

Carl was all set to return to the flea market and unload his Drake receiver. He even told me he had planned to panel his room to give the listening shack a comfortable air, but now he wouldn't. "Now just a minute, before you quit," I said to Carl, "let's give it a try." We swiped his wife's kitchen roll of wrapping aluminum and hung it on the wall with masking tape. Two walls were outside walls, so this is where we placed the foil. Figure 3-30 shows what we did. It looked kind of silly until we attached a clip lead from the foil to the antenna post of the receiver. Wow! Carl practically cried as he tuned the bands. His wife practically cried too when she saw the wall but calmed down once she realized that wall panels were going up.

The last I heard from Carl was he was planning to move to the suburbs where he had purchased an old homestead on six acres. I wonder what he had in mind.

FOX HUNT TRANSMITTER

Ever been to a radio fox hunt? Everyone brings a portable radio and a very directional antenna and tries to find where a small transmitter has been hidden. First one to find it wins. And here's just the transmitter to bring this old ham radio game to the rest of us. Transistor Q1 acts as a crystal oscillator in the new 49 MHz walkie-talkie band. The output of this oscillator is very low, and no license is required if you keep your antenna down to just a few inches in length.

Refer to Fig. 3-31.

Trimmer capacitor C2 lets you tweak the frequency of this transmitter right into the middle of the channel. Use a walkie-talkie and listen for carrier; when you hear it best, you're on frequency. This same circuit can be used as a wireless mike. Connect a carbon microphone, like an old telephone handset mike, in series with R2 and ground.

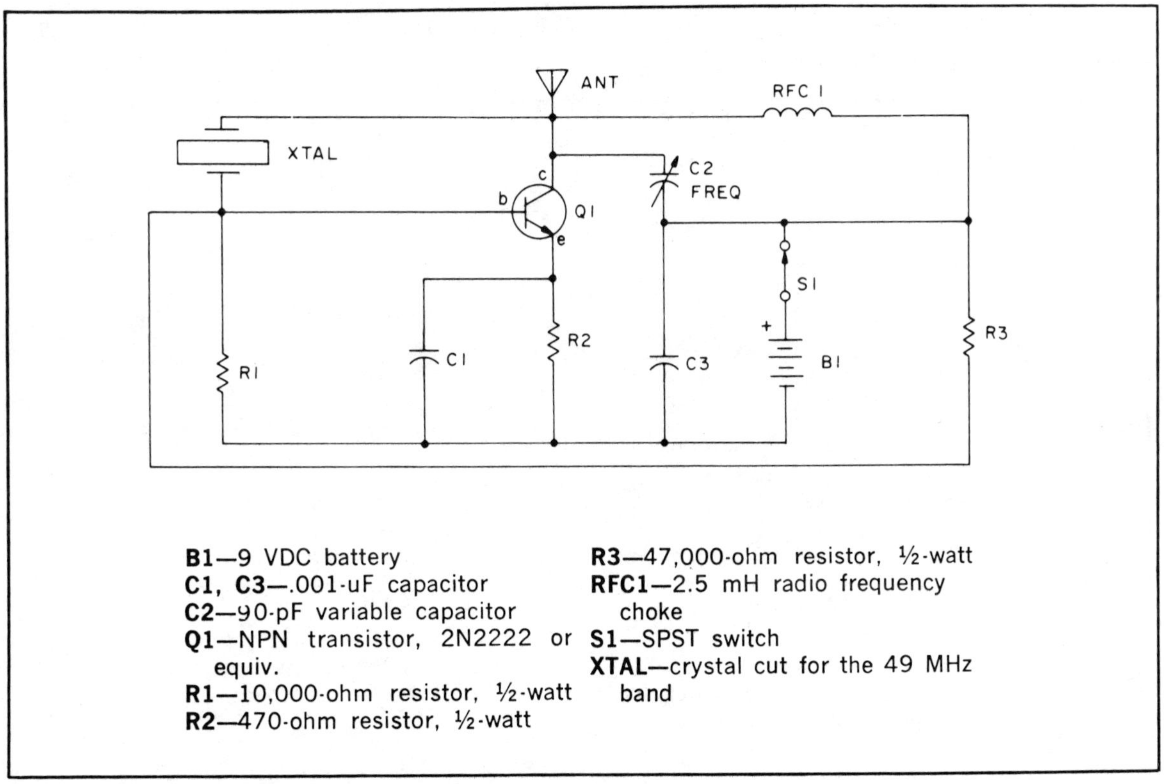

Fig. 3-31. Fox Hunt Transmitter schematic and parts list.

HIGH-PERFORMANCE TRANSISTOR RADIO

Here's a neat way to update your crystal set, assuming you can still find it. Or use these few inexpensive parts to build from scratch. Instead of using a cat's whisker or a diode, this radio uses the very sensitive junction of a junction FET as its detector. This makes it a very "hot," very sensitive high impedance detector. Then the JFET does double duty by converting the high input impedance to a lower output impedance—low enough and with enough drive to power a set of high impedance headphones or a high impedance earphone (about 1000 ohms or so).

Refer to Fig. 3-32.

The antenna coil is one of those simple loopsticks you've seen at the parts stores. (Or you might want to wind your own on an oatmeal box.) The broadcast variable capacitor is one of the tuning capacitors taken from an old, defunct radio. You can use any long wire for the antenna, but if you string it outdoors, be sure to use a lightning arrestor. You can also clip an alligator clip to your bedspring, a windowscreen, or the metal part of a telephone.

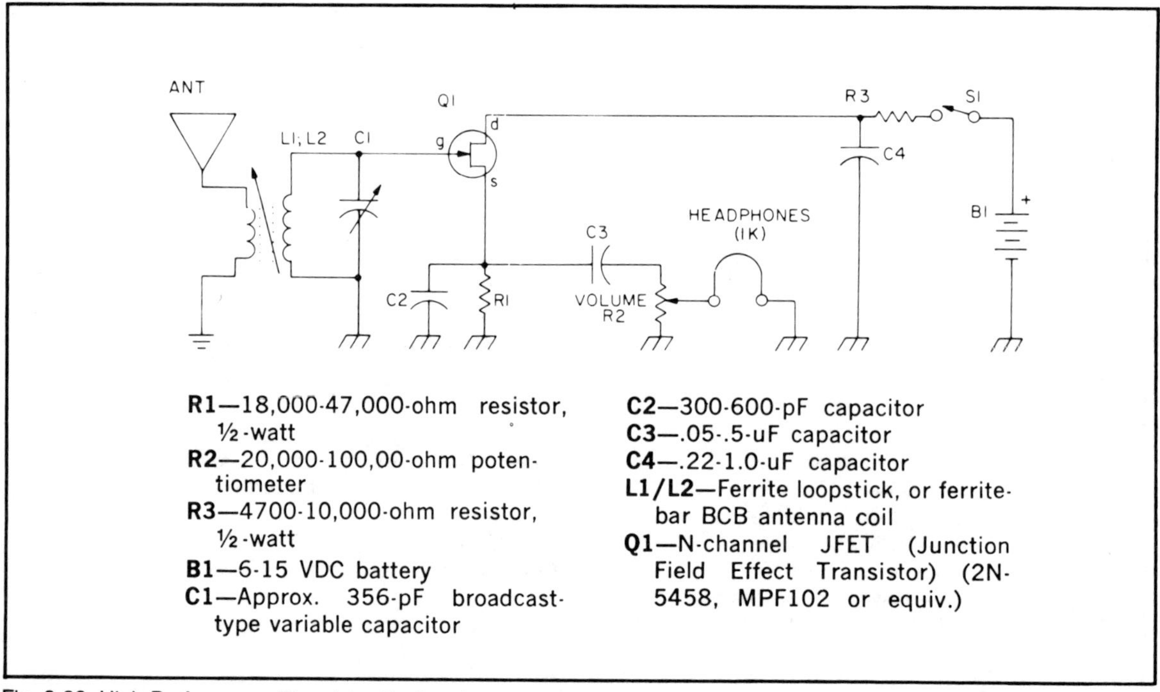

R1—18,000-47,000-ohm resistor, ½-watt

R2—20,000-100,00-ohm potentiometer

R3—4700-10,000-ohm resistor, ½-watt

B1—6-15 VDC battery

C1—Approx. 356-pF broadcast-type variable capacitor

C2—300-600-pF capacitor

C3—.05-.5-uF capacitor

C4—.22-1.0-uF capacitor

L1/L2—Ferrite loopstick, or ferrite-bar BCB antenna coil

Q1—N-channel JFET (Junction Field Effect Transistor) (2N-5458, MPF102 or equiv.)

Fig. 3-32. High-Performance Transistor Radio schematic and parts list.

SWL'S LOW-BAND CONVERTER

Ever listen in on the long waves, from 25-500 kHz? It's easy with this simple converter. It'll put those long waves between 3.5 and 4.0 MHz on your SWL receiver.

Refer to Fig. 3-33.

Q1 acts as a 3.5 MHz crystal oscillator, mixing the crystal frequency with the long wave input from the antenna and forwarding the mix to your receiver.

L1 is a standard broadcast loopstick antenna coil. The crystal is available from many companies by mail order, or is likely to be at a ham radio store near you. You could also use a 3.58 MHz TV color crystal.

Adjust the slug of L1 for your best signal after tuning to a strong station.

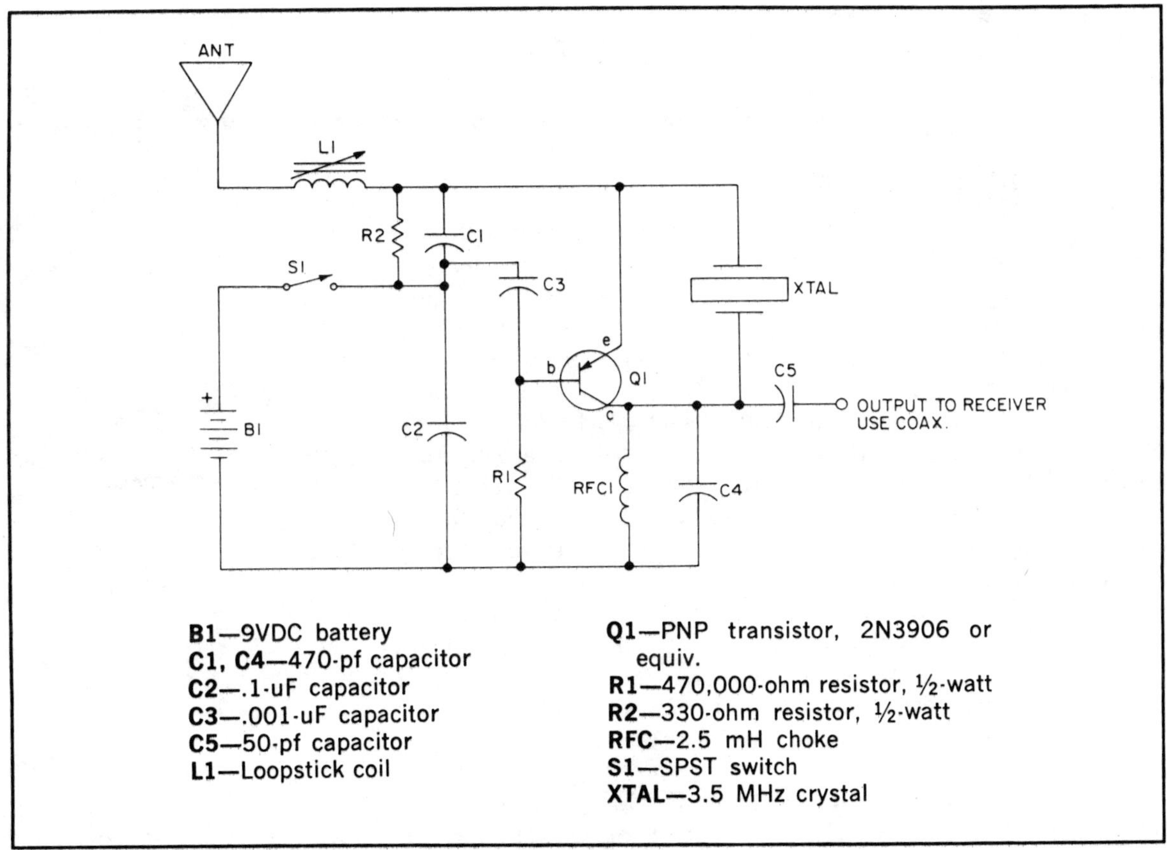

B1—9VDC battery
C1, C4—470-pf capacitor
C2—.1-uF capacitor
C3—.001-uF capacitor
C5—50-pf capacitor
L1—Loopstick coil
Q1—PNP transistor, 2N3906 or equiv.
R1—470,000-ohm resistor, ½-watt
R2—330-ohm resistor, ½-watt
RFC—2.5 mH choke
S1—SPST switch
XTAL—3.5 MHz crystal

Fig. 3-33. SWL's Low-Band Converter schematic and parts list.

Chapter 4

Projects for the Electronics Workbench

LONG-DELAY TIMER

By now, most electronics hobbyists are quite familiar with timers, such as the 555. Projects built with these chips can serve a multitude of purposes, ranging from sirens to triggers for digital counters. However, in straight timing applications, the relatively high frequency at which these chips operate dictates the necessity for a lot of additional circuitry in the construction of long-range timer circuits.

This limitation has been solved with the introduction of the EXAR XR-2242 timer, which allows the user to set delays ranging from microseconds to (with the use of another 2242) virtually a year. So if you want to build a once-a-year alarm clock to nab Punxatawney Phil on Groundhog Day, or whether you need a repeating clock to trigger a light display every few seconds, try our Long-Delay Timer.

Refer to Figs. 4-1 through 4-4.

Heart of the Matter. The Exar XR-2242 integrated circuit is a long-delay timer and very low-frequency oscillator. The functional plan of the 8-pin chip is shown in Fig. 4-1. It consists of a time base oscillator, the output of which is supplied to an 8-bit counter. There is a total count of 128. A control circuit is included to initiate and reset the time delay activity.

A simplified schematic of the timer IC, Fig. 4-2, shows the timebase circuit at the left. The external resistor and capacitor are connected among pins 1, 7, and 8. the time pulse generated by the

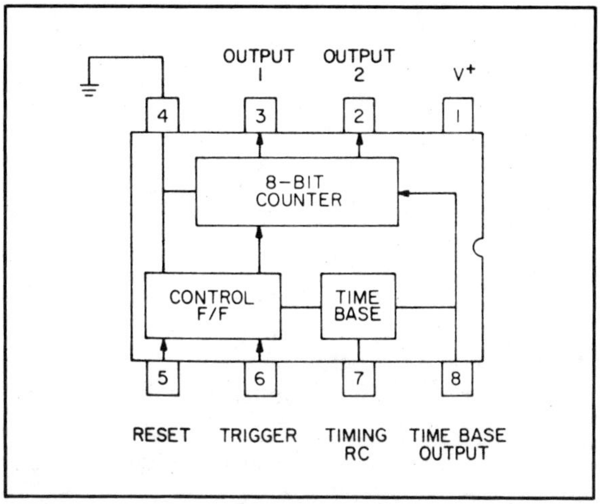

Fig. 4-1. Block diagram of the EXAR XR-2242, the heart of the timer.

time-base oscillator has a period equal to 1RC. The output of the time-base oscillator is applied to a counter chain that provides a total count of 128. Therefore the delay time at the output is equal to 128RC.

$$Td = 128To = 128RC$$

This output is made available at pin 3. An output is also derived from the input counter and is in the form of a square wave with a period of 2RC. This output is derived at pin 2.

The recommended range of operation for the time constant circuit is between 1000-ohms and 10-megohms for the resistance R1 and between 0.01 and 1000-μF for capacitor C1. The control logic provides

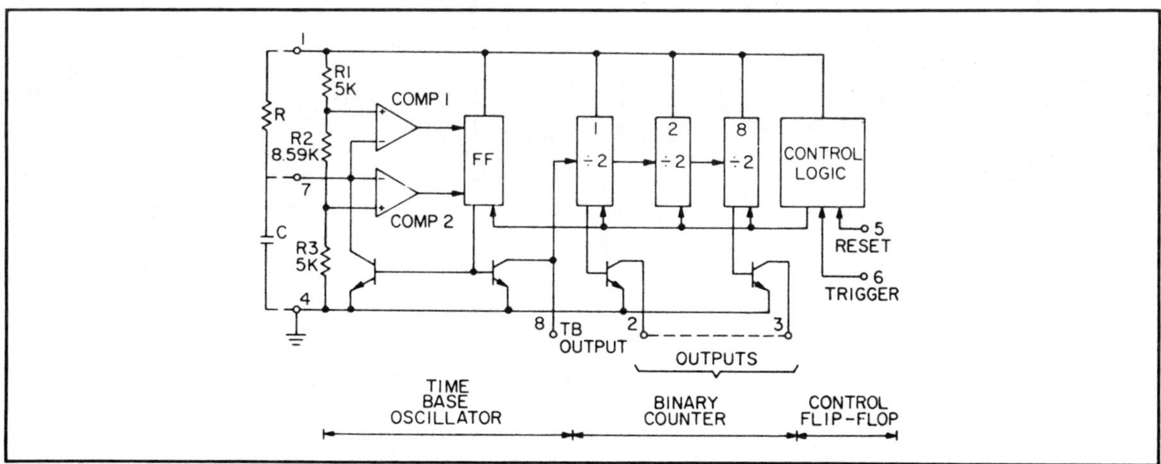

Fig. 4-2. This function diagram of the workings of the XR-2242 indicates just what is happening inside the unit. The three major segments of the chip are detailed in the text, but a quick glance at this diagram when experimenting or troubleshooting will save you time.

a trigger input at pin 6. A positive edge initiates the delay interval. After the delay interval is initiated, it will continue for a time period of 128RC.

Circuit Function. A positive pulse applied to the trigger input, pin 6, will start the time-base oscillator. At the moment it is triggered, it generates a very short duration negative pulse. Capacitor C1 charges through resistor R1 forming a ramp voltage with a period equal to the RC product (time constant). At the conclusion of the charging period (ramp), a second sharp negative pulse is generated. This train of negative pulses at pin 8 is applied to the input of the counter chain. The trigger time-base output and output of the first binary counter are shown in Fig. 4-3. The output of the last counter is shown, but not in the same scale.

The output swings negative to logic 0 when the positive trigger pulse is applied. The output remains negative for a time interval equal to 128RC. After this interval of time, the output again returns to the logic 1 position and awaits the arrival of a trigger pulse at pin 6. Any trigger pulses that arrive during the time delay period (T) do not influence the operation of the circuit. However, the arrival of the first trigger after the delay period initiates a new time-delay interval. If the

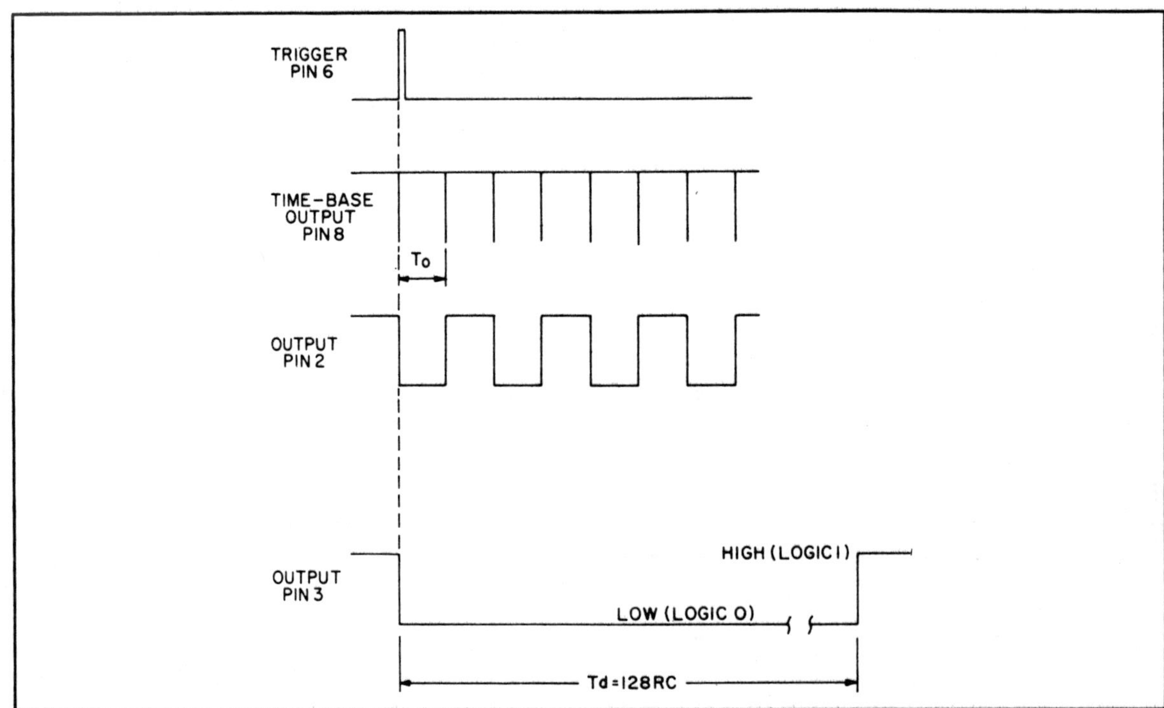

Fig. 4-3. This diagram depicts the output waveforms of the trigger, time base and of the first binary counter (pin 2). The second binary counter's output (at pin 3) is also shown for reference, but not in the same time scale as the other three; its own scale is shown beneath it (Td=128RC).

time-delay interval is to be stopped or reset it can be done by applying a positive pulse trigger to pin 5. This will reset the circuit, making it ready for the arrival of the very next trigger at pin 6.

Operating Modes. The external circuit can be arranged for three basic modes of operation. These are: monostable, astable trigger and reset operation, or astable free-running self-triggered operation. In monostable operation, the timer awaits the arrival of a trigger pulse. When it does arrive, the circuit will go through one time-delay operation and stop. It will remain inoperable until the next trigger arrives.

In triggered astable operation, the arrival of the external trigger will start the circuit in operation. However, at the conclusion of the first delay interval, it will recycle itself and continue operation in this manner even though there is no arriving trigger pulse. The free-running circuit provides continuous operation. The circuit self-triggers as soon as power is supplied and there is no necessity for re-applying a pulse.

Experimental Circuit. The circuit in Fig. 4-4A shows the connection arrangement for using the timer for either monostable or triggered bistable operation. The time period equals the product of resistor R1 and capacitor C1:

$$Td = 39K \times 2 \times 10^{-6} = 0.078 \text{ seconds}$$

This delay is multiplied by the counter chain to a value of:

$$Td = 128To = 128 \times 0.078 = 9.984 \text{ seconds}$$

Thus the delay time at the output should be approximately 10 seconds.

The actual delay time depends very much on the true values of resistor R1 and capacitor C1. If you wish to adjust the circuit for some precise time interval, use a fixed resistor and a potentiometer for resistor R1 as shown in the optional circuit of Fig. 4-4A. This arrangement permits you to adjust the time-base oscillator frequency precisely to obtain an exact 10-second delay interval. A stop-watch can be used to make the potentiometer adjustment.

Testing and Operation. Close switch S3 for monostable operation. Open switches S1 and S2. Connect the LED indicator circuit to the output and apply power. The LED indicator should light indicating that the output is high (logic 1). *Momentarily* close switch S1. Since S2 connects to the supply voltage, this is the same as applying a positive trigger. At the moment switch S1 is closed, the LED will be unlit indicating that the delay interval has been initiated. Output is low (logic 0). For the circuit of Fig. 4-4 it should remain unlit for approximately 10 seconds. After this interval the LED will again light and remain so until switch S1 is momentarily closed again.

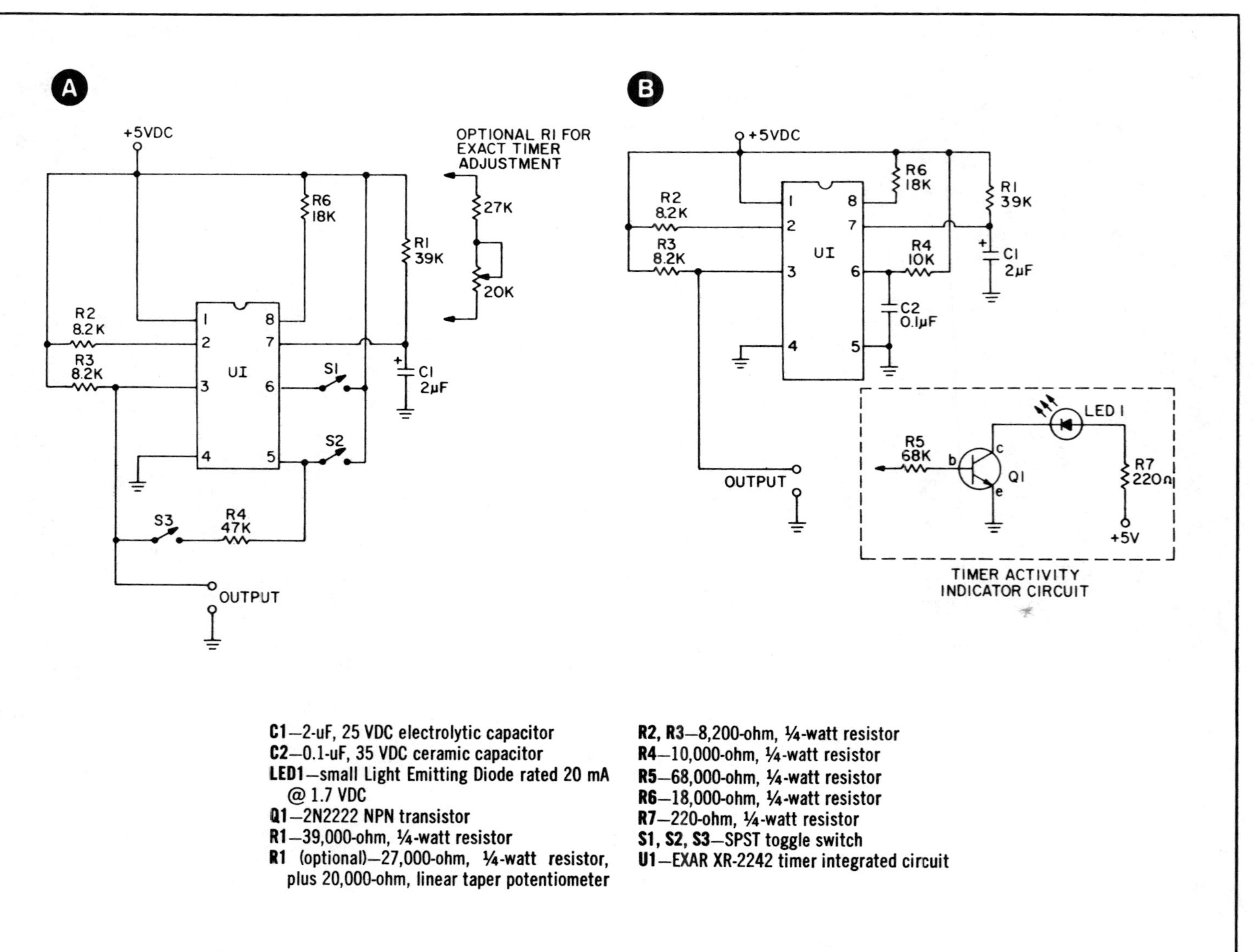

C1—2-uF, 25 VDC electrolytic capacitor
C2—0.1-uF, 35 VDC ceramic capacitor
LED1—small Light Emitting Diode rated 20 mA @ 1.7 VDC
Q1—2N2222 NPN transistor
R1—39,000-ohm, ¼-watt resistor
R1 (optional)—27,000-ohm, ¼-watt resistor, plus 20,000-ohm, linear taper potentiometer
R2, R3—8,200-ohm, ¼-watt resistor
R4—10,000-ohm, ¼-watt resistor
R5—68,000-ohm, ¼-watt resistor
R6—18,000-ohm, ¼-watt resistor
R7—220-ohm, ¼-watt resistor
S1, S2, S3—SPST toggle switch
U1—EXAR XR-2242 timer integrated circuit

Fig. 4-4. Long-Delay Timer schematic and parts list.

Close switch S1 momentarily to initiate a new cycle. Then very quickly close S2 momentarily. Note that the LED will light immediately after the closing of switch S2. This activity is the same as applying a positive reset trigger pulse to pin 5.

Triggered astable operation is obtained by opening switch S3. Momentarily close switch S1 to initiate a delay interval. (The LED will be unlit.) After the delay interval, the output will switch to high and the LED will light. It will remain lit for approximately 10 seconds and then it will go out indicating the initiation of a new delay interval. This time, the new cycle was initiated even though switch S1 was not closed momentarily. This activity will repeat itself in an astable manner until the reset switch (S2) is momentarily closed during one of the time delay intervals. The circuit will not recycle itself again until the trigger switch S1 is again momentarily closed.

Free-Running Oscillator Operation. To operate the timer chip as a low-frequency oscillator, connect the circuit of Fig. 4-4B. Note in this arrangement that pin 5 is grounded and that the trigger (pin 6) is connected permanently to the supply voltage through a resistor. This means that a fixed supply voltage is supplied to the trigger circuit which will insure self-triggering operation.

Rearrange the circuit for oscillator operation. Supply power to the circuit. Note that the LED will be lit for ten seconds and then off for ten seconds and so on. This indicates that a very low-frequency square wave (one cycle each twenty seconds) is being generated at the output, the output swinging between high and low and back to high again during a single pulse period.

Connect the LED indicator to pin 2. The LED will flicker, indicating the much higher frequency square wave that is to be found in this initial stage of the counter chain. Return the LED indicator circuit to the pin 3 output.

If you wish, you can now experiment with other components for resistor R1 and capacitor C1. A 10-μF capacitor substituted for C1 will result in a delay of approximately 50 seconds. The LED will be off for 50 seconds and on for 50 seconds, etc. The use of a 1-megohm resistor and a 100-μF capacitor would result in a delay time in excess of 3 hours. A 3900-ohm resistor and a 2-μF capacitor would cut the delay time down to approximately 1 second.

You have learned how the XR-2242 operates as a long-delay timer and a very low-frequency oscillator. A more elaborate circuit could be added to the output, operating a mechanical or solid state relay, the activities of which could be regulated by the time-constant combination resistor R1 and capacitor C1. Other applications may come to mind for this versatile circuit as you continue to experiment.

MINI-REG

Here's a low-cost precision regulated dc power supply which is sure to be a welcome addition to any workbench—provided some family member doesn't appropriate the power supply for use as a universal ac adapter! Compactly assembled in an eye-catching low profile, the Mini-Reg is continuously adjustable from 3.4 volts to 15 volts dc and delivers up to 500 milliamperes, enough for just about any job. Using the HEP C6049R precision monolithic IC regulator, the Mini-Reg effects 0.01 percent regulation with the line voltage variations, 0.05 percent regulation for load variations, and its output impedance is a mere 35 milliohms. Short-circuit proofed, the Mini-Reg also features adjustable current limiting which greatly reduces the chances of damaging valuable components in the circuits you are working on. You can also use the Mini-Reg as a constant-current source and recharge nicad batteries.

Refer to Figs. 4-5 through 4-9.

Circuit Operation. The HEP C6049R is actually a dc regulator within a regulator which accounts for its high performance. As shown in the block diagram, a very stable reference voltage (*V*r) is applied to the non-inverting or voltage follower input of an op-amp which serves as the first regulator and dc level shift amplifier. The output voltage of this stage can be varied from 3.4 volts to 15 volts by varying pot R11. This voltage is applied to the noninverting input of the second op-amp which is capable of supplying up to 5000 milliamperes current to the load. This stage has unity voltage gain wherein V-out follows the input voltage to this stage. This double regulator arrangement fully isolates the dc level shift amplifier and results in very close regulation. Capacitor C4 provides frequency compensation and precludes possible circuit oscillation.

External components consisting of transistor Q1 and selectable resistor Rsc provide constant-current limiting should the supply be short-circuited. When the load current passing through Rsc becomes sufficiently high, the base of Q1 becomes forward biased causing Q1 to conduct. When Q1 conducts, the voltage regulator delivers an essentially constant current to the load at a level depending on the value of Rsc. In the schematic diagram, resistor R3 places a minimum load on the regulator. Switch S3 selects the desired current limit. Jacks J1 and J2 permit insertion of a milliammeter to read load current but without impairing regulation. Diode D2 provides meter protection and diode D1 provides reverse voltage protection.

Construction. Assemble the Mini-Reg in an aluminum case or in

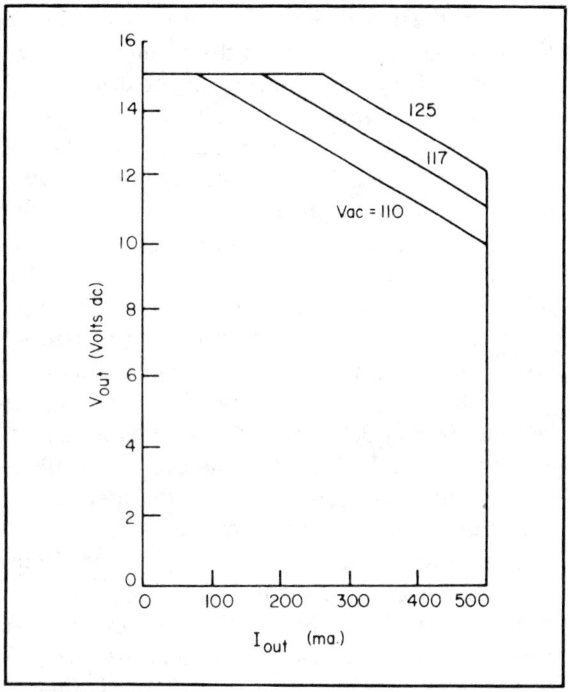

Fig. 4-5. This chart shows the operating range of the Mini-Reg at various line voltages. The full 15 Vdc is only available at lower currents, but few IC projects ever require that much voltage or current supply.

a plastic case with aluminum cover plate. Select a case which will accommodate the particular meter and transformer you plan to use. Plan the layout allowing room for the PC board assembly when the cover plate is secured.

Begin by laying out and drilling mounting holes for IC1 in the heatsink. Drill a 7/16-inch diameter hole in the heatsink to pass the lead wires of IC1. File off drill burrs and ridges so that IC1 mates perfectly on the heatsink. Drill matching holes in the cover plate. For ventilation, drill a number of holes in the cover plate and on the bottom of the case.

Make the PC board using the circuit pattern shown, taking care to locate pads for IC1 just right. Push IC1 into the drilled board and mark and drill the mounting holes. For easier mating, countersink the lead holes for IC1 on the insulation side of the board by twirling a small drill bit.

Install and solder the jumper on the insulation side of the board and install the solder T42-1 micro-clips on the copper side at all resistor and board take-off terminals. Clip a small heatsink and on the leads of Q1 when soldering. Install remaining circuit board components excepting trim resistor R5. Using 6-32 machine screws, bolt IC1 and the heatsink to the cover plate, Place a lock washer and two 6-32 nuts on each mounting bolt. *Omit the mica washer between IC1 and the heatsink* and apply a bit of silicone heatsink grease between IC1 and the heatsink. Coil a ¼- by 1½-inch strip of fishpaper insulation and slip it down into the hole in the cover plate around the IC lead wires. Push the PC board assembly down on the mounting screws and mate with the protruding IC leads and secure. If you can't install the assembly, look for bent pins or reversed installation of IC1.

Install switches S1 and S2 along with jacks J1 and J2 on the left side of the case. Install diode D2 an capacitor C7 on switches S2. Secure two solder lugs on each binding post and install diode D1 and capacitor C6 on the binding posts. Pass the ac line cord through the left side of the case and knot the cord for strain relief. Install resistors R6 through R10 on switch S1. Depending on the base-emitter characteristics of Q1, the specified values of current limit resistors R6 through R10 may differ somewhat in your power supply. This is why trim resistor R5 was included to properly trim the 500 mA current limit. For this reason, you may defer installation of R6 through R9 but *do* install R10.

Place RECT-1, R1, C5, and C8 on a small piece of perfboard and situate this sub-assembly behind the meter. Connect meter M1 directly to binding posts BP1 and BP2. Use No. 20 stranded wire for connections to the PC board. Connect a wire from board pin G to BP2. Run a wire from board pin E to the rotor lug of S3. Connect a wire from board pin D to resistors on S3. Run a wire from board pin F directly to BP1. Run a pair of wires from pot R11 to board pins B and C. Connect a wire from *V-in minus* directly to BP2. Do not make the connection from *V-in plus* to board pin A at this time. You may omit the double-fused plug and provide but one fuse in the primary side of tranformer T1. Carefully check all wiring and solder connections.

Checking It Out. We intentionally deferred installation of several components and some wiring, so that you can perform a few simple tests which preclude damage to circuit components. Connect a voltmeter across R1 and verify that V-in plus is nineteen volts dc. Connect a milliammeter and 100-ohm resistor in series from V-in plus to board pin A. Set S3 to pick up R10 and set R11 to minimum resistance. Turn

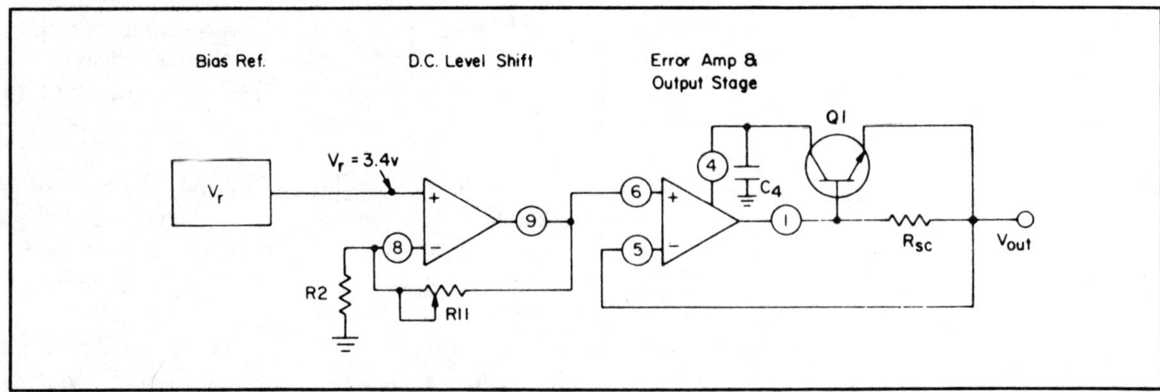

Fig. 4-6. Simplified block diagram of the C6049R regulator chip, which is the heart of the Mini-Reg power supply. Thanks to such ICs, construction projects are easy to build.

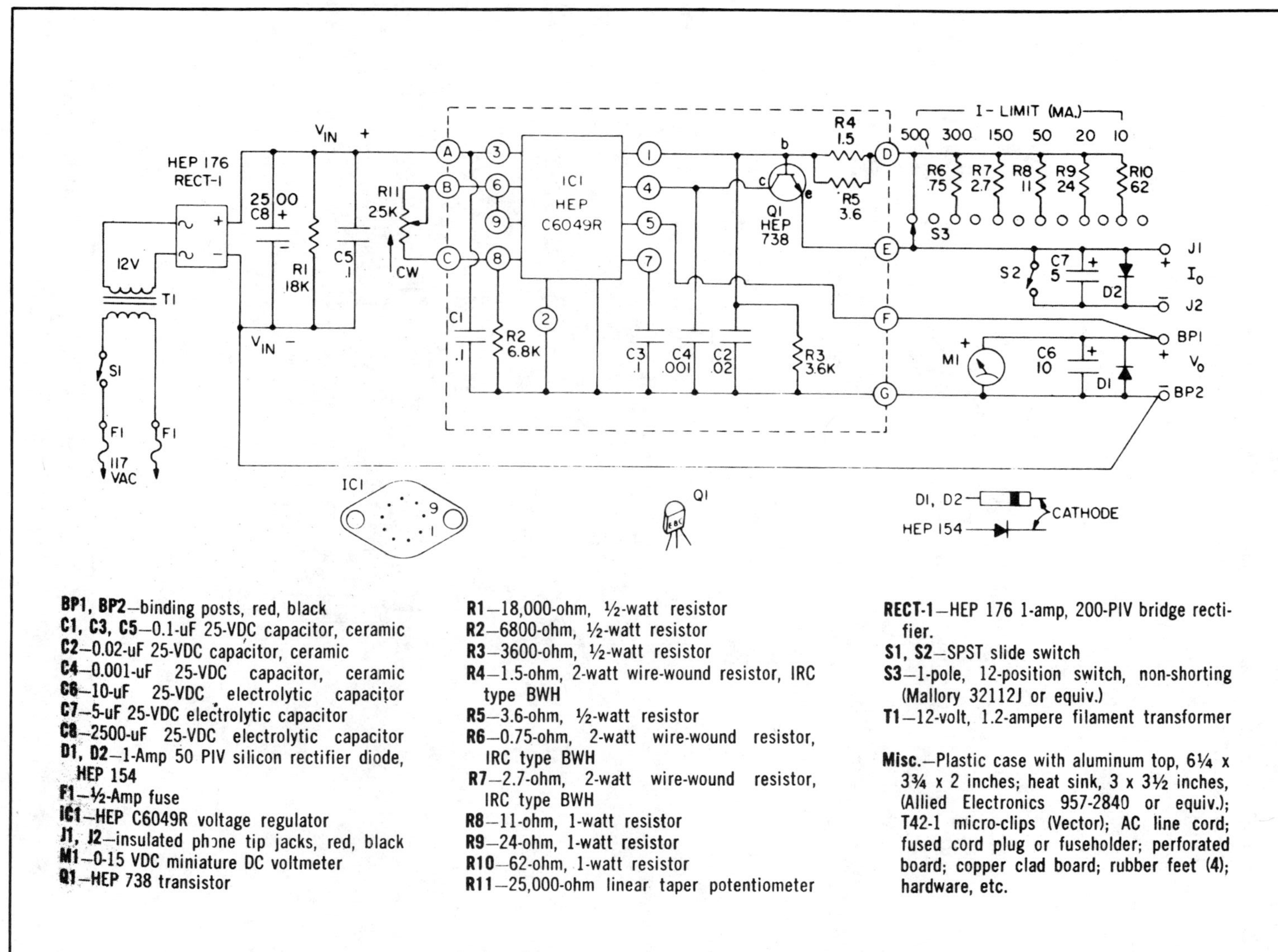

BP1, BP2—binding posts, red, black
C1, C3, C5—0.1-uF 25-VDC capacitor, ceramic
C2—0.02-uF 25-VDC capacitor, ceramic
C4—0.001-uF 25-VDC capacitor, ceramic
C6—10-uF 25-VDC electrolytic capacitor
C7—5-uF 25-VDC electrolytic capacitor
C8—2500-uF 25-VDC electrolytic capacitor
D1, D2—1-Amp 50 PIV silicon rectifier diode, HEP 154
F1—½-Amp fuse
IC1—HEP C6049R voltage regulator
J1, J2—insulated phone tip jacks, red, black
M1—0-15 VDC miniature DC voltmeter
Q1—HEP 738 transistor
R1—18,000-ohm, ½-watt resistor
R2—6800-ohm, ½-watt resistor
R3—3600-ohm, ½-watt resistor
R4—1.5-ohm, 2-watt wire-wound resistor, IRC type BWH
R5—3.6-ohm, ½-watt resistor
R6—0.75-ohm, 2-watt wire-wound resistor, IRC type BWH
R7—2.7-ohm, 2-watt wire-wound resistor, IRC type BWH
R8—11-ohm, 1-watt resistor
R9—24-ohm, 1-watt resistor
R10—62-ohm, 1-watt resistor
R11—25,000-ohm linear taper potentiometer
RECT-1—HEP 176 1-amp, 200-PIV bridge rectifier.
S1, S2—SPST slide switch
S3—1-pole, 12-position switch, non-shorting (Mallory 32112J or equiv.)
T1—12-volt, 1.2-ampere filament transformer

Misc.—Plastic case with aluminum top, 6¼ x 3¾ x 2 inches; heat sink, 3 x 3½ inches, (Allied Electronics 957-2840 or equiv.); T42-1 micro-clips (Vector); AC line cord; fused cord plug or fuseholder; perforated board; copper clad board; rubber feet (4); hardware, etc.

Fig. 4-7. Mini-Reg schematic and parts list.

S1 on and observe about five milliamperes current on the milliammeter and 3.4 volts on meter M1. Advance R11 and observe a voltage increase up to fifteen volts dc. If the output voltage is less than fifteen volts, the value of R11 may be too small or R2 may be too large. Having verified the above, you may now install the wire from V-in plus to PC board pin A.

Plug the milliammeter into jacks J1 and J2 and open S2 (Meter In). Adjust R11 for ten volts output and set S3 to ten milliamperes current limit. Then, connect a 500-ohm ½-watt resistor across the output terminals. If current limiting action is taking place, the milliammeter should indicate roughly ten milliamperes and the output voltage should drop to about five volts. If much higher values are observed, current limiting is not taking place. Look for a defective or improperly installed Q1. If your current limit is, say seven milliamperes, you can bring it up to ten by using a smaller value for R10 or by connecting a suitably larger value resistor across R10.

Only after you have verified current limiting action at low current, set S3, to pick up R4 (500 mA setting) and set the VOM accordingly. You will need either a 50-ohm 10-watt rheostat or adjustable power resistor to gradually load the supply. Or, you can use a number of

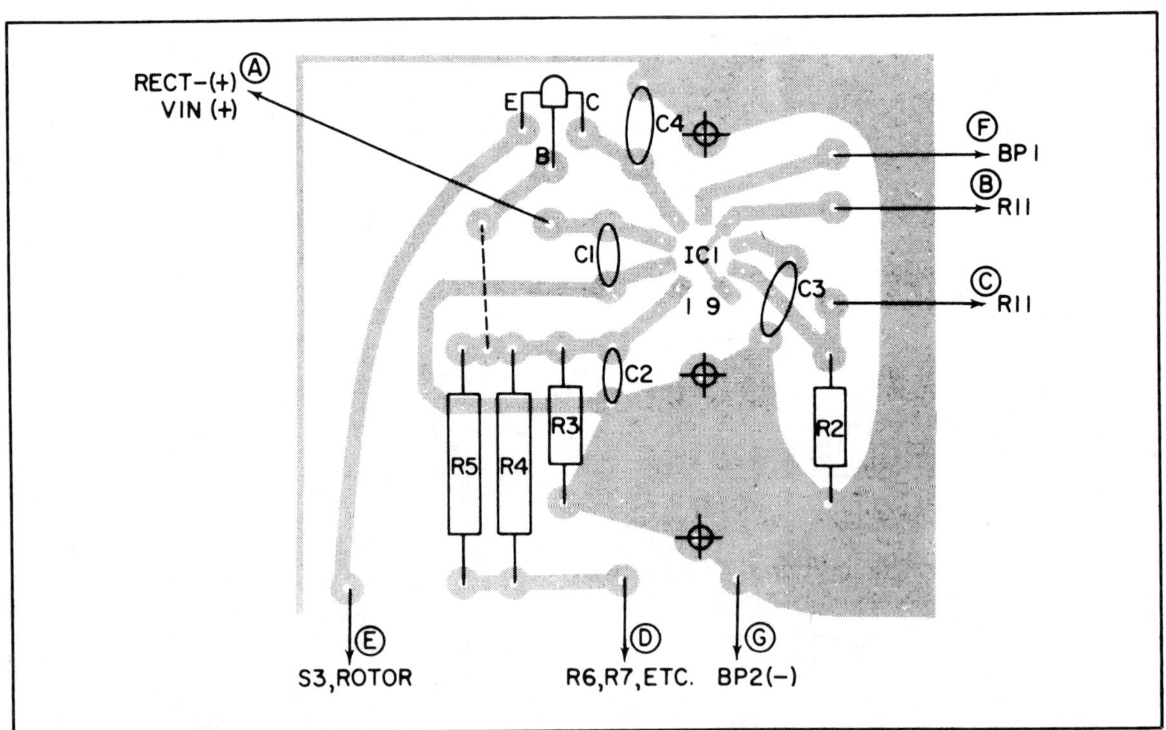

Fig. 4-8. The parts should be placed according to this diagram. Note the location of the three drill holes for securing the IC and the board to the chassis. Locate the IC mounting holes very carefully so that everything mates snugly. This will help keep the chip cool.

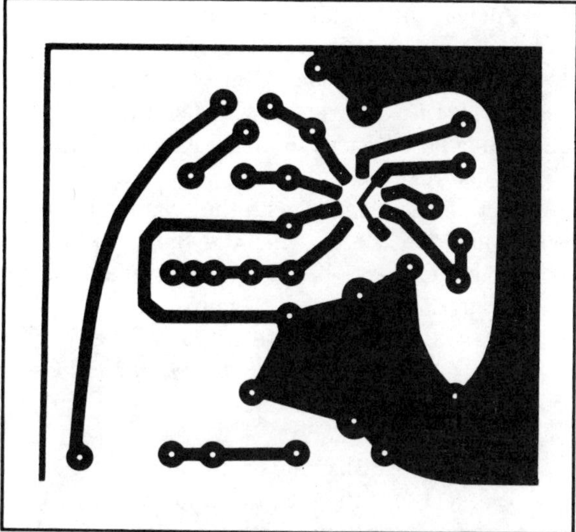

Fig. 4-9. This is an exact-scale printed circuit board pattern showing the foil side of the board. Contrary to normal this side is where the components are to be mounted. Only the jumper and the IC chip are mounted on the other side. Be careful to keep the foil-side component bodies off the metal surface to avoid shorts. Be especially careful with resistors R4 and R5 and capacitor C3.

small-valued power resistors. Set the rheostat to maximum resistance and connect it across the output terminals. Gradually reduce load resistance while observing output voltage and current. Current limiting should occur at below 500 mA. To increase the limit to 500 milliamperes, select and install a suitable resistor for R5. Proceed similarly to size or trim resistors R6 through R10. You can easily include other current limits in the spare positions on S3 to match the charging currents of your nicad batteries. Do not exceed 500 milliamperes or else IC1 will be damaged.

Application. The operating range of the Mini-Reg for several line voltages is shown. The supply "drop-out" shown in the upper right hand corner of this chart is due to an insufficient difference between V-in plus and V-out which in turn depends on transformer T1 voltage. When you are not using a meter at jacks J1 and J2, close S2.

The adjustable current limiting feature of the Mini-Reg greatly reduces the chances of damaging circuit components of the circuit powered by the supply. Suppose you are experimenting with a transistorized circuit drawing five milliamperes at five volts. You would then set S2 to 10 mA. At these settings, the maximum power the supply can deliver is but a mere fifty milliwatts.

If you plug a transistor in backwards, the most it can draw is fifty milliwatts, probably much less; hence, the device will survive the error. However, certain semiconductors can be damaged with but microwatts of power. Nevertheless, you are far better off using current limiting supplies. If your experimental circuit draws 400 mA at five volts, set S3 to 500 mA limiting the power to 2.5 watts. This power level is more than enough to zap many devices if you make an error. If you have another five volt supply, split the circuit supply lines and protect those devices you cannot spare with the Mini-Reg.

Almost any circuit operating off three volts can safely operate at 3.4 volts. The output voltage can be further reduced by connecting a low-voltage zener diode in series with the plus lead to the load and monitoring the load voltage with a voltmeter. In this case, load voltage regulation now depends on zener diode characteristics.

When recharging batteries with the Mini-Reg, connect a silicon rectifier diode in series with the plus lead going to the battery. This eliminates "back-leak" when the supply is turned off with battery yet connected. Observe battery polarity when making connections. Circuits using op amps usually require a dual or split supply. To provide a dual six-volt supply, set the output voltage to fifteen volts, set S3 to 100 mA, and connect two 6-volt zener diodes in series across the output terminals. Then, connect a 100 μF 25V electrolytic capacitor across each zener diode.

The Mini-Reg handily checks and sorts zener diodes of 15 volts or less. Set R11 for fifteen volts output and set S3 to 10 mA. Connect the diode across the output terminals with plus lead wire to BP1. Observe zener diode voltage on M1. Advance S3 to high currents but do not exceed rated current of the diode. The better the quality of the diode, the less increase in voltage observed on M1.

When you operate radio or audio equipment from the Mini-Reg, set S3 to a current level which supplies peak currents on audio peaks. Otherwise, you will notice audio distortion on audio peaks. With some radio and audio equipment, operations off an ac adapter of the Mini-Reg may introduce an ac hum. Reversing the ac plug usually remedies the problem. If not, connect a ground wire to either the plus or minus terminal of the Mini-Reg, whichever proves most effective. In addition to its use as a universal ac adapter, the Mini-Reg serves as an excellent power supply when servicing battery operated transistorized equipment. You'll wonder how you ever solved your power supply problems before you discovered Mini-Reg.

SUPERCHARGER

Proliferation of portable electronic gadgets, such as calculators, tape recorders, walkie-talkies, and radios, gave a big boost to sales of rechargeable batteries. This article should bring your knowledge on the rechargeable battery up-to-date and tell you about a truly universal charging circuit with an electronic timer which you can build.

Rechargeable sealed batteries, besides many other advantages, make the operation of portable equipment quite inexpensive. Do you still remember the high cost of B and filament batteries for portable tube radios? But even with transistorized equipment, the cost of "cheap" throw-away batteries may be quite high. For example, a portable calculator or a radio using four AA throw-away cells needs battery replacement about once a week if it is used for 2 to 3 hours each day. This comes to about $50 per year. A set of four rechargeable AA-size nickel-cadmium (NiCad) batteries costs around $8 and with proper care should last three to five years or more. The cost of electricity used for recharging comes to only 10 cents per year. Quite a difference in cost!

Refer to Figs. 4-10 and 4-11.

What Proper Care? We mentioned that a rechargeable Ni-Cad battery will last for many years if proper care is exercised. Our charger described in this article will give your rechargeable batteries such proper care. There are three rules to observe when handling rechargeable batteries. They are all expressed in terms of battery capacity in milliampere hours (mAh). This value is usually given by the manufacturer on the battery label. If no battery capacity is given, some common values are shown in this table. However, watch for the figures given by the manufacturer. For example, you may find a sub C cell in a D cell package.

Battery Size	Capacity (mAh)	10-Hour Rate (mA)	5-Hour Rate (mA)
AA	450	45	90
sub C	1000	100	200
C	1500	150	300
D	3500	350	700

Rule 1. Do not discharge continuously at more than the hourly rate (450 mA for AA cells). Whether this rule is satisfied depends on the kind of equipment you are using. This rule will seldom be violated. Just don't try to run your electric power mower on a bunch of AA cells!

Rule 2. Do not continue discharging when the battery voltage is 0 volt (cell reversal). If you have several batteries in series, one will always have slightly smaller capacity than the others. When that battery is completely discharged, the other batteries will still pump current through it. The only way to avoid this condition is to turn off your appliance immediately when the total series battery voltage drops significantly (by more than 1 volt). You will notice it when, for example, your radio starts distorting. Turn it off immediately.

Rule 3. Do not charge at more than the 10-hour rate (45 mA for AA cells) and do not continue charging at that rate beyond full capacity for more than a few hours. Slightly higher charging rates of up to the 5-hour rate are permissible as long as the battery is still discharged. To satisfy this rule, you need to control the charging current and the charging time as is provided by this charger. Some so-called universal battery chargers put either a too-high or a too-low current into your batteries. As a result either the battery will be damaged and its life shortened or it will not get fully charged in a reasonable amount of time.

These are general and safe rules. Specially-constructed batteries (for example, the so-called quick-charge batteries) may let you break one or more of these without causing permanent damage. However, unless the battery manufacturer assures you to the contrary you better stick with our three rules; otherwise permanent damage may result. Either the battery will fail (go dead) immediately or its life-span and capacity will be shortened.

Battery Charger. This charger is capable of charging one to six cells from AA to D size. It lets you control the charging current and the charging time. You turn the charger on, set the current to the *10-hour rate* for a full charge or 5-hour rate for a quick boost, and forget it. After 14 hours (or 1¾ hours for a quick boost) the charger will turn itself off. In other words, we pump in *140 percent of battery capacity* to charge it fully (40 percent is the typical loss in the charging process). For a quick boost of 1¾ hours when the battery is completely or partially discharged, we can go up to the 5-hour rate to obtain about one-quarter full battery capacity. For special quick-charge batteries follow manufacturer's recommendations.

The charger makes use of a newly-developed integrated circuit which combines a built-in oscillator (similar to the 555-type) and a frequency divider of up to 65,536 (2^{16}). This way we can choose a basic oscillator frequency of 0.77 Hz which can be obtained with reasonable resistance and capacitance values and divide it by 2^{16} to obtain timing values of up to 14 hours. The basic frequency, f, is determined by C2, R3, and R4. The frequency

$$f = \frac{1}{2 \times R3 \times C2}$$

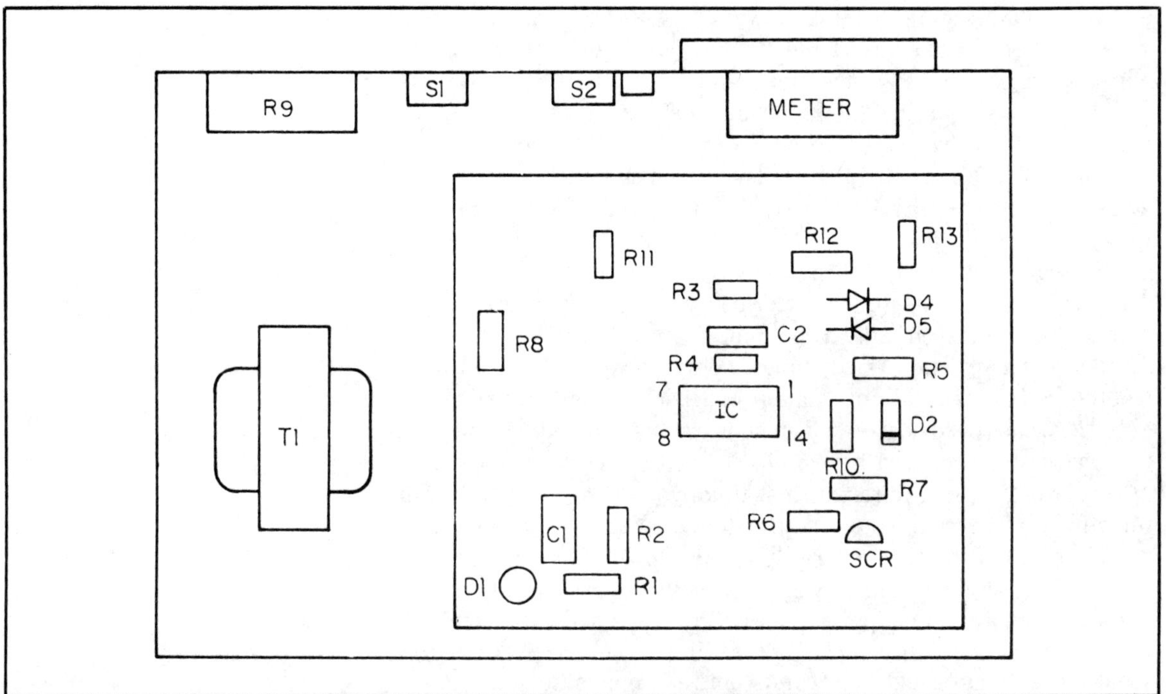

Fig. 4-10. You should have little difficulty locating parts on your supercharger perfboard. While it is possible to build this unit in a much smaller area if you wish, beginners will find the extra room a benefit.

where R4 = 2 × R3. The IC is connected in such a way that the timer resets itself when the circuit is first turned on. When its timing interval is up, it will turn the SCR off permanently until the circuit is first removed from, then connected to the power line again. The rest of the circuit is straightforward. The output of the IC (pin 8) controls the gate of the SCR and lights up the LED. The charging current is controlled by the variable resistor R9. The current range with the values shown in between approximately 40 and 500 mA for up to 6 cells. Switch S1 selects the IC divider output of either 2^{16} or 2^{13}.

The lowest divider ratio the IC is capable of, 256, is particularly useful during the charger calibration. To select this counting/dividing mode, disconnect pins 12 and 13 from S1 and temporarily connect pin 12 to pin 14 and pin 13 to pin 5. When you have finished the test, reconnect pins 12 and 13 to S1 after removing your temporary connection. In this mode the timer should turn itself off after 3 minutes 17 seconds plus or minus 10 seconds. The meter M1 is used as a volt meter (0 to 10 volts) across the batteries or as a charging current milliamp meter of 0 to 500 mA. Its function is selectable with S2. The diodes D1 and D2 protect the meter from overload.

Put It Together. You can mount all components on a perfboard. The wiring is not critical. The MOS integrated circuit is internally protected against static charges, however we still recommend using a 14-pin socket. Do not insert the IC until you are finished with the wiring, have checked all connections, and make sure the power is off.

If you plan to charge the batteries outside your equipment, then you must provide battery holders for various size batteries which you want to connect to the charger. Under certain conditions, you may be able to connect the charger directly to your appliance without removing the batteries, usually via the "adapter" jack. You may have to look at the schematic of your radio or walkie-talkie to find out if the "adapter" jack is connected to batteries when a plug is inserted. If so, you can charge the NiCads in the unit.

Once construction is complete, apply power and check to see whether or not the LED pilot lamp is on. If so, it should remain on for either one and three quarters of an hour or fourteen hours, whichever time you have selected with the *time select* switch. To check the correct operation of the timing circuit in less time, you can make the following temporary connections to enable the divide by 256 function. Connect pin 12 and 13 of the IC temporarily to pins 14 and 15 respectively to select the 256 divider ratio. Try different values of capacitor C2 till you get a timing interval of approximately 3 minutes and 17 seconds. Of course, this is not a critical parameter, but it should be accurate to at least three minutes and 17 seconds plus and minus 30 seconds.

More Savings. Besides rechargeable batteries, regular throw-away, zinc-carbon batteries can also be recharged under certain conditions. Those conditions follow.

- Battery should not be completely discharged (battery voltage should stay about 1 volt).
- Battery should not be leaking.
- Battery should be used soon after being recharged.

Other popular throw-away batteries are alkaline and mercury batteries. Mercury batteries are used where high energy concentration in low volume is required. A camera or a hearing aid is a prime example of such an application. The mercury cell has three to five times the capacity of a carbon-zinc cell of the same size but it costs five to ten times as much.

Non-rechargeable alkaline batteries have about twice the capacity of a comparable carbon-zinc cell at approximately three times the price. Mercury and alkaline cells have similar nearly constant discharge voltage and low internal resistance characteristics as the NiCad cells. However, they are not leakproof and should be removed from equipment if not in use. We strongly discourage you from trying to recharge mercury or non-rechargeable alkaline batteries. Gases generated by the recharging process in the sealed cell may cause an explosion and spread the caustic electrolyte.

You may also run across rechargeable alkaline batteries. They are not as popular as NiCad batteries, but are slightly cheaper and have

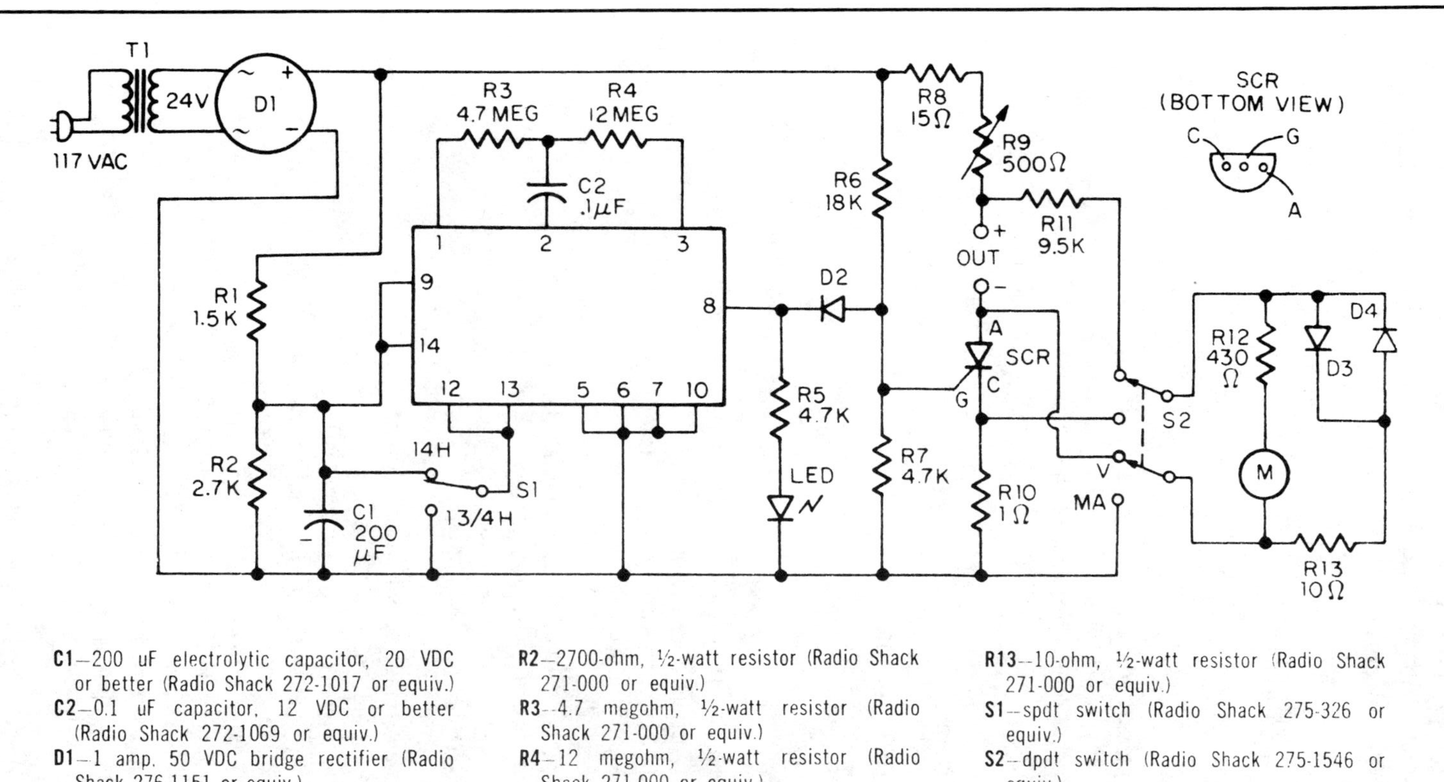

C1—200 uF electrolytic capacitor, 20 VDC or better (Radio Shack 272-1017 or equiv.)
C2—0.1 uF capacitor, 12 VDC or better (Radio Shack 272-1069 or equiv.)
D1—1 amp, 50 VDC bridge rectifier (Radio Shack 276-1151 or equiv.)
D2—general purpose germanium diode such as 1N34A
D3, D4—general purpose silicon diode such as 1N914
IC1—oscillator-timer integrated circuit, Motorola MC14541CP
LED—light emitting diode, red, 20 mA (Radio Shack 276-041 or equiv.)
M1—0 to 1 mA panel meter (Radio Shack 22-052 or equiv.)
R1—1500-ohm, ½-watt resistor (Radio Shack 271-000 or equiv.)
R2—2700-ohm, ½-watt resistor (Radio Shack 271-000 or equiv.)
R3—4.7 megohm, ½-watt resistor (Radio Shack 271-000 or equiv.)
R4—12 megohm, ½-watt resistor (Radio Shack 271-000 or equiv.)
R5, R7—4700-ohm, ½-watt resistor (Radio Shack 271-000 or equiv.)
R6—18,000-ohm, ½-watt resistor (Radio Shack 271-000 or equiv.)
R8—15-ohm, 3-watt or better resistor
Note—You can use two 7½ ohm resistors in series such as Radio Shack 271-147.
R9—500-ohm wire-wound potentiometer (Allied Electronics 875-4041 or equiv.)
R10—1-ohm, ½-watt resistor
R11—9500-ohm, ½-watt resistor, 5%
R12—430-ohm, ½-watt resistor, 5%
R13—10-ohm, ½-watt resistor (Radio Shack 271-000 or equiv.)
S1—spdt switch (Radio Shack 275-326 or equiv.)
S2—dpdt switch (Radio Shack 275-1546 or equiv.)
SCR—0.8 to 1 amp, 100 volt silicon controlled rectifier, G.E. C103 (Radio Shack 276-1059 or equiv.)
T1—power transformer, 117 V primary to 24 V secondary @ 1 amp (Radio Shack 273-1480 or equiv.)
Misc.—perf board, hardware, push-in clips, case approx. 6 x 4 x 3-in. (Radio Shack 270-252 or equiv.), 14-pin IC socket, output terminals such as Radio Shack 274-724 phone tip jacks, wire, solder, etc.

Fig. 4-11. Supercharger schematic and parts list.

similar characteristics to NiCad batteries. They are not, however, as long-lived. Many other excellent types of batteries are used in military and commercial applications. They did not yet find their way to the consumer market because of high cost.

From this short description, you may deduce that the NiCad battery is the most cost-effective battery in many applications where the appliance is in *frequent* use.

On the Inside. A NiCad battery consists of layers of sintered cadmium and sintered nickel separated by fiber soaked in potassium hydroxide electrolyte.

Sintering consists of baking a powdered metal to the consistency of a solid. A sintered material is highly porous. Its active area is several hundred times larger than that of a solid plate of the same dimension. The basic chemical reaction in a NiCad battery is as follows:

Charged

Cathode Electrolyte Anode

$Cd(OH)_2 + 2KOH + 2NiO$

Discharged

Cathode Electrolyte Anode

$Cd + 2KOH + 2Ni00H$

This reaction does not generate any gases. However, during the latter part of the charging cycle, during overcharging and during high discharge, hydrogen, oxygen and electrolyte fumes are being generated. These gases will normally reach an equilibrium condition reacting with each other and with the porous electrodes. Sealed cells also have a safety venting mechanism (activated above 100 PSI) assuring that the cell will not rupture under extreme conditions. Repeated venting however, causes loss of the electrolyte and subsequent battery deterioration. For this reason controlled charging is beneficial to NiCad batteries.

Other Advantages. A major advantage of NiCad cells, in particular when used for portable radios and walkie-talkies, is a nearly constant voltage during the discharging cycle. Regular zinc-carbon batteries lose their voltage at a fairly constant rate and thus affect the performance of the equipment they are powering; however, rechargeable batteries keep their voltage nearly constant until they nearly completely discharge. For example, the voltage of a carbon zinc battery drops by approximately 0.3 volts per cell when it is 50 percent discharged. The voltage of a NiCad battery drops by only 0.1 volt during the same period. Another important feature of NiCad batteries is the low internal resistance on the order of about 30 milliohm (AA cells)—about ten times less than for a comparable zinc carbon battery. This feature is particularly important for class B type audio circuits which require more power during peaks of speech or music. Batteries with a low internal resistance can supply the sudden surges of power required for good, low distortion sound. Another important feature of NiCad batteries, as compared to zinc carbon, is that they can be stored in a charged or discharged state and are virtually leakproof.

From flashlight to photoflood, from toys to two-way radio, NiCads are in widespread use. Everyone is ready to save a buck these days; from a money-saving standpoint, NiCad batteries have some definite advantages. Maybe, if you are a heavy battery user, NiCad rechargeable batteries can help you.

THE JUNK BOX SPECIAL

Between 555 timers, TTL, CMOS, op amps and run-of-the-mill transistor projects, the average experimenter is often faced with the need for a regulated power supply with a range of about 5 to 15 volts—just to try out a breadboard project. If you've priced any regulated supplies lately you know they don't come cheap. Maybe, just maybe, you might get one for $30 or $35.

With a little careful shopping, a reasonably stocked junk box and one or two "brand new" components, you can throw together a regulated supply costing less than $10 that will handle most of your experimenter power supply requirements. One of these Junk Box Specials is shown here. The range of this model is 5 to 15 volts dc at currents up to 1 ampere. One of the common, three-terminal regulators which are now flooding the surplus market provides everything in the way of regulation. Depending on the source, the regulator will cost you from $1 to $2.50; the higher prices often include an insulated mounting kit (worth about 25 cents).

Refer to Fig. 4-12.

If regulator IC1's collector terminal is connected to a voltage divider across the output—R1 and R2—the output voltage will be that at the junction plus the voltage rating of the regulator, which in this instance if 5 volts. So, when potentiometer R2 is adjusted so its wiper is grounded the power supply's output is that of the regulator, 5 volts—perfect for TTL projects. As R2 is advanced, increasing the resistance from IC1's collector to ground, the voltage output increases.

Getting the Parts. There are plenty of parts around to build this supply for under $10. If you go out and round up "all new" components the cost is likely to go well over $30, so forget about new parts. Power transformer T1 can be 18 volts at 1 ampere (or rated at higher current, though the supply's maximum output is 1 ampere), or 36 volts center-tapped at 1 ampere or more. Both the 18-volt and 36-volt transformers are glutting the surplus market. If you get an 18-volt transformer use the bridge rectifier shown in the schematic. If you get a 36-volt C.T. transformer use the full-wave rectifier shown below the schematic. The diode rectifiers, SR1 through SR4, are type 1N4001, 1N4002, 1N4003, or 1N4004, which are also glutting the surplus market. Just to show you the savings possible, at the time this article is being prepared you can buy fifteen surplus 1N4001s for $1. Just one single "general replacement" for the 1N4001 from a national supplier is selling for over 40 cents.

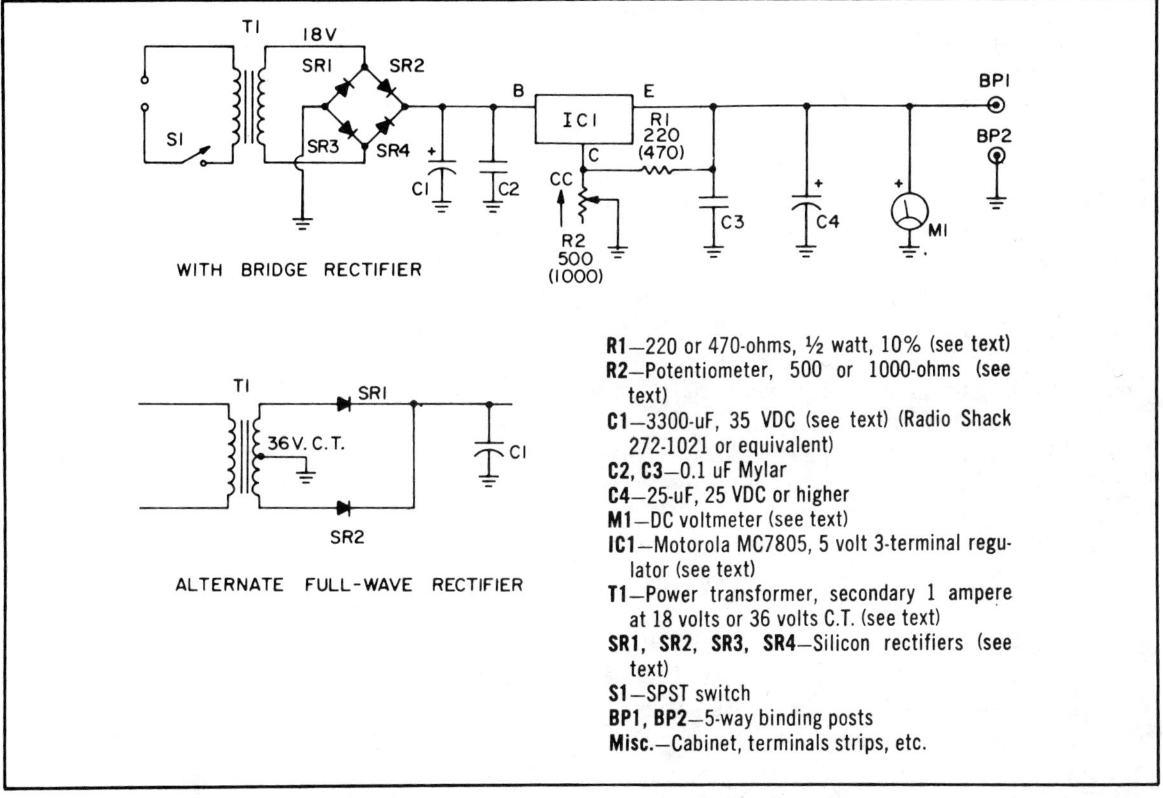

Fig. 4-12. Junk-Box Special schematic and parts list. Virtually any dc voltmeter that can display the range of 0 to 15 volts can be used. The surplus market is loaded with less-than-$5 meters that are suitable if you don't mind a little extra scale coverage.

Capacitor C1 can be anything from 2000 to 4000 μF at 25 volts or higher. Look for an outfit selling surplus computer capacitors. If worse comes to worse you can get the value specified in the parts list in a Radio Shack store.

The 3-terminal, 5-volt regulator is another item easily found on the surplus market. With an adequate heatsink—such as the cabinet itself—the device can safely deliver 1 ampere. The terminals B, C, and E are indicated directly on the device or on the terminals—where they join the case. The collector (C) lead is connected to the IC's metal tab, and is normally grounded. Note that in this project, however, the collector terminal, and therefore the tab, is not grounded. You must use an insulated mounting kit consisting of a mica insulator and a shoulder washer. Place the insulator between the IC's body and the cabinet, or the tab and the cabinet, and slip the shoulder washer into the opening (hole) in the body or tab. Pass the mounting screw from outside the cabinet through the mica washer, through the IC, and through the shoulder washer. Secre with a ¼-inch (or smaller, not larger) nut hand-tightened against the shoulder washer. Before going any further check with an ohmmeter to be certain the collector terminal is insulated from the cabinet.

Connecting wires are soldered directly to IC1's terminal leads; use a heatsink such as an alligator clip on each terminal if you have a large (greater than 40 watts) iron. Since the layout is not important, we suggest positioning IC1 between two mounting strips so R1 can span across the strips and be soldered to IC1's collector terminal.

Finally, we come to the meter, a device that has become slightly more expensive than a barrel of Arabian oil. Any meter that can indicate at least the range of 0 to 15 Vdc is adequate. You might not end up with a meter case the looks suitable for NASA, but the output voltage doesn't care two hoots whether the meter is a modern $25 model or a surplus-special for a buck ninety-nine.

Power switch S1 can be a separate SPST as shown in our project, or it can be part of R2. But keep in mind that a separate S1 allows you to turn the supply on and off without affecting voltage control R2's adjustment.

Finally, we come to R1 and R2. You will note that the schematic shows two values for each. One value for each resistor is in brackets (parenthesis). You can use either set of values as long as they are matched. If R2 is 500 ohms R1 is 220 ohms; if R2 is 1000 ohms R1 is 470 ohms. The reason we show both sets of values is because 500 and 1000 ohm potentiometers appear on the surplus market from time to time, but usually not together. This way, you can use whatever is available at low cost.

Checkout. Set potentiometer R2 so the wiper shorts to the end connected to IC1's collector terminal, thereby connecting the collector directly to ground. If you wired R2 correctly it should be full counterclockwise. Then set S1 to *on*. The meter should rise instantly to 5 volts dc. As R2 is adjusted clockwise the output voltage should increase to 15 Vdc or slightly higher. If R2 can adjust the output voltage only over the range or approximately 12 to 15 Vdc, or 12 to 15+ Vdc, IC1 is defective, or has been damaged.

A HEAT CONTROLLER FOR YOUR SOLDERING IRON

One trick that old timers have used for years is to connect a diode in series with a medium-to-heavy duty soldering iron. This halves the value of the iron's wattage rating, making it especially useful for soldering transistors, integrated circuits and low-wattage resistors.

But this arrangement limits the versatility of the iron, since there are times when one may wish to solder to a metal chassis or make other heavy-duty type connections. The soldering iron Heat Controller described here provides low/high-wattage versatility in a compact case, with a convenience outlet.

Refer to Figs. 4-13 and 4-14.

How It Works. Diode D1 shown in Fig. 4-13 provides half-wave rectification of the ac line voltage when switch S1 is in the *Low* position. Throwing switch S1 to the *High* position allows full line voltage to be applied to the soldering iron receptacle. The center *Off* position removes power from the outlet. Fuse F1 prevents any harmful effect if the iron's element should become short-circuited. Indicator lamps I1 and I2 add a professional touch to the equipment, and also act as on-off pilot lights.

Construction. All components except diode D1 are mounted on the case. Diode D1 is soldered to a terminal strip, which also provides

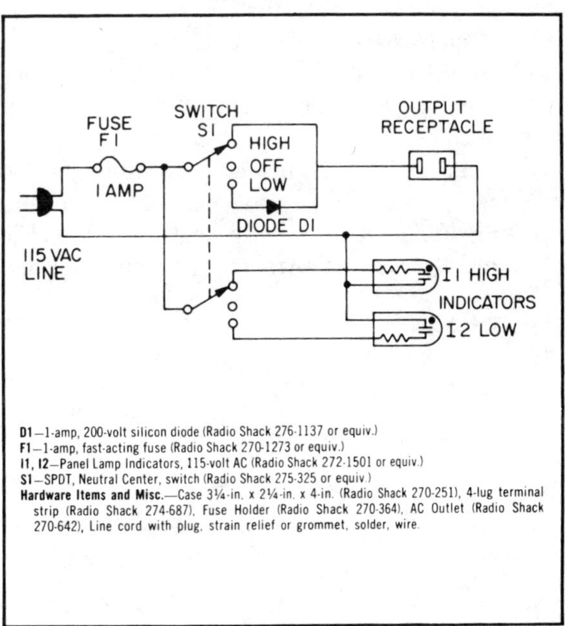

D1—1-amp, 200-volt silicon diode (Radio Shack 276-1137 or equiv.)
F1—1-amp, fast-acting fuse (Radio Shack 270-1273 or equiv.)
I1, I2—Panel Lamp Indicators, 115-volt AC (Radio Shack 272-1501 or equiv.)
S1—SPDT, Neutral Center, switch (Radio Shack 275-325 or equiv.)
Hardware Items and Misc.—Case 3¼-in. x 2¼-in. x 4-in. (Radio Shack 270-251), 4-lug terminal strip (Radio Shack 274-687), Fuse Holder (Radio Shack 270-364), AC Outlet (Radio Shack 270-642), Line cord with plug, strain relief or grommet, solder, wire.

Fig. 4-13. A Heat Controller for Your Soldering Iron parts list. Schematic diagram of two-position heat controller for 50-watt soldering iron. It's easy to wire-up—should take about an hour for an entire project.

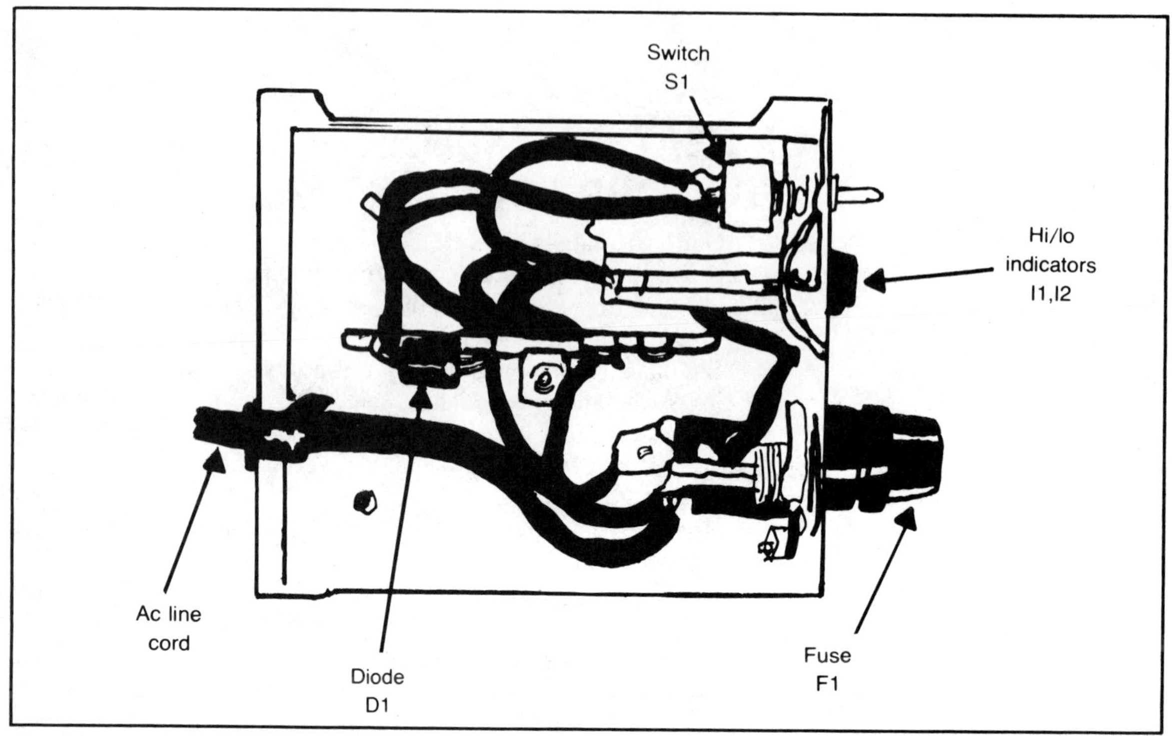

Fig. 4-14. Mechanical layout of heat controller. Putting this project together is not only good practice for beginners, but it also gives you a versatile tool that will make future projects easier.

terminal points for the various interconnecting wires. Wiring is point-to-point and not critical. Transfer letters are used for the individual panel markings. This is a very easy project, and the hour or so it takes to assemble it (once you've got the parts together) will be quickly repaid by the added convenience of having two different iron heats to work with.

Using the Heat Controller with a 50-watt soldering iron that takes various tips will handle about 95 percent of all your soldering iron work. It's only the very occasional super-heavy job that will require anything else, and that would require a much bigger iron anyhow.

A DESIGNER'S ±15V POWER SUPPLY

With very few exceptions, equipment and projects using operational amplifiers require a bipolar power supply, that is, a power source with both positive and negative voltage outputs in relation to ground. While it is generally possible to power an op amp from a single-ended power source, this technique requires a lot of extra hardware and filtering, and more often than not causes more problems than it's worth. The best way to power an op amp or a project using an op amp is with a dual tracking bipolar supply. Now, thanks to the latest in IC technology, you can build a ±15V dual tracking supply for well under $20.

Refer to Figs. 4-15 and 4-16.

It's Short Proof. The bipolar supply shown puts out up to 100 mA with full overload protection; in the event of a short circuit, the supply automatically shuts down before it's damaged. The really big plus in operation is *dual tracking*, which requires an explanation. The usual way to obtain a dual voltage output is shown in Fig. 4-15, a center-tapped transformer, bridge rectifier and zener diode regulators do a mighty fine job in non-critical circuits. This circuit is usually used for audio preamplifiers. However, suppose you power not only an op amp but a relay amplifier connected across half the output; the op amp might pull 10 mA while the relay amp pulls 60 mA from one half the supply. This will usually cause a voltage drop on one-half of the supply. If the op amp is in a critical circuit, the voltage unbalance between each side of the supply can cause improper circuit operation. Further, internal power supply heating can change the charactertics of the zener diodes, again causing voltage mismatch.

A dual-tracking supply, on the other hand, does just what it says, tracking one side against the other. Any voltage change on one side of the circuit automatically corrects the voltage on the other side to match. Heating effects are also compensated for on the opposite side, so that regardless of load current, ambient heat or whatever, the voltages on both sides of the supply track together. In the model shown, the worst-case mismatch is 0.15 volts (150 millivolts). If one side is −15.00 volts, the other side can be, worst-case, +15.15 volts.

Two more big pluses for the bipolar supply are voltage regulation and extra, electronic filtering. Within the power line range of 100 to 135 volts, the bipolar supply output will not vary more than 0.004 volt (typical), nor will current output changes from zero to full load (100 mA) cause more than a 0.005 volt typical variation. There is better than 120 dB filtering, with the ripple components less than 10 μV.

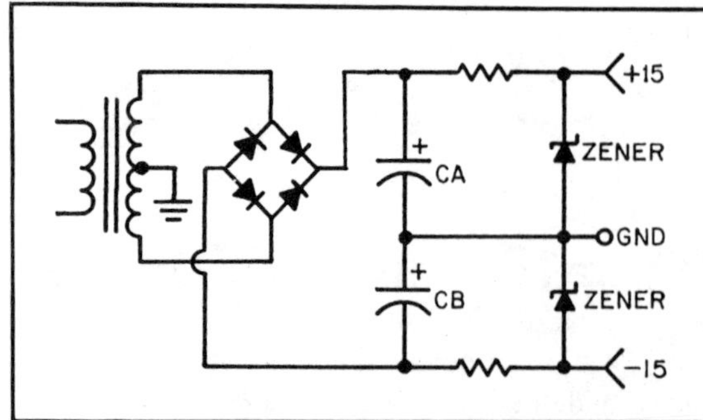

Fig. 4-15. Simple zener-regulated shunt-type supply lacks regulation, tracking ability, and stability necessary for precision use.

Smaller Caps. In short, this is an ideal supply for the lab and experimenter as it eliminates any worry about the power supply; you can spend your time checking and developing the circuit rather than fussing with power supply regulation and filtering. Actually, the circuit is not much more expensive than a cheap diode-type supply because it doesn't require brute-force filter capacitors which often cost more than an integrated circuit regulator.

The heart of the bipolar supply is IC1, a complete integrated regulator that replaces 21 transistors, seven zener diodes, and 11 resistors. Until this IC was made available a dual tracking bipolar supply took all these parts. In addition to regulation, IC1 also features current limiting; just two ½-watt resistors protect IC1 against damage if the output current attempts to exceed 100 mA. Though the worst-case regulation is specified as 1 percent, in actual tests on complete supplies the regulation was better than 0.5 percent.

Construction. The entire power supply is built on a 2⅞-inch × 4-inch printed circuit board which can be mounted inside any cabinet along with the equipment it is powering. Alternately, it can be installed in a small cabinet for use as a bench or test supply.

If you use the template available in the parts list to lay out your PC board, we suggest you use the Radio Shack parts specified in the parts list, because lead spacing matches the printed circuit board component mounting holes. No components other than IC1, however, are critical, and substitutions can be made.

First step is to prepare a printed circuit board. Cut a section of copper-clad board to size and scrub the copper clean with steel wool or a strong household cleanser. Place a piece of carbon paper, carbon side toward the copper, on the board and tape the board under the full-scale template provided and secure the board in position with a few strips of masking tape.

Using a sharp pointed tool, such as a scribe, indent the copper at each mounting hole by pressing the point of the tool firmly through the template and into the foil. The indents will provide the markings for the component mounting holes. Using a ball point pen and a lot of pressure, trace the foil outlines on the template. Remove the copper-clad board from under the template and using a resist ink pen trace the foil outline on the board: then fill in the areas with resist. No foil should be less than 1/16 inch thick as undercutting by the etchant will produce a slightly thinner foil.

Pour enough etchant into a container slightly larger than the PC board so there's at least ¼-inch depth. Then float the PC board on top of the etchant with the copper side down into the etchant. Every few minutes agitate the etchant container to speed up removal of the undesired copper. After all the undesired copper has been removed (about 20 minutes) rinse the board under running water and remove the resist with a small rag soaked with resist remover, resist solvent or acetone.

Holes and Such. All components mounting holes are drilled with a No. 55, 56 or 57 bit. The board's corner mounting holes should clear a No. 4 screw; the transformer's and IC1's mounting holes should clear a No. 6 screw. Do not make the holes larger than needed.

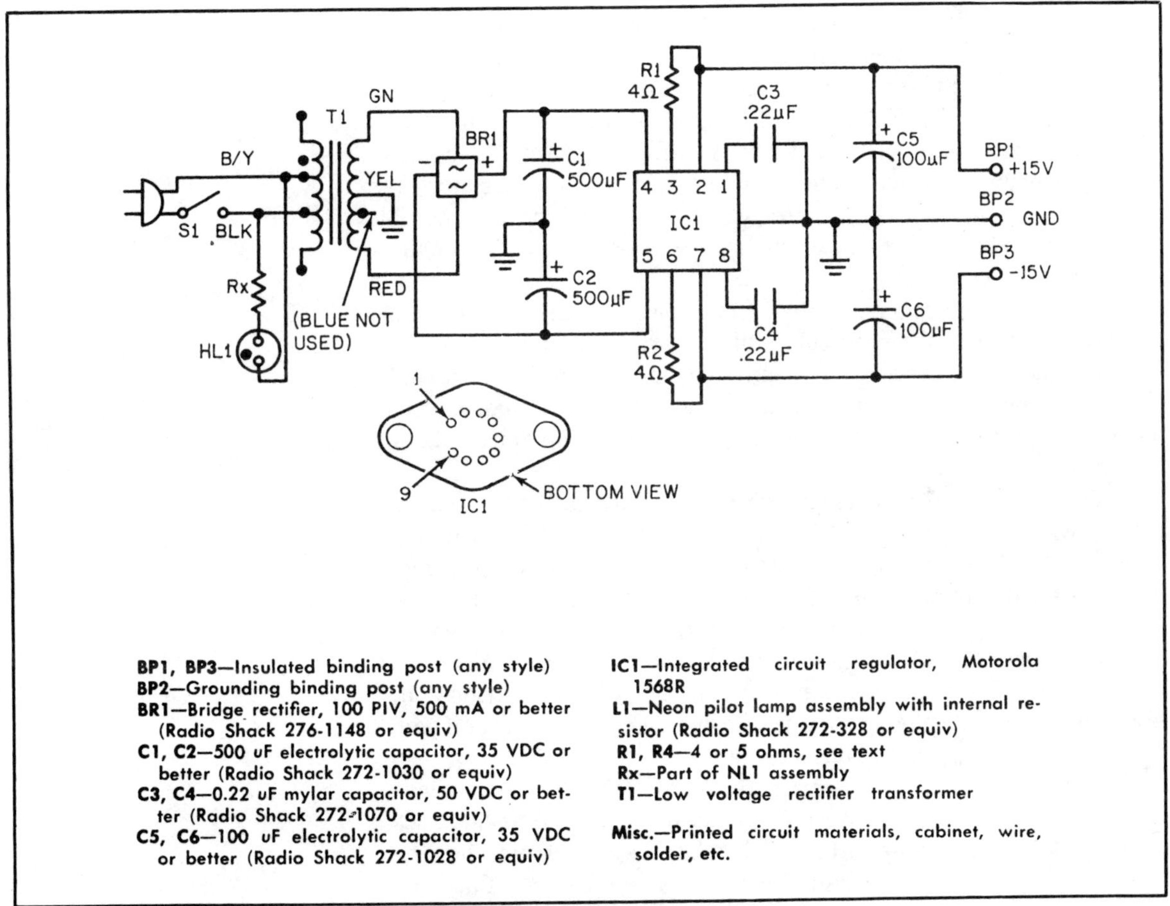

BP1, BP3—Insulated binding post (any style)
BP2—Grounding binding post (any style)
BR1—Bridge rectifier, 100 PIV, 500 mA or better (Radio Shack 276-1148 or equiv)
C1, C2—500 uF electrolytic capacitor, 35 VDC or better (Radio Shack 272-1030 or equiv)
C3, C4—0.22 uF mylar capacitor, 50 VDC or better (Radio Shack 272-1070 or equiv)
C5, C6—100 uF electrolytic capacitor, 35 VDC or better (Radio Shack 272-1028 or equiv)
IC1—Integrated circuit regulator, Motorola 1568R
L1—Neon pilot lamp assembly with internal resistor (Radio Shack 272-328 or equiv)
R1, R4—4 or 5 ohms, see text
Rx—Part of NL1 assembly
T1—Low voltage rectifier transformer
Misc.—Printed circuit materials, cabinet, wire, solder, etc.

Fig. 4-16. A Designer's ± 15V Power Supply schematic and parts list.

Install bridge rectifier BR1 first, using extra care in orientation. The usual terminal arrangement of the low cost "surplus" bridges available to the experimenter usually have a similar lead arrangement. Note that one terminal is marked +, indicating the positive dc output. Two other terminals are generally marked with a "~" (sine wave) or the letters ac indicating the connections from the power transformer. The fourth lead, the one diagonally opposite from the + lead, is the negative dc output. Make certain your bridge rectifier has this terminal arrangement. If it has any other lead arrangement you must modify the PC board template (if you use it) to correspond to the difference. Install BR1 so that when the leads pass through the board to the foil side the + lead comes through the "plus" hole.

Allow about ¼-inch space between the bottom of BR1 and the PC board and solder the leads to the foil. Then install and solder IC1. IC1 can fit the board both right and wrong. Make certain the space between the pins faces the nearest edge of the PC board, then secure IC1 with two short No. 6 screws. Install power transformer T1 and all other components. T1 has several unused leads, simply cut them off at the transformer with diagonal cutters.

Current limiting resistors R1 and R2 should be 4 ohms, a somewhat expensive and difficult value to obtain. You can substitute two parallel 10 ohm, 10 percent, ½-watt resistors; the board already has the extra mounting holes for two resistors at R1 and two resistors at R2. The difference between the required 4 ohms and the resultant 5 ohms from the parallel resistors will have no appreciable affect on the power supply operation.

To make a bench supply, the PC assembly can be installed in a metal cabinet. The three output terminals are 5-way binding posts. Two outside posts, used for the + and – voltage outputs, are insulated from the cabinet. The center terminal, the common ground, is connected to the cabinet by its own mounting nut (do not place an insulator under the nut).

Power switch S1 mounts on the rear apron and can be any type of SPST switch. To prevent the foil on the underside of the PC board from shorting to the cabinet use a ¼-inch metal spacer or stack of washers between the board and cabinet at each mounting hole. If desired, a neon pilot lamp can be installed. Make certain the pilot lamp is the type with a built in current limiting resistors of 56,000 to 200,000 ohms.

Service Note. If the power supply fails to operate properly, if, for example, there is voltage on one side but not on the other, or if both voltages are extremely low, it is most likely that the polarity of C1, C2, C5 or C6 is reversed. Take extra care to check that C2 and C6 have their *positive* terminal connected to ground. If you use the Radio Shack capacitors specified in the parts list, all polarities are correct when the vertical arrow on each capacitor faces the same direction. But don't take chances. The capacitors you get might have the arrow misprinted or used to denote another polarity. Double check that the polarity is correct.

If the capacitor polarities are correct and the output voltage is nearly correct but tends to wander, if the output voltage can't settle down to a rock steady value when current is drawn, it is most likely that you have used the wrong primary connections on T1. The proper wires are color-coded. Use a secondary output voltage of 40 volts rms center-tapped (20-0-20).

CUSTOM SWITCHES YOU COULDN'T AFFORD TO BUY

How often have you searched fruitlessly for a special switch? Probably dozens of times—if you're at all an active builder. The next time this happens, consider custom-building your own complex switches using inexpensive magnetic reed switches and small ceramic magnets. Such do-it-yourself switches offer several advantages. They are relatively inexpensive, silent, and long lasting, as there are no rubbing contacts and the reed contacts are sealed in glass, away from corrosive atmospheric gases.

The several custom-built switches shown in this article only hint at the virtually limitless design combinations that are possible. Study the drawings to learn how magnet orientation and direction of travel past the reed switches affect switching action.

Refer to Figs. 4-17 through 4-20.

Carpetak Tape, a cloth tape with adhesive on both sides that is used to hold down carpets, is excellent for mounting the reed switches to panels. The switches adhere firmly, yet can be removed without damage. For greater permanence, you may wish to use epoxy cement for mounting once you know exactly where to locate the reed switches. Generally, it is best to locate the magnet and reed switches on the same side of the panel; however, it is also possible to put the switches on one side of a non-magnetic panel and orient the magnet on the other side. The ceramic magnets are of extremely hard material, and you may have poor success if you try to hacksaw them smaller. Try breaking the magnet by clamping in a vise and striking with a chisel; it may not break cleanly across, but grinding on an emery wheel may be practical. When possible, just use the magnets as they are. Mount them in aluminum holders as shown here, or glue to support arms with epoxy adhesive.

The following brief descriptions of various switch types should help clarify the principles of building switches:

Single-throw, multi-pole. These can be constructed simply by mounting reed switches in parallel, and passing the *edge* of a vertically-mounted magnet over them to trip all switches simultaneously. If you need a sequential switching action, just angle the magnet about 30 degrees so that the parallel reed switches are tripped in 1, 2, 3, 4 order. If the magnet movement continues in the same direction, the switches will go off in the same 1, 2, 3, 4 sequence, on the other hand, if magnet movement is reversed when all switches are on, the switches will go off in the reverse 4, 3, 2, 1 order.

Multi-position, single-pole. Arrange the reed switches one

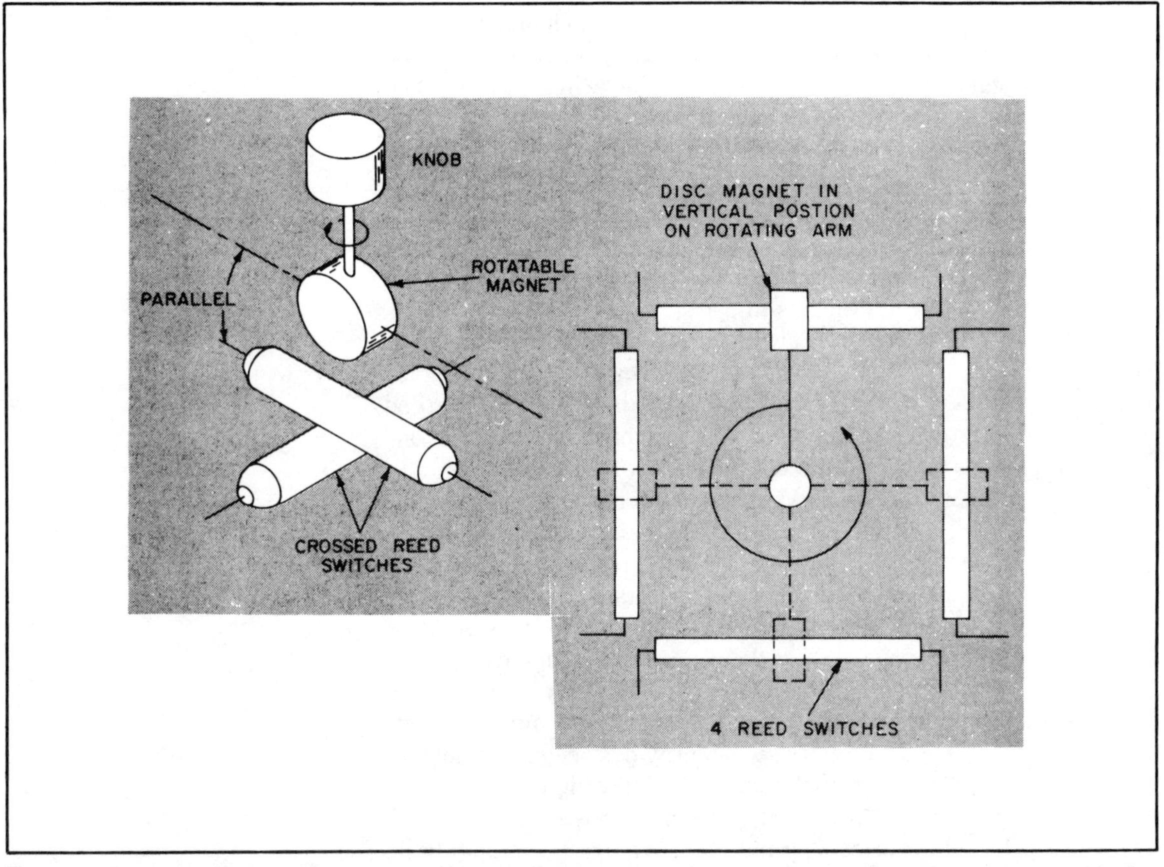

Fig. 4-17. Make a single-pole, two-position, on-before-off rotary switch with two crossed-reed switches and a rotatable disc magnet positioned above them. When the axis of the magnet is parallel with a switch's long dimension, that switch remains off; the other switch is simultaneously on. In the intermediate, 45-degree position, both switches are on. Use this arrangement when the second switch must come on before the first switch turns off. A single-pole, four-position switch is accomplished by a vertically-oriented disc magnet passing over four reed switches, which are arranged in a square.

after the other, like cars of a train. You can keep the switch smaller by using two lines of staggered switches, as shown. As the vertically-oriented disc or rectangular magnet passes over the switches, each "on" switch goes off before the next switch comes on.

A photograph shows a sliding switch of this general type, but one made to function as a double-pole, single-pole, double-pole sequencer. A simple locking device consisting of a lock washer under the knob on the other side of the panel permits locking the movement at any desired position. Note the "guide" strip near the slot; a square nut that holds the magnet support arm on the knob shaft bears against this guide to keep the magnet properly aligned over the reed switches.

Rotary switches. These are easier to build than slide switches, and there are many ways to achieve special switching characteristics. Note that when the edge (diameter axis) of a disc magnet is aligned with the long axis of a reed switch there is no switching action.

Thus if you mount several reed switches next to each other, and rotate the magnet directly over the center of the switches, you obtain more or less simultaneous on-off action.

If two switches are crossed a vertically-mounted rotating magnet will turn one switch on and the other off when the magnet axis is parallel to the long axis of one reed switch. In the intermediate position, both switches are on; thus you can have on-before-off action with a very simple physical arrangement. To make a double-pole version, cross four reed switches in pairs.

A 4-pole, 4-position rotary switch can be made by arranging four reed switches in a "square" and adding a vertically-oriented disc magnet so that it can be swung in a circle over the centers of the reed switches. This provides an off-before-on switching action, each "on" switch first going off before the next one comes on.

Some strangely useful things begin to happen if you mount the disc magnet horizontally instead of vertically. As the magnet passes over a corner of the square of reed switches so that it is partly over the ends of two adjacent switches, both switches go on. Rotate the magnet a little further, so that it is over just one switch and that switch stays on while the other goes off. Curiously, when the *horizontally* mounted magnet is over the center of a switch, that switch goes off. This is exactly the opposite of the on-action caused by a vertically-mounted magnet. Consequently, this type of rotary switch provides sequential double-pole and single-pole action, with four fully off positions.

Multi-pole Rotary. Such switches can be constructed by stacking additional reed switches atop the first four that make up a basic square. Mount the rotatable magnet inside the "box" formed from the stacked reed switches. When the long dimension of the magnet is perpendicular to stacks on opposite sides of the box, all of those switches will go on; other stacks at 90 degrees to these will remain off because the magnet axis is parallel to them. By using one of the larger rectangular magnets, you can easily stack at least a half dozen switches on a side, for a total of 24 switches; 12 would be on at any one time, 12 off. When the magnet is in the intermediate, 45-degree position, all switches are off.

Linear Off-Before-On. Mount parallel switches far enough apart so that, as a vertically positioned magnet approaches the switches from one side, the first switch will go on and off before the next switch is affected. Here's a handy trick that enables you to pull the parallel reed switches much closer together to form a more compact switch: Position a small disc magnet over the center of the first switch so that its long axis (diameter) is parallel to the long axis of the switch. If you have been paying attention, you already know that in this position the magnet has no effect on the switch. Now slowly rotate the switch away from this parallel orientation until you hear the switch

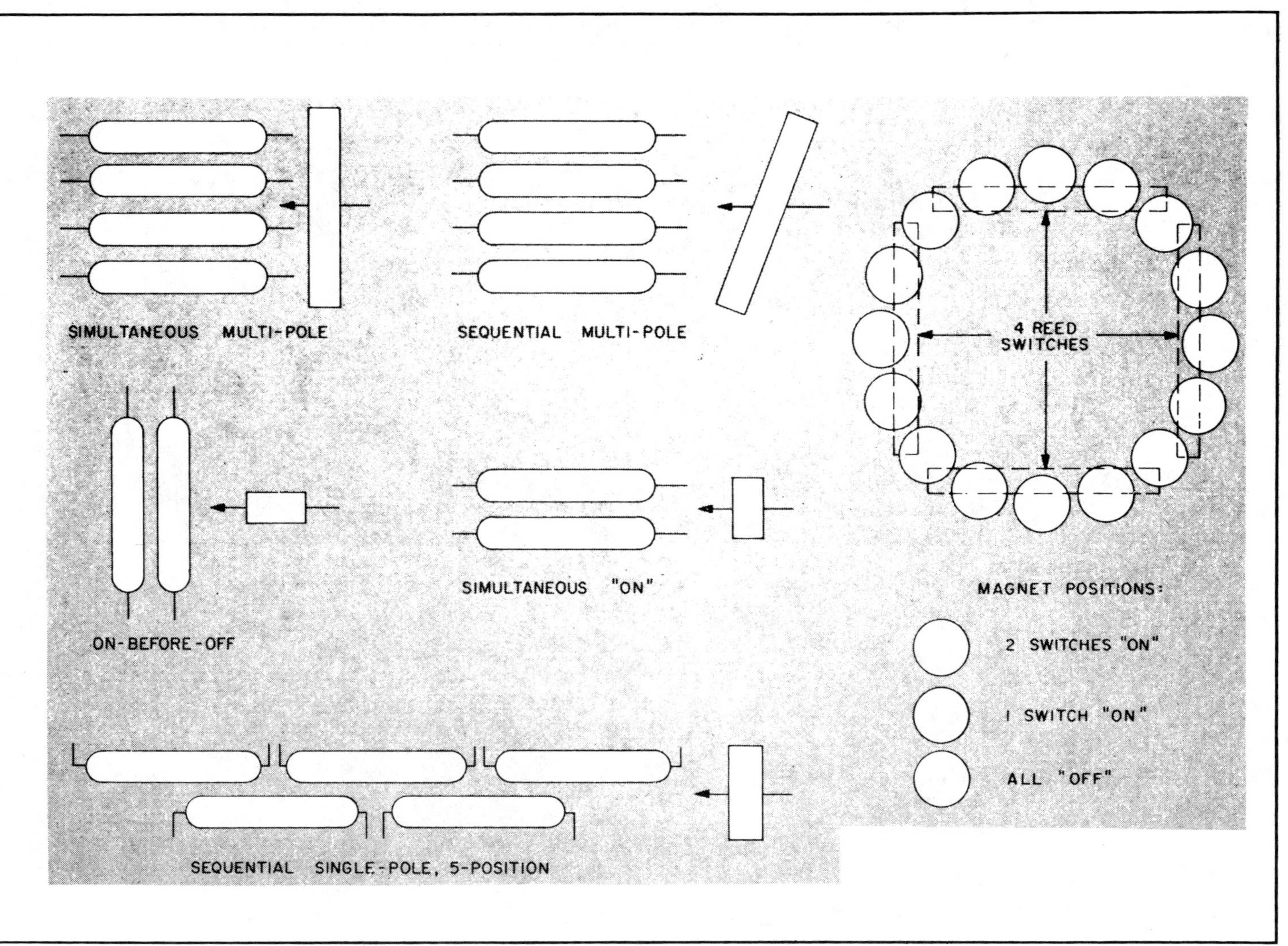

Fig. 4-18. Here are various magnet and reed switch orientations all of which achieve different switching actions. Position the magnets and reed switches exactly as shown, and as further discussed in the text, for best results. A complex switching method is achieved by passing a horizontally-oriented magnet over a square of four reed switches. When the magnet is in a corner, two adjacent switches are on; in other positions, only one switch is on.

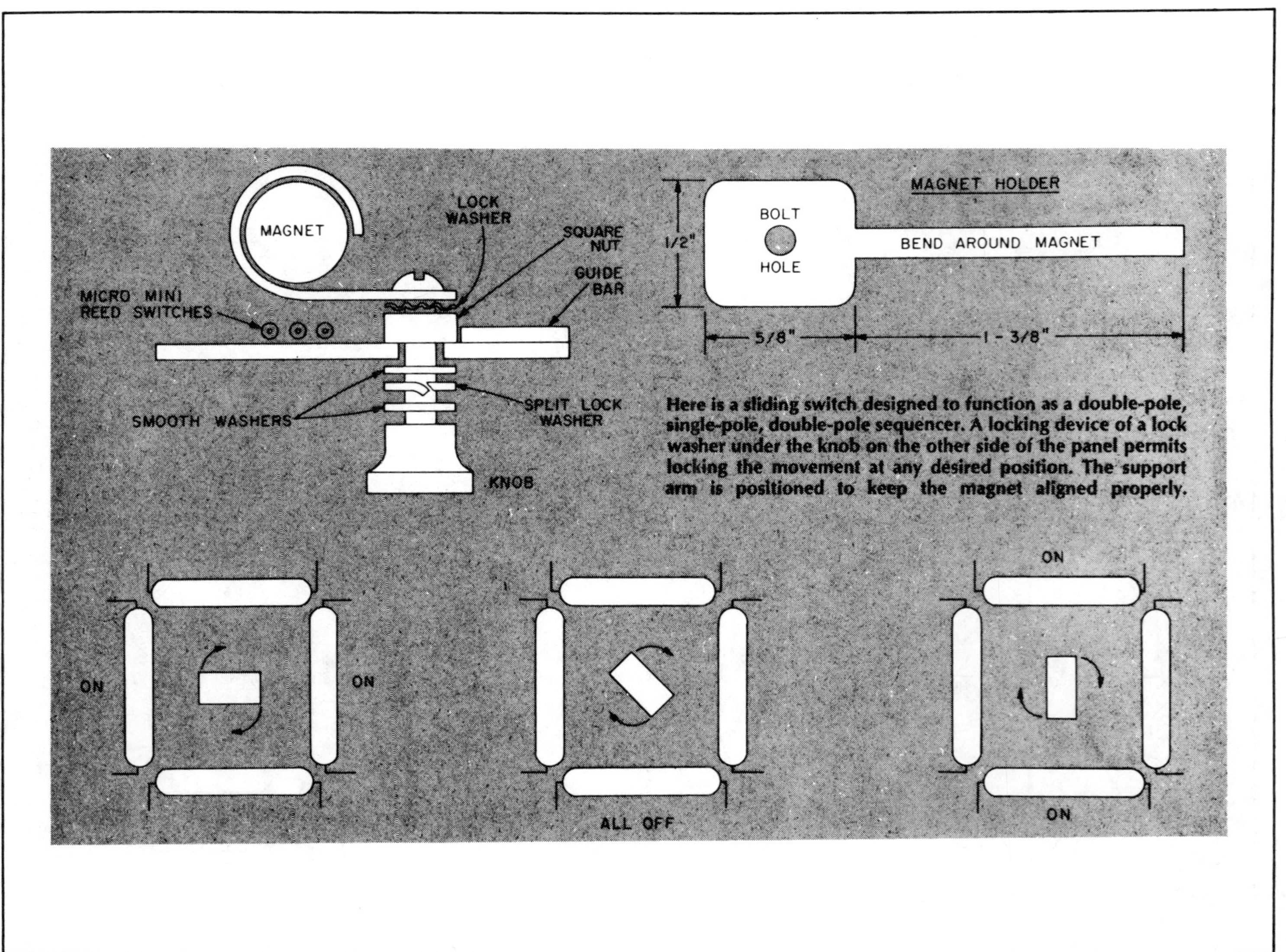

Fig. 4-19. By positioning a rotatable magnet in the center of a square of four reed switches you have a double-pole rotary switch. By stacking magnets on top of the basic four, and using a larger magnet, as many as 12 on-positions are possible at each active orientation.

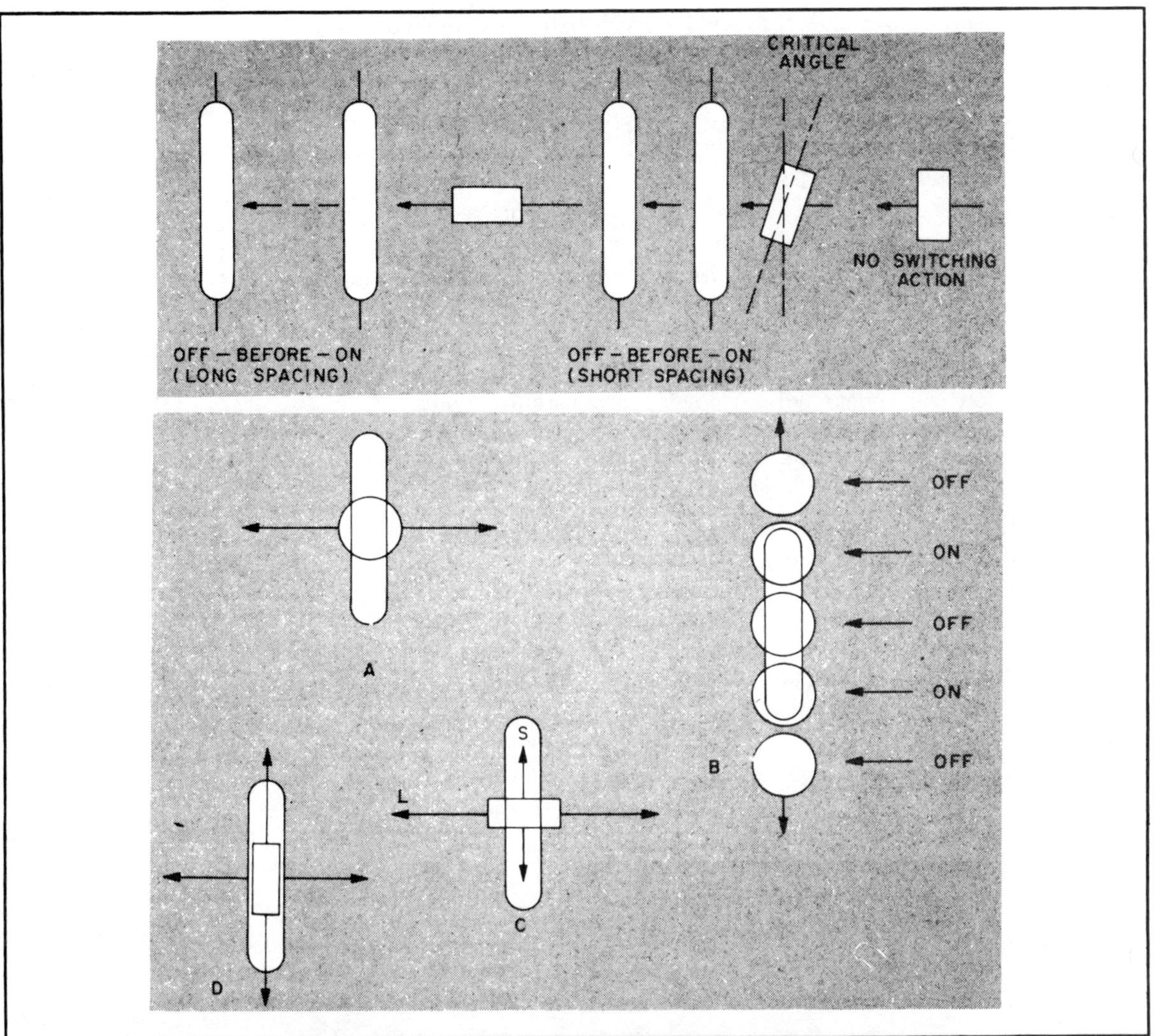

Fig. 4-20. Spacing of adjacent reed switches is important if the first switch must go off before the next turns on. When the magnet is in its strongest orientation (left), the reed switches must be far apart. If the magnet is positioned at an angle slightly removed from where it does not switch, the two reed switches can be closer together and retain the on-before-off action. Orientation and direction of movement of a disc magnet influence switching action. No switching occurs as a horizontal magnet moves across the reed switch as at A. If the magnet passes along the length of the reed (B), switching to on occurs when the magnet is near either end of the reed, but not when it is directly over the contact points in the center of the reed. Switching occurs if the magnet is vertically oriented as at D, but over a shorter range if the magnet moves along the "S" path than along the "L" path. If the magnet is turned 90 degrees (C), no switching occurs when it is moved along either of the indicated paths.

click on. Mount the magnet in its sliding holder so that it passes over the center of the reed switch in this slightly angled position. This deliberately weakened magnetic action permits location of the next switch much closer to the first—as close as about ⅜ inch—and still obtain off-before-on switching.

There's no problem, of course, if you want on-before-off action because then you can mount the reed switches as close to each other as you like. A magnet approaching from the side will turn the first on, then the second, before turning the first off. Bear in mind that if the magnet approaches the pair of switches from the end directions, you obtain approximately simultaneous double-pole switching.

If you have but one reed switch, and the vertically-oriented magnet approaches from one side of the reed switch, it has a longer "reach" and switching action occurs while it is relatively far from the center of the switch. This relative sensitivity, relating to direction of magnet travel, could be an important factor in some switch-design problems.

Special Designs. As you play around with your reed switches and magnets you will undoubtedly discover many variations on the ideas given here. For example, suppose you wanted a sliding multi-pole switch that always switches the reed switches on in the same sequence. As the magnet travels over the parallel reeds, the switches come on in 1, 2, 3, 4 order. But as you slide the magnet back to its original starting position, the reeds would go on in reverse 4, 3, 2, 1 sequence—which is what you do *not* want. So what's the answer? Simply mount the magnet on a holder that permits it to be turned 90 degrees at the end of each sweep. This way the magnet can be in its "active" orientation going one way, and in its "dead" orientation going the other way. Thus it is possible to return the magnet to the starting position, for subsequent normal switching order, without affecting the switches on the return sweep.

It seems likely that you could construct such truly off-beat switches as, for example, a level-indicating switch by suspending the magnet on a short pendulum so that it will swing close to either of two parallel magnets to electrically signal tilting. And it may be feasible to create a vibration-detector in much the same manner, but mounting the magnet on the end of a flat or coil spring so that vibrations will swing it in-and-out of the switching range.

THE SMART POWER SUPPLY

When working with various electronic projects, it's easy to get carried away with too many current-eating components, which can overload a power supply. Our Smart Power Supply solves this problem with its built-in LED ammeter, which always tells you what the current draw is.

Refer to Fig. 4-21.

The supply delivers a regulated 5-volt and 8-volt output at up to 1-amp, and you'll never be in the dark as to how much current is being drawn. 4 LEDs display the amount of current being utilized by the load. Each LED lights respectively to show the level of current being drawn. For example, if ¾ of an amp (.75) is being used, the first three LEDs (".25, " ".50," and ".75") will all glow to show that a current of at least .75 A is flowing. Best of all, the current measuring resistance is an unprecedented 0.1-ohm! What's more, the cost for the ammeter portion of the circuit is only about $5. That's way less than you'd pay for a good mechanical meter. The 5-volt output is ideal for all of your TTL IC projects, while the 8-volt output may be selected for CMOS circuits, and other higher-power requirements. The total cost for the whole supply, including the bargraph ammeter, is about $15-20, depending on your buying habits and choice of parts suppliers.

How it Works. IC4 is supplied by an accurate reference voltage of 5-volts by IC3. IC4 is a quad op amp used in a quad comparator configuration. The four op amps (comparators) in IC4 are each fed a separate reference voltage by the divider network made up of R1-R4 and R5-R8. These comparators in IC4 are very sensitive, and they can detect extremely small voltage differences and compare them.

Let's take the first op amp comparator as an example. Its inputs are pins 2 and 3, and its output is pin 1. The reference voltage appearing at pin 3 is compared to the voltage coming into the first comparator at pin 2. When ¼ of an amp or more is flowing through R10, .025-volts or more (0.1-ohms times 0.25A = .025V) appears across R10, which is enough voltage to equal pin 3's reference voltage, thus turning on the first op amp. The output of this op amp is at pin 1, so LED1 turns on to signify that at least ¼ of an amp is being drawn. In a like manner, the other LEDs turn on or off with the changing current. The rest of the circuitry makes up a basic voltage-regulated power supply.

Construction. All of the circuitry, except ICs 1 and 2, can be mounted on a small piece of perfboard. These two ICs must be mounted to the cabinet. In operation, IC1 and IC2 will get hot when the

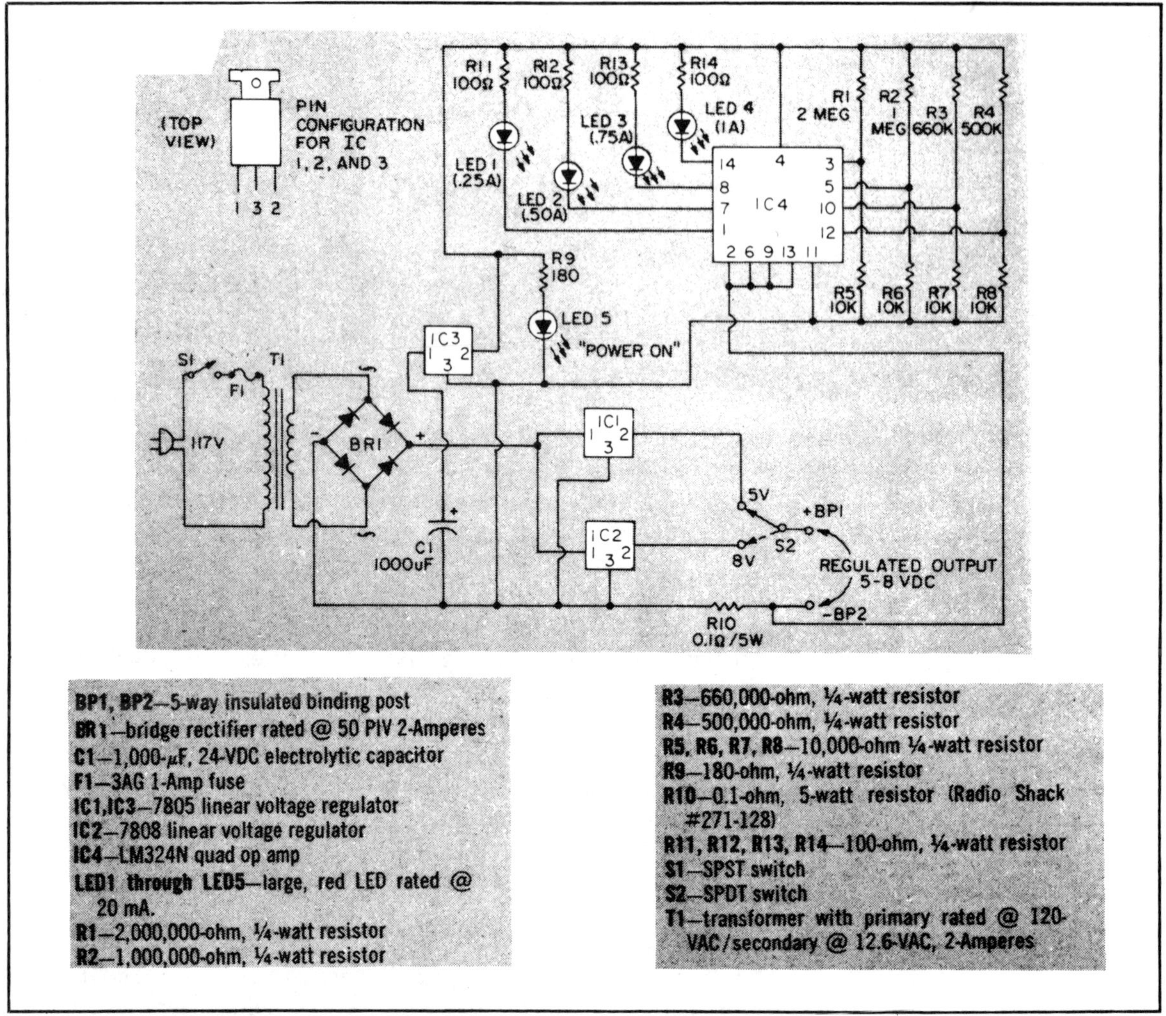

BP1, BP2—5-way insulated binding post
BR1—bridge rectifier rated @ 50 PIV 2-Amperes
C1—1,000-μF, 24-VDC electrolytic capacitor
F1—3AG 1-Amp fuse
IC1,IC3—7805 linear voltage regulator
IC2—7808 linear voltage regulator
IC4—LM324N quad op amp
LED1 through LED5—large, red LED rated @ 20 mA.
R1—2,000,000-ohm, ¼-watt resistor
R2—1,000,000-ohm, ¼-watt resistor
R3—660,000-ohm, ¼-watt resistor
R4—500,000-ohm, ¼-watt resistor
R5, R6, R7, R8—10,000-ohm ¼-watt resistor
R9—180-ohm, ¼-watt resistor
R10—0.1-ohm, 5-watt resistor (Radio Shack #271-128)
R11, R12, R13, R14—100-ohm, ¼-watt resistor
S1—SPST switch
S2—SPDT switch
T1—transformer with primary rated @ 120-VAC/secondary @ 12.6-VAC, 2-Amperes

Fig. 4-21. Smart Power Supply schematic and parts list.

supply is run at higher currents, and they may shut down if the heat is not carried away. The back of the cabinet is the best place to mount ICs 1 and 2, for it allows a large heat dissipating area, while keeping the rest of the cabinet cool to the touch. When mounting ICs 1 and 2, smear heatsink grease between the IC cases and the cabinet, then bolt the ICs down tightly. Connect three long wires to IC1 and 2. These will be connected to the main circuit board later.

If the transformer that you wish to use has a center tap, cut it off or tuck it away. You won't need it. Bolt T1 down to the cabinet. Use heavy gauge (No. 16) wire for all line voltage connections, and carefully wrap all ac line connections with electrical tape. Use a grommet around the line cord exit hole in the chassis to protect the cord from the heat that will be there due to ICs 1 and 2. Tie a knot in the line cord just inside the cabinet hole to prevent it from being pulled out.

IC3, unlike ICs 1 and 2, can be mounted on the perfboard because it will not get hot in operation. You should use a 14-pin socket for IC4. Install IC4 only after all of your wiring to the socket is complete.

Be careful not to make any solder "bridges" between socket pins, as they are close together. When you install IC4 in its socket, make sure that you observe the correct orientation with regard to pin 1.

After you've installed the circuit board, attach the wires from ICs 1 and 2 to their proper places on the board. Connect the wires to the display LEDs last, and make sure that you observe polarity on each LED. Be careful not to let the LED leads short against the metal cabinet.

Operation. Carefully inspect your wiring on the circuit board, especially the wiring to IC4's pins. This is a very important step, as one misplaced wire here can produce some real odd-ball systems. If everything appears to be in order, turn the unit on. The "power" LED (LED5) should glow.

Connect a voltmeter to the output jacks. Depending on what position switch S1 is in, the voltmeter will read 5 or 8 volts. Throwing S2 to its other position should cause the voltmeter to read the other of the two voltages that the supply delivers.

To test the ammeter section, connect a circuit to the output jacks. With the supply set for 5-volts, a TTL IC circuit would be good for this test.

If the circuit that you hooked up draws more than .25 A, then one or more of the display LEDs will go on to show you how much current is being drawn.

Conclusion. You shouldn't worry about overloading the power supply, as fuse F1 will limit current draw to a peak of about 1.3 A momentarily, before acting, and we deliberately overloaded several times in a row, with no damage occurring to the circuitry.

You might wish to attach a solderless breadboard to the top of the cabinet, to act as a permanently-powered breadboard for your experiments, or to construct an output voltage switcher for powering several projects alternately.

Chapter 5
Automotive Projects

AUTOMATIC HEADLIGHTER

The night may be dark and dismal when you pull in your driveway, but thanks to your Automatic Headlighter you'll have no trouble seeing the way to your doorstep. This automatic headlight timer is a two-transistor circuit that keeps your headlights illuminated for 60 seconds after you leave your car. A simple ten-dollar project (all new parts), it's connected in parallel with your existing auto headlight switch and is operated by a single pushbutton. The timer has been field-tested for over a year and has been proven extremely reliable. It is for installation in a 12-volt negative ground electrical system, but installing the unit in positive ground systems is a snap for hobbyists.

Refer to Figs. 5-1 through 5-3.

The How Of It. When the driver depresses pushbutton switch S1, timing capacitor C1 charges to 12 volts and turns on transistor Q1, which drives power transistor Q2 into conduction. This, in turn, energizes the relay which has its contacts connected in parallel with the headlight switch. The relay will stay energized until C1 discharges to the Q1 turn-off level. The lights-on period is determined by the value of C1, R1, and the characteristics of transistor Q1. With values shown on the schematic, about 60 lights-on seconds are provided.

What To Do. Construct the timer in a small 3¼-inch × 2⅛-inch × 1⅝ inch mini-box where all components except the relay and pushbutton switch S1 are mounted on a 1⅞ inch × 1⅜ inch printed circuit card. The circuit card may be made by any of the conventional methods.

Special attention must be given in mounting some of the components on the printed circuit card. Capacitor C1 must be mounted with its + terminal connected to R1. Also, special attention must be given to spacing the transistors at least ⅛ inch away from the circuit card to avoid overheating during soldering. Two insulated washers are used to mount power transistor Q2 to the printed circuit board.

Interconnecting wires from the circuit board should each be at least 4 inches long. They can be cut to proper length just before wiring in the mini-box. Mount RY1 and S1 in the mini-box, with heavy gauge (No. 14) wires soldered to the relay terminals as shown on the schematic. Make these wires about 24 inches long and mark "+" and "Lights" for easy identification during installation. The last step is to mount and connect the printed circuit card in the mini-box.

Under the Dash. The protruding mounting screw of RY1 may be used to mount the timer under your dash, or make a bracket of your own design for your particular automobile and installation. Just make

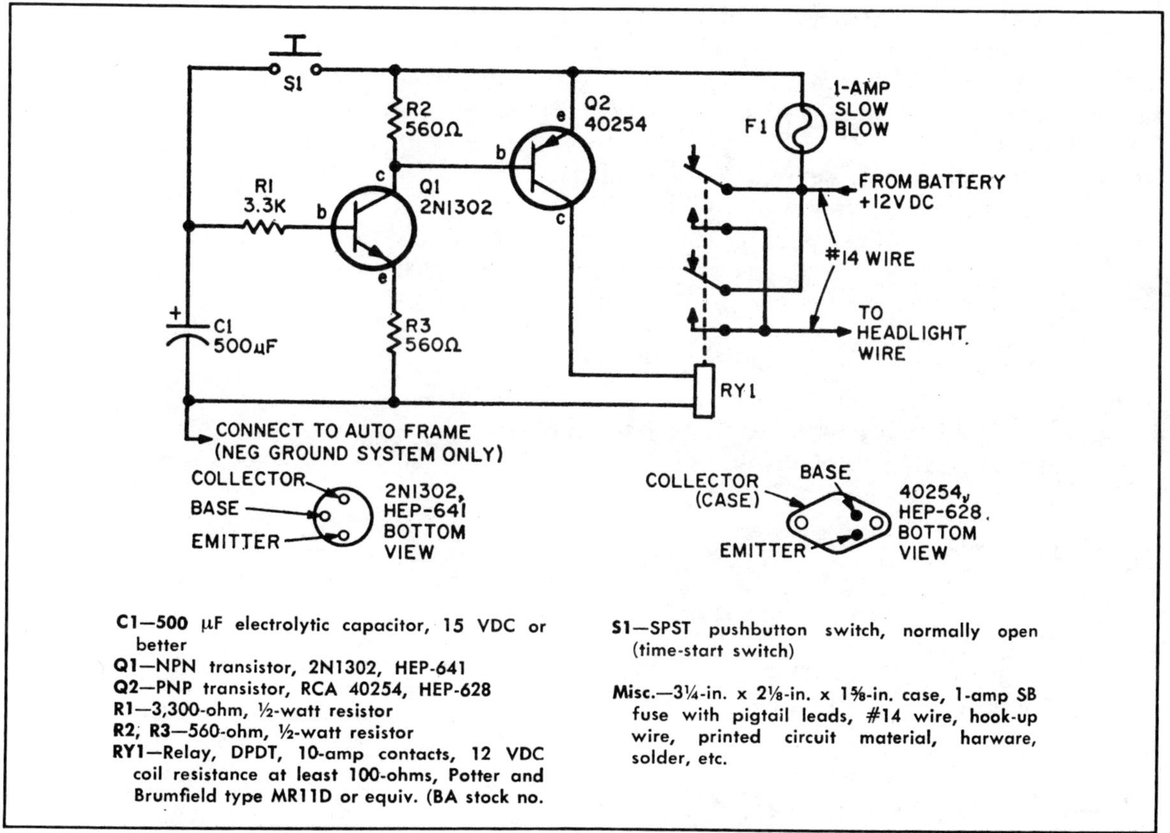

Fig. 5-1. Automatic Headlighter schematic and parts list.

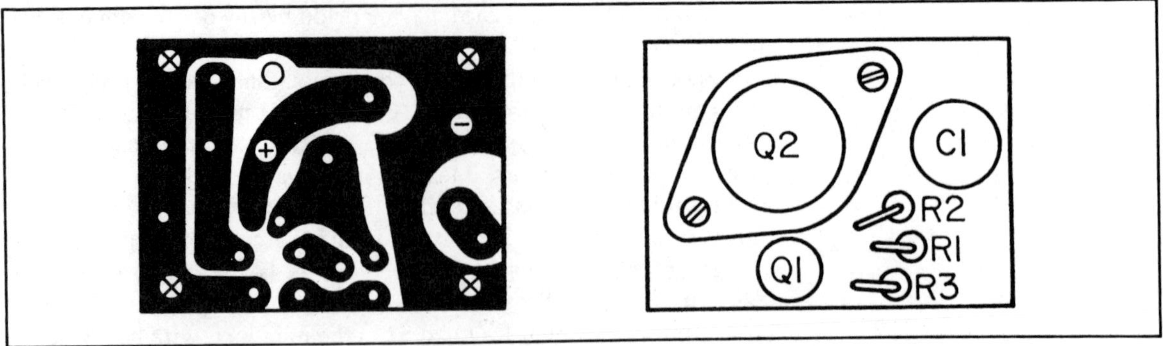

Fig. 5-2. Actual size copper (bottom) side of the board on left and component (top) side on right.

sure the timer chassis is well grounded to the dashboard frame before proceeding.

At the headlight switch, find the wire coming from the headlamps and select a convenient (probably a wire right on the switch) to tap into the 12-volt battery power. You can locate these two wires by checking for voltages with a VOM at the headlight switch. A +12-volt battery wire will be the only wire that will read +12 volts with the headlight switch in the *off* position. The wire reads +12 volts only when the headlights are *on* is the other correct one; it normally runs directly to the foot dimmer switch. After locating these wires and marking them "+" and "Lights," disconnect the battery cable at the battery so there will be no accidental shorts. Remove ½-inch insulation from the two underdash headlamp and battery wires you have located. Now connect each timer wire to the corresponding automobile headlamp and battery wire. Wrap electrical tape over the exposed wires and arrange them in an out-of-the-way spot under the dash. Reconnect the + battery cable and the timer is ready for use.

Modifications Anyone? If more than 60 seconds delay is desired, a larger capacitor can be installed in place of C1. For example, a 1000 μF capacitor will give two minutes or more of time delay. If you wish, a 10K trimmer resistor in series with R1 will allow exact adjustment of the time interval.

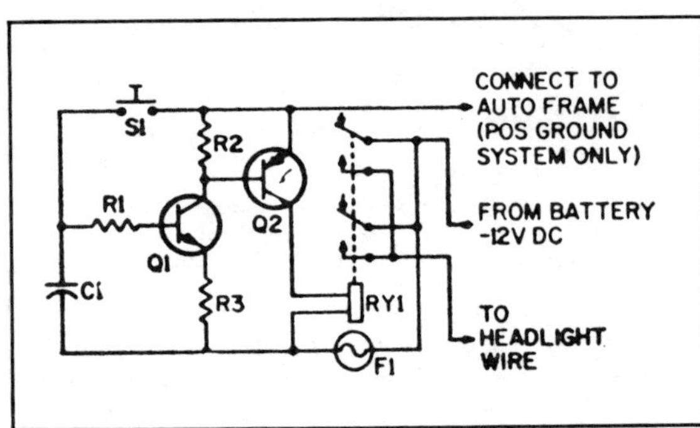

Fig. 5-3. Use this schematic in *positive* ground cars only. There are no connections made to the normally open contacts of the relay. Use the same printed circuit board. All of the necessary changes are made around a board.

Several fixed timing periods such as 60, 30 and 15 seconds can be chosen with a switch wired to select different discharge resistors connected in parallel across capacitor C1. All of these modifications combined would give a longer delay period and a variable delay with your choice of lights-on times.

Positive Ground? To modify the basic timer for a 12-volt positive ground system, do not ground the printed circuit board to the chassis. Instead, use insulated standoffs when mounting the circuit board in the mini-box. Ground the "+" point on the *circuit card* to the chassis using a solder lug mounted under one standoff. Run a wire from the circuit card negative points to the relay. Operation will be identical to negative ground systems.

ALTERNATOR TESTER

Automobiles have been coming off the production lines with alternators instead of generators for many years now, and these units have proven to be reliable and superior to the ones they replace. Being alternating current machines, they are inherently more complicated than generators and require slightly more sophisticated testing procedures to indicate their condition. This problem is brought about by the fact that automotive alternators are three phase machines, with full-wave rectification of the output to produce direct current as required by the automotive and its battery. The schematic shows a typical automotive alternator connected to its three-phase full-wave rectifier circuit.

Refer to Figs. 5-4 through 5-9.

Rectification is accomplished by six high-current silicon diodes in the alternator, and this is where the problem comes in. Many of the troubles encountered with automotive alternators are due to failure of one or more of the diodes, either by opening or shorting. Neither of these conditions will result in an inoperative alternator, and no doubt some of the automobiles on the road today have just such a problem. A shorted diode is the more serious of the two conditions, since it will result in the loss of about 50 percent of the output capability of the alternator. Such a condition is easily detected by an ordinary output test on the alternator. However, an open diode is another matter. This condition will result in loss of only a few amperes of output capability of the alternator due to the fact that only one half of one phase of the machine is disabled. Some of this lost capacity is carried by the other two phases, which will be overloaded when the alternator is required to produce full output as demanded by the automotive electrical system. Such a condition may well result in further failure of more diodes. An ordinary output test of an alternator with an open diode generally will not detect any malfunction. Because of those testing problems, another test method to determine the condition of alternators has been developed, and the construction of the Alternator Tester is the subject of this article.

The ability of Alternator Tester to detect defective diodes, both open and shorted, depends on the fact that the output ripple voltage of an alternator with a defective diode rises dramatically higher than that produced by a normally-operating alternator. When the pulsating dc waveform output voltage of an automobile alternator is measured the magnitude of the ripple voltage is about 0.2 to 0.5 volts, peak-to-peak. When one of the diodes in the alternator fails the ripple voltage

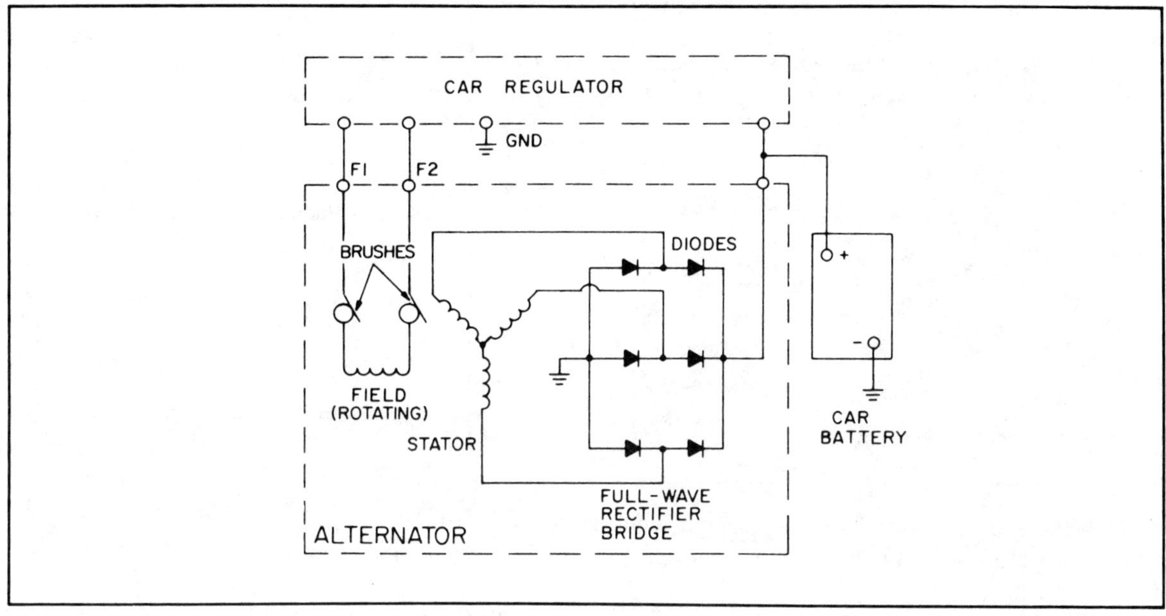

Fig. 5-4. Typical automotive charging circuit. Latest model alternators have solid-state regulator circuit built into alternator frame.

increases to 1-volt peak-to-peak or more. The Alternator Tester measures the peak-to-peak ripple voltage so that the condition of the alternator can be determined.

Construction Details. In order to keep construction costs low, the Alternator Tester was designed to be used with an ordinary VOM or VTVM as the indicating device. Since the output impedance of the test instrument is close to zero, any meter of at least 1000-ohms-per-volt sensitivity can be used. The circuit is constructed on a small printed circuit board and fitted into a metal or plastic cabinet. Two tip jacks are mounted in the cabinet which serve as the connection to the VOM. A pair of test leads is brought out through a grommet and these

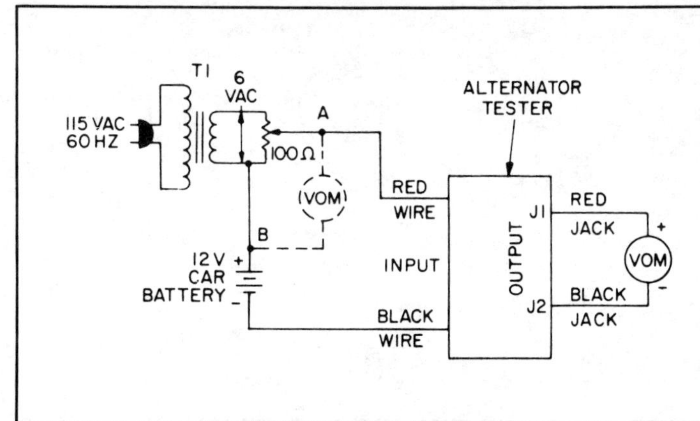

Fig. 5-5. Calibration of the Alternator Tester.

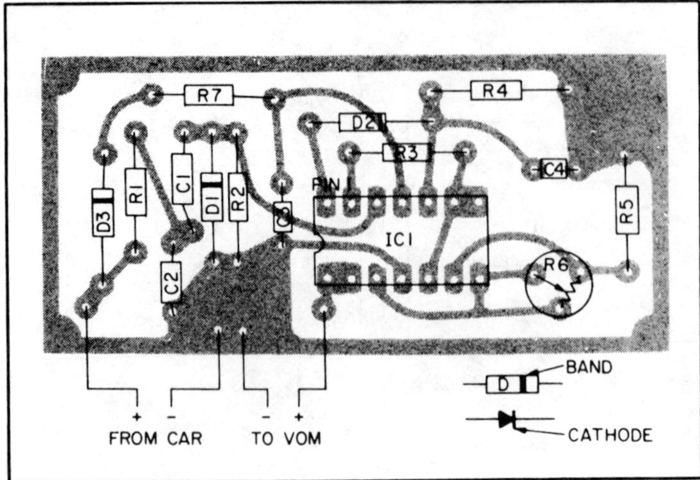

Fig. 5-6. Parts placement on the printed circuit board (shown larger than actual size for clarity).

provide the dc power to operate the circuit as well as the connection to the alternator output (battery) terminal where the ripple measurement is to be made.

About the Circuit. The Alternator Tester is basically a peak detector circuit which responds to the peak-to-peak value of an ac voltage fed to its input terminal. Power to operate the circuit is derived from the output of the alternator on the same lead which feeds the ripple voltage to the input of the peak detector. The dc output of the alternator is blocked by C1, which allows only the ripple voltage to pass through.

Operational amplifier IC1A and IC1B are connected together to form a peak detector circuit. The ripple voltage from the output terminal of the alternator is fed to the positive input of IC1A after the dc

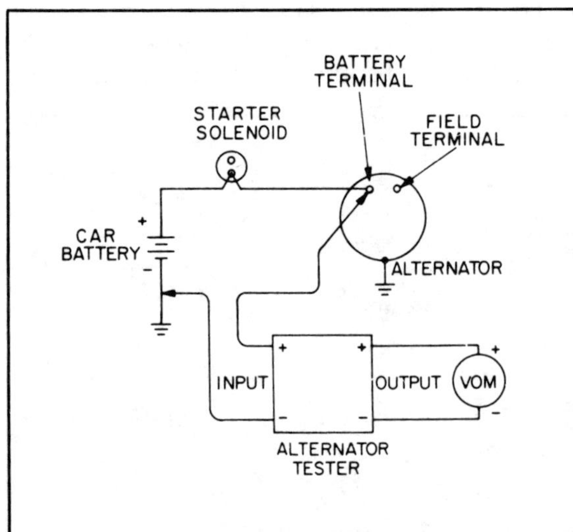

Fig. 5-7. With the three-phase output of automobile alternators, which is rectified by a six-diode full-wave rectifier, it is possible for the output of the system to appear normal even though one diode is open. With this circuit, a mechanic can test the rectifier output and discover the increase in the ripple voltage that would be caused by such a failure.

voltage of the alternator is blocked by C1. D1 clamps the ripple voltage to ground, so that it varies between zero and some positive value. Op amp IC1A charges C4 to the peak value of the ripple voltage. Op amp IC1B is a voltage follower which feeds back the peak value of the ripple voltage to the negative input of IC1A. This stabilizes the circuit so that the voltage appearing at the output of IC1B holds to the peak-to-peak value of the ripple voltage fed to the input of IC1A. Capacitor C4 is prevented from discharging through IC1A by D2, and can discharge only through R4 at a rate much slower than the ripple frequency of the alternator. This holds the meter reading constant between voltage peaks of the alternator. Amplifier IC1C has an adjustable gain of slightly more than unity to compensate for the slight error (loss) caused by D2, as well as providing a means for calibration of the instrument. Voltage follower IC1D provides an extremely low output impedance to drive any meter of 1000-ohms-per-volt or more. Power for the circuit, about 2 mA, is taken directly from the alternator output terminal. Diode D3 prevents damage to the circuit in the event of any reverse polarity connections.

Calibration of the Instrument. Calibration of the Alternator Tester is accomplished by feeding an ac voltage of known amplitude between the input terminal and ground, and adjusting R6 for the correct meter reading. The calibrating ac voltage input can be measured by the ac voltmeter function of the VOM, which reads RMS volts. To convert RMS to peak-to-peak voltage multiply the value by 2.83. The calibration circuit uses a 6-volt filament transformer and potentiometer as a source of low voltage ac. To calibrate the instrument connect the filament transformer, potentiometer, and alternator test circuit as shown, using any twelve volt dc supply for power. (Be sure there is no ripple voltage on the output of the supply, since this will cause an error

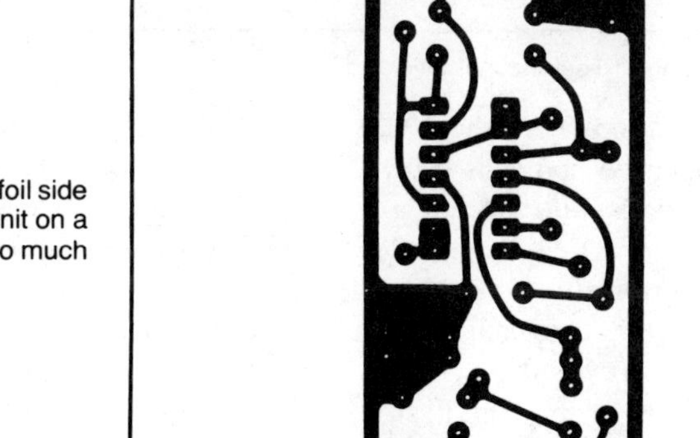

Fig. 5-8. This pattern shows the printed circuit board (foil side up) for the Alternator Tester. You can construct the unit on a perfboard if printed circuit board fabrication seems too much trouble.

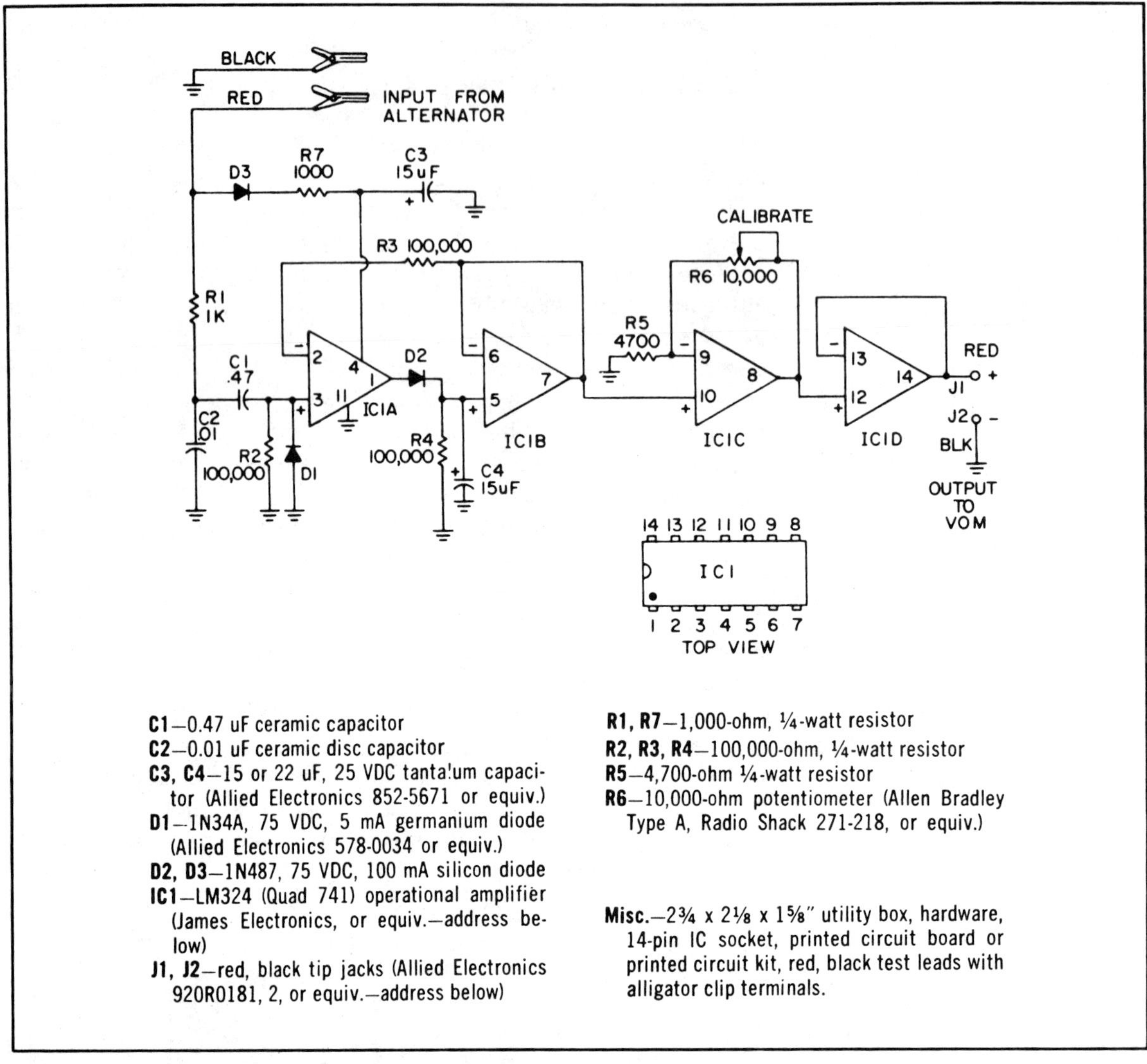

Fig. 5-9. Alternator Tester schematic and parts list.

in the calibration.) Set the VOM to read ac volts, and connect it between points A and B as shown. Set the potentiometer so that the VOM reads 0.35 volts RMS. This is equivalent to 1 volt peak-to-peak. Disconnect the VOM, set it to a 1.5 to 3 volts dc scale, and connect it to the output terminals of the Alternator Tester Calibrate potentiometer reading of 1 volt. This completes calibration of the Alternator Tester.

Alternator Testing. The testing of an automotive alternator consists of two parts. The first test is the output test, which determines if the alternator can deliver the full current that it was designed to produce. Bear in mind that the following procedure tests both the alternator and the voltage regulator at the same time, and failure of the alternator to deliver rated output also may be caused by a defective voltage regulator. Before making the following tests inspect the connections to the alternator and battery to be sure they are tight. A loose or bad connection between the alternator and the battery may cause an excessive ripple measurement even though there are no defective diodes in the alternator.

The alternator output test requires the use of only the VOM which is set to read dc volts on a 0 to 15 volts or greater scale. Connect the VOM directly across the battery, observing correct polarity. Start the engine and turn on the headlights (high beam), windshield wiper, blower motor (high speed), and radio. Race the engine to a moderate speed (about 2000 RPM) and note the reading of the meter. A properly operating charging system should maintain at least 13.5 volts and not more than 15 volts across the battery. Voltage readings below 13.5 indicate a defective alternator or voltage regulator. Voltage readings above 15 indicate a defective voltage regulator. Some automobiles have voltage regulators which can be adjusted. Refer to the service manual for your car for voltage regulator tests and adjustments. If the above test indicates satisfactory performance proceed to the ripple voltage test, using the connections shown in the testing diagram. Note that the positive lead of the Alternator Tester is connected directly to the battery terminal of the alternator. The reason for this is that the ripple measurement depends upon the small, but finite, resistance between the alternator and battery.

In order for the ripple test to be accurate, the alternator must be delivering a sizeable current. This is accomplished by slightly discharging the battery. Before starting the test, shut the engine off and turn on the car headlights for about ten minutes. During this time you can connect the Alternator Tester to the car. Leave the headlights on while making the test. Start the engine and bring the RPM up to about 2000. Note the reading of the meter. An alternator in proper operating condition will have a ripple voltage somewhere between 0.2 and 0.5 volts peak-to-peak. Should one or more of the diodes be defective the ripple voltage will increase to 1 volt peak-to-peak, or more. If this is the case you will have to remove the alternator from the car, disassemble it and locate the defective diode.

AUTOMOTIVE EXHAUST EMISSIONS ANALYZER

If the car you drive was manufactured after 1963, it has an emissions control system. As exhaust emission requirements become more strict (some states have annual inspections that include an emissions test) with each passing year, it has become difficult to properly adjust the car engine.

Gone are the days when the idle mixture screws of a carburetor were purposely set to less than optimum condition to assure meeting permitted emission levels. Some tune-up manuals provide a technique for adjusting the carburetor using a tachometer to measure engine RPM.

Refer to Figs. 5-10 through 5-15.

This method is fine, if all engine systems are operating properly. But what happens if system malfunction causes excessive emissions?

CO Measurement. This is where the Automobile Exhaust Emissions Analyzer comes in. This easy to build instrument provides a visual indication of an exhaust component—carbon monoxide (CO). The level of CO in automobile exhaust is inversely proportional to the air-fuel ratio and combustion efficiency. The air-fuel ratio is varied when the idle mixture screws of a carburetor are adjusted. (Some carburetors have just one idle mixture adjustment screw; others have two.)

The CO level is low when the air-fuel ratio is 14.7. Modern automobile engines are designed to idle with air-fuel mixtures near this theoretically perfect air-fuel ratio.

The Automotive Exhaust Emissions Analyzer contains a sensing element (thermistor) that is exposed to automobile exhaust. A meter readout permits measurement of the air-fuel ratio.

Since one other exhaust pollutant, hydrocarbons, are also controlled by the idle mixture screws, the exhaust analyzer will help to minimize this component even though it is not responsive to it. This is because the optimum setting for minimum CO is close to the optimum setting for hydrocarbon emissions.

Good Mileage. Thus, by adjusting your carburetor with this instrument, you will help reduce air pollution. A properly adjusted carburetor will also help provide better gasoline mileage.

The heart of the instrument is a bridge circuit that contains two 100-ohm resistors (R3 and R4), and two thermistors (T1 & T2). A thermistor is a special resistor having a large negative temperature coefficient of resistance. Its resistance value decreases as the resistors temperature increases.

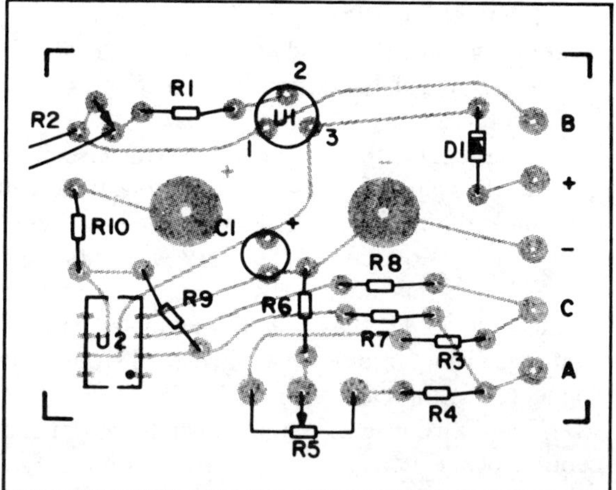

Fig. 5-10. When you install the various components onto the PC board, make sure that you install the ICs properly. Use IC sockets for the two integrated circuits. If you have a bad IC, removal is easier.

At room temperature the resistance of T1 & T2 is about 2000-ohms. When they are each heated to 150 degrees centigrade by a 10 mA current, the resistance value decreases to 100-ohms. Thus, the four elements comprise a bridge circuit.

A characteristic of CO is that it conducts heat away from a thermistor at a different rate than air. One thermistor, T1, is exposed to the automobile exhaust while the other, T2, is isolated in a pure air environment. The difference in thermal conduction unbalances the bridge.

A voltage difference is caused between points A and C. A differential amplifier, U1, amplifies this difference and drives the meter with sufficient current to read out the percentage of CO and the air-fuel ratio.

A front panel balance control, R5, balances the bridge and calibrates the instrument. Calibration is performed when both thermistors are exposed to the outside air.

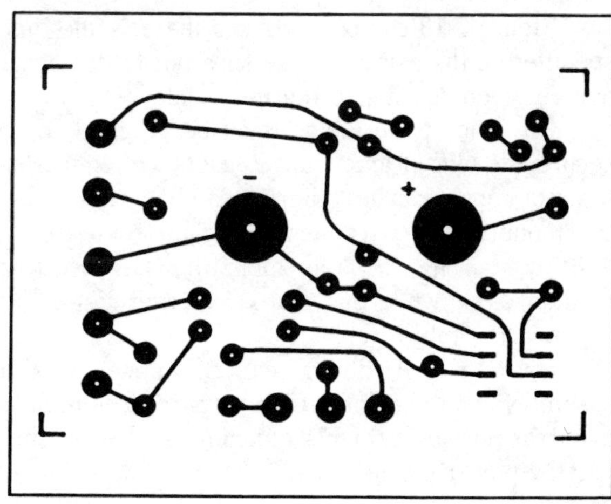

Fig. 5-11. When you construct this PC board, be very careful when you etch the various circuits. Make sure that there are no solder bridges and that all of the connections are good. Use an ohmmeter to check out the whole board.

Accurate Calibration. To assure accuracy, U1 is used as a current regulator that is adjusted to deliver 20 mA into the bridge. U1 is a voltage regulator designed to maintain 4 Vdc between terminals two and three.

Since most of the current fed into the circuit must pass through R1 and R2, the value of this series resistor string will control the output current of the regulator. The current is set by adjusting R2 and will hold its adjusted value over a very wide range of input voltages.

The circuit is powered by the car's battery and generating system. A silicon diode, D1, has been placed in series with the positive power lead to prevent circuit damage in case the power leads are improperly connected to the car battery.

The entire circuit of the exhaust system analyzer, with the exception of the sensing and reference elements, is on a PC board, mounted directly to the rear of the meter. The two large circular pads are designed for the meter studs and are properly spaced for the 0-1 mA meter that is specified in the parts list.

Check the center to center distance of the meter studs and modify the foil layout if you use a different meter. Extra large pads have been provided to accommodate the external wires that connect to the PC board.

Use a socket for all integrated circuits. Once a multipin integrated circuit has been soldered directly into a printed circuit board, it is difficult to remove an IC without destroying it or the printed circuit.

Follow correct orientation for the integrated circuits, electrolytic capacitor, and diode as illustrated. These parts will not work if they are incorrectly placed. Pin 1 of U2 is identified in the layouts by a small dot. On the IC itself, pin 1 is identified also by a small dot or a semicircle that is molded into the plastic.

Construction of the sensing assembly is illustrated in Fig. 5-12. Use any material for the tubing. Aluminum is ideal because of its light weight and non-rusting quality.

Figure 5-13 can be used as a meter scale that will fit the meter specified in the parts list. Be sure not to damage the delicate meter needle when you install the new scale.

One end of the tubing must be plugged. Epoxy can be used to secure the plug if aluminum is used. Other materials such as steel, brass or copper can be soldered. As shown in Fig. 5-12, drill and tap the two mounting holes for the sensing unit for 4-40 machine screws. Use 4-40 by ¾-inch screws for mounting. Be sure to orient the sensing element as shown so that the exhaust thermistor will be exposed to the gases in the tube.

Use a No. 20 flexible wire for the connections between the sensing assembly and the printed circuit board. Use different colored wire to avoid the possibility of misconnections. Use red and black wires for the battery power connections.

Check the Length. Before making the PC board connections, be sure that the wire between the sensing assembly and unit will be placed somewhere near the engine compartment so that the meter can be monitored during carburetor adjustments.

The bridge current must be adjusted to 20 mA by means of R2. To make this adjustment you need a 12-Vdc source and an accurate dc voltmeter. Use a car battery for the power source if you do not have a line operated dc supply.

Connect the positive and negative power leads of the analyzer to the power source observing correct polarity. Set the front panel balance control to midposition. Measure the voltage across R6 and adjust R2 for 2 volts.

Make sure that the control will adjust the meter reading to both sides of center scale. If the unit performs this test, the adjustment procedure is completed. If you are unable to get the proper meter readings, recheck the wiring and the printed circuit board.

The analyzer is designed only for gasoline engines. Do not use it on diesel engines or automobiles with heavy smoke exhaust. To do so may coat the sensing element.

For cars that have catalytic converters, the sensing assembly should be exposed to the exhaust gases before they pass through the converter. Some cars have a plug in the exhaust system for this purpose. On other cars, remove the exhaust gas recirculating (EGR) valve and examine the gases through it.

Balance Meter. Before inserting the sensing assembly into the tail pipe, connect the power leads to the car battery and slowly adjust the balance control for a center scale reading of the meter at the point marked Bal. There will be a time lag between adjustment of the control

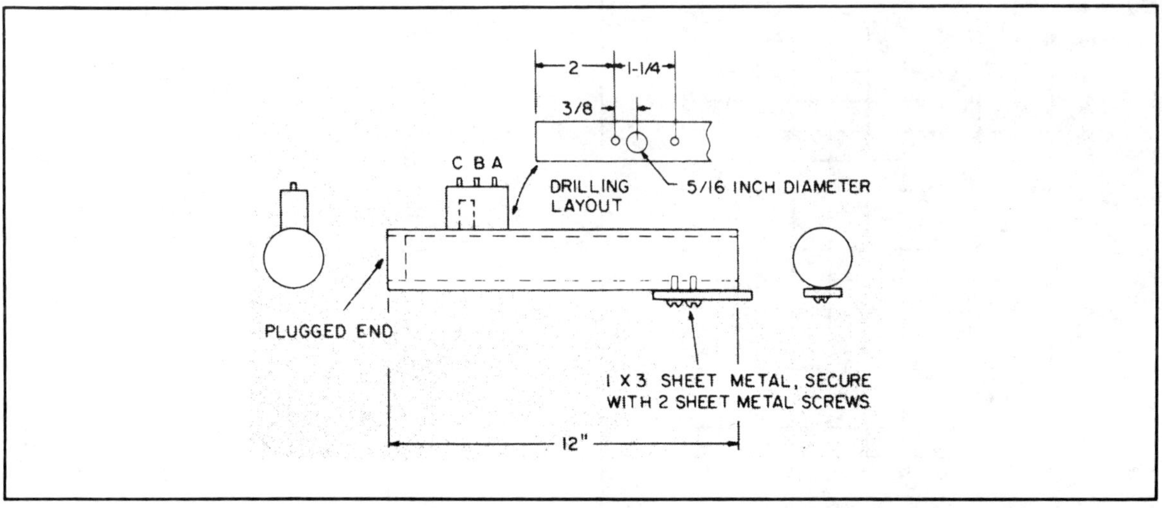

Fig. 5-12. The sensing unit for the Auto Exhaust Emissions Analyzer is constructed as shown here. Make sure that one end is plugged when constructing.

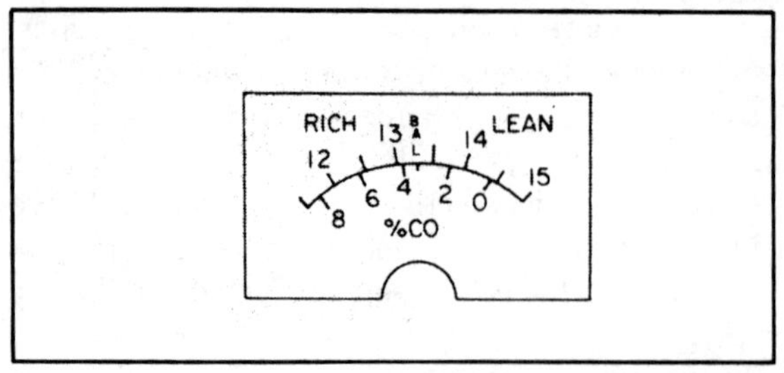

Fig. 5-13. This is a scale template to use for your Auto Exhaust Analyzer.

and response of the meter. Be sure the engine is up to normal operating temperature and the choke is fully open.

Place a long flexible tube into the tail pipe and the sensing assembly into the other tube end with the thermistor assembly facing upward. The tube should be long enough to eliminate any temperature difference between the air and CO. Note the reading of the meter.

Readings to the left of center indicate an exhaust containing an excessive amount of carbon monoxide by today's standards. Cars with emission control systems should have meter readings to the right of center, indicating a lean mixture and a low concentration of carbon monoxide.

If your car is a recent model and indicates excessive exhaust emissions, you may reduce these emissions by turning the idle mixture screws counterclockwise to lean out the mixture. When this is done, the idle speed of the engine may be reduced. It is best to follow the

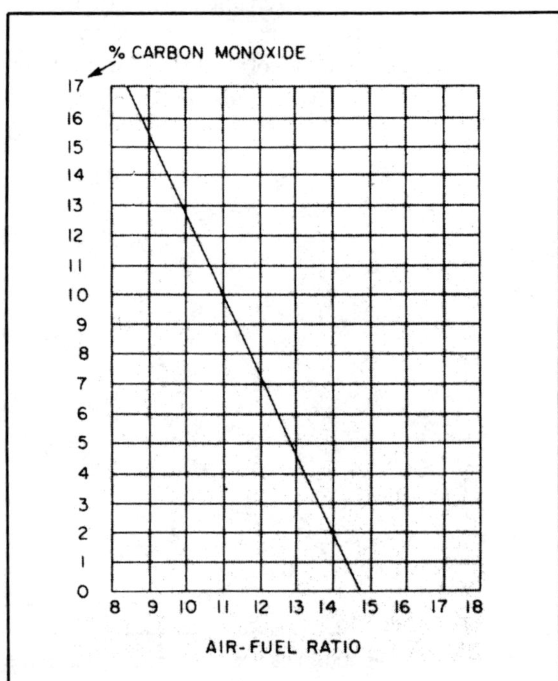

Fig. 5-14. The idealized air to fuel ratio is 14.7. Keeping it as close to that ratio will maximize your car's performance.

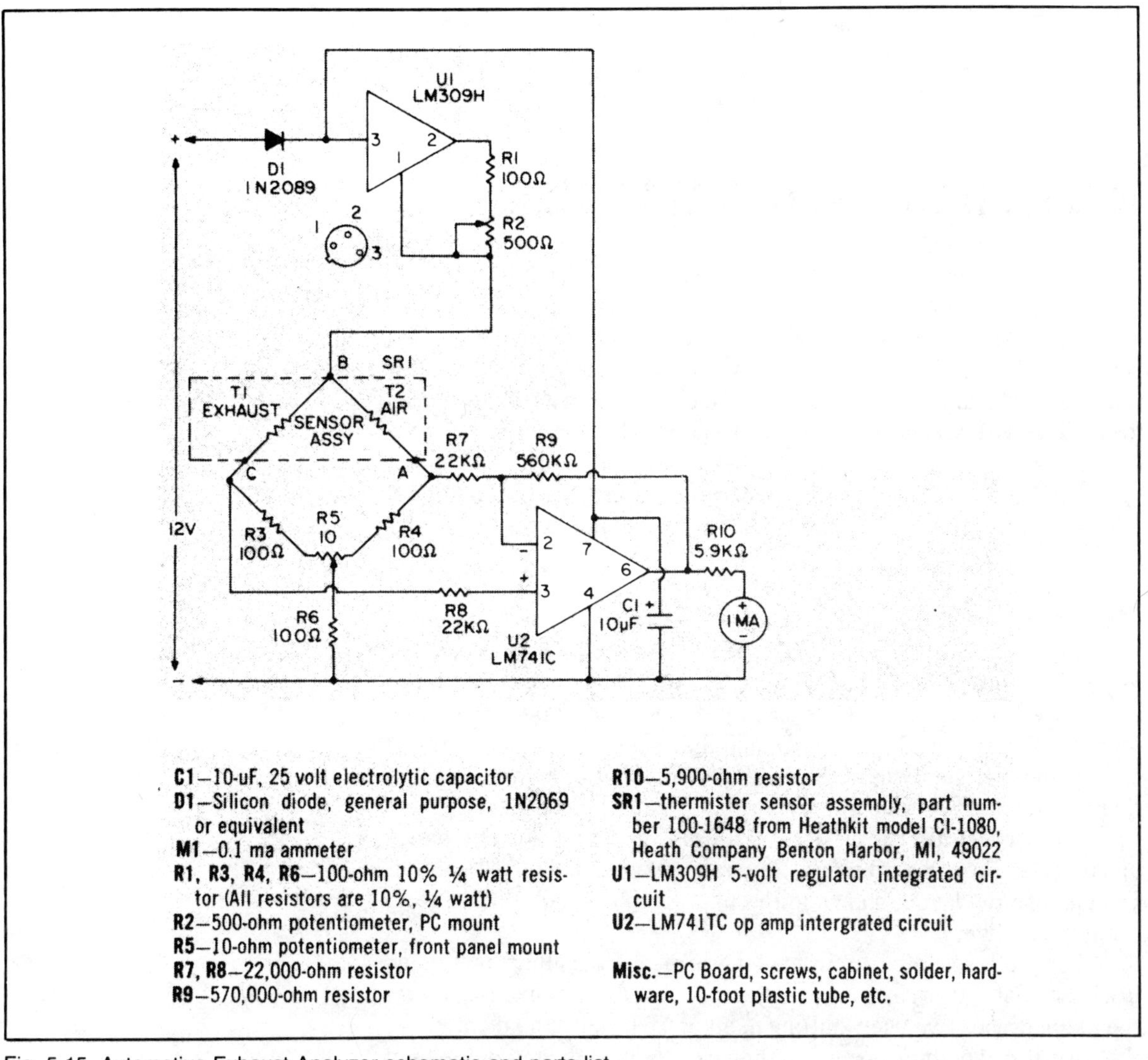

C1—10-uF, 25 volt electrolytic capacitor
D1—Silicon diode, general purpose, 1N2069 or equivalent
M1—0.1 ma ammeter
R1, R3, R4, R6—100-ohm 10% ¼ watt resistor (All resistors are 10%, ¼ watt)
R2—500-ohm potentiometer, PC mount
R5—10-ohm potentiometer, front panel mount
R7, R8—22,000-ohm resistor
R9—570,000-ohm resistor
R10—5,900-ohm resistor
SR1—thermister sensor assembly, part number 100-1648 from Heathkit model CI-1080, Heath Company Benton Harbor, MI, 49022
U1—LM309H 5-volt regulator integrated circuit
U2—LM741TC op amp intergrated circuit

Misc.—PC Board, screws, cabinet, solder, hardware, 10-foot plastic tube, etc.

Fig. 5-15. Automotive Exhaust Analyzer schematic and parts list.

manufacturer's instructions (usually located on a decal in the engine compartment) as to the proper setting for idle speed.

When you are finished with the test, allow the sensing assembly to purge itself of moisture and exhaust gas before putting it away or using it again. Do not use compressed air to dry out the assembly.

WINDSHIELD WASHER WATCHER

Ever had the annoying experience of having the car in front of yours on the highway begin to kick up dirty water on your windshield, and when you went for the washer button, nothing happened? Well, chin up, bunkie, the Washer Watcher is just for you. It not only warns you when the tank is empty, it warns you when you're nearing refill time, *before* it's too late. This handy device can also be an engine saver for those of you who have water-injected turbocharger setups on your car.

The heart of the unit is National Semiconductor's LM1830 fluid detector, which responds to the conductivity of fluids across its probe leads.

Refer to Fig. 5-16.

How It Works. The LM1830 generates an ac oscillator signal (in order to prevent electrolytic coating of the probes, thereby reducing their efficiency) which is coupled to the probes by a 0.05-μF capacitor. When the conductive fluid in the tank reaches a low enough level, the resistance between the probes rises past the 13,000-ohm reference level (set internally within the IC), and the oscillator signal is coupled to the amplifier segment of the IC. The amplified 6,000 Hz signal is then fed to an LED which indicates the low fluid condition.

Construction. The circuit can be assembled quickly and easily on solderless breadboard stock. Component layout is not critical, but you can use the layout shown in the photograph if you're not feeling terribly creative.

The probe assembly in our model was made by drilling three holes in a triangular pattern in a large rubber stopper. After doing this, insert the metal probes (we used knitting needles). Measure the diameter of the narrow end of the stopper and then cut a hole in the top of the fluid reservoir just slightly larger than the stopper's narrowest diameter, thus allowing a snug fit. Solder a wire to each probe, and connect them to the appropriate pins on the IC. Do not cement the probes into position on the stopper yet, because you still have to calibrate.

Calibration and Operation. The longest probe, which connects to a ground anywhere in the engine compartment that's handy, should be pushed down through the stopper so that it's just touching the bottom of the reservoir.

The "refill" probe may be inserted to any length, depending upon how much margin you wish to have between the first and last warnings.

Remember, water varies from area to area, and the type of solvent you use in your washer reservoir, and the resulting mixture, can affect

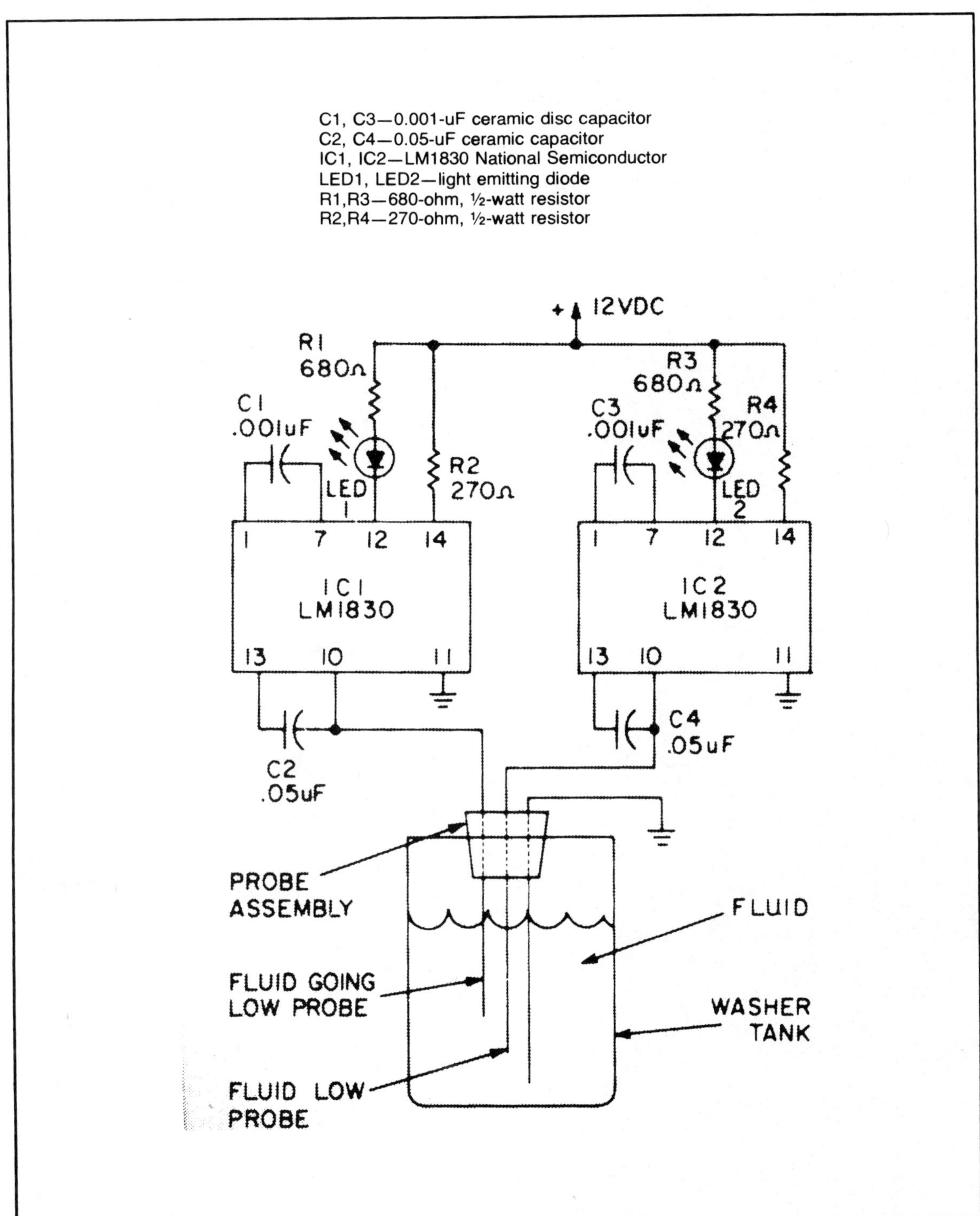

Fig. 5-16. Windshield Washer Watcher schematic and parts list.

the calibration. Before you cement the probes and stopper in place, check the operation of the unit by referencing the warning lights with visual observations. Try to keep the solvent/water mixture the same each time you refill, in order to keep the unit calibrated.

When you're satisfied with the unit's calibration, cement the probes into the stopper—both on the top and bottom—and then cement the stopper into the hole in the reservoir. Silicon bathtub sealer should do the job very nicely. The next time you take a trip, let the Washer Watcher take some of the grief out of driving for you.

DWELL/TACHOMETER

Outside of a good set of wrenches, the most commonly called for automotive tool is the dwell/tachometer. When tuning up an engine, there's no substitute for the kind of accuracy a dwell/tach can bring to your engine adjustments. A commercial version of this apparatus might run as high as $25. With some judicious parts buying, you should be able to do the job for roughly half that much. In addition, our dwell/tachometer gives you an additional feature not found on any but the most expensive commercial units: a dc voltmeter, which is highly useful in checking a car's electrical system and, in particular, the ignition.

Refer to Figs. 5-17 through 5-24.

Most of the parts used in the construction of this instrument will probably be found in the electronics hobbyist junk box. The meter is a common 1 mA dc movement. If desired, other meter movements may be used by simply changing circuit values to accommodate a more or less sensitive meter.

The Circuit. In order to best understand the operation of the dwell/tachometer circuit, it is necessary for the reader to be familiar with the voltage waveform appearing at the primary terminal of the ignition coil. This is shown in Fig. 5-17. The basic voltage waveform is a rectangular pulse with a considerable amount of ringing on the rising edge of the pulse. This ringing is caused by the sudden cut-off of current in the ignition coil, and results in the high-voltage generation which fires the spark plugs. The duty cycle of the rectangular pulse is determined by the dwell angle of the ignition points (or solid state circuit in electronic systems), and must fall within specified limits for proper engine performance.

The dwell meter section of the instrument is composed of Q2, Q3, and associated components. Q3 is connected as a constant current generator with 8 volts impressed upon the base and the meter placed in the collector circuit by the FUNCTION switch. The value of resistance placed in the emitter of Q3 determines the collector current of the transistor, and is adjusted so that the meter reads the full dwell angle (45 or 60 degrees) when the sensing lead of the instrument is shorted to ground. Q2 acts as a switch that controls the base of Q3, and causes Q3 to either be on or off, depending upon the state of the ignition points. When the points close, Q2 is cut off and Q3 passes its calibrated constant current through the meter. When the points open, Q2 is forward biased and saturated by the voltage appearing across the points. This cuts off Q3 and the meter current is zero. This action takes

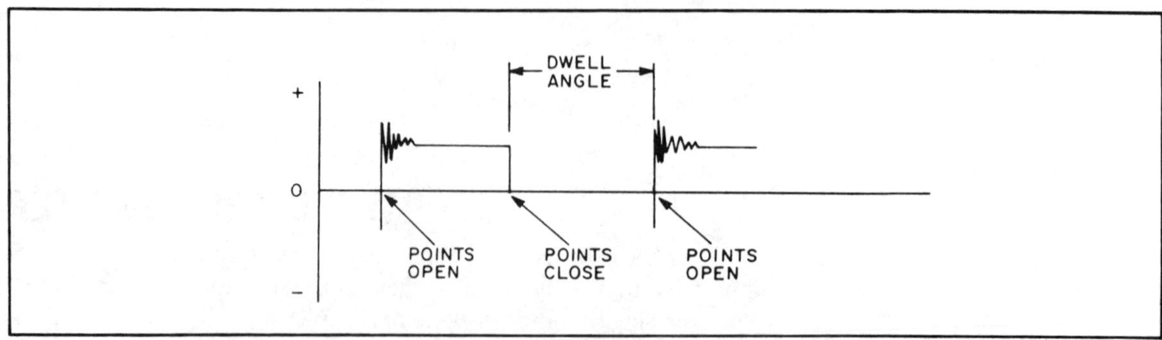

Fig. 5-17. A waveform diagram of the voltage across the ignition points shows exactly what happens during operation and what it is that you're measuring when you use the instrument.

place much faster than the meter needle can follow, so that meter reading is the average of the two conditions and is the actual dwell angle measurement.

The tachometer section of the instrument makes use of the fact that the meter of spark plug firings per second is directly related to the

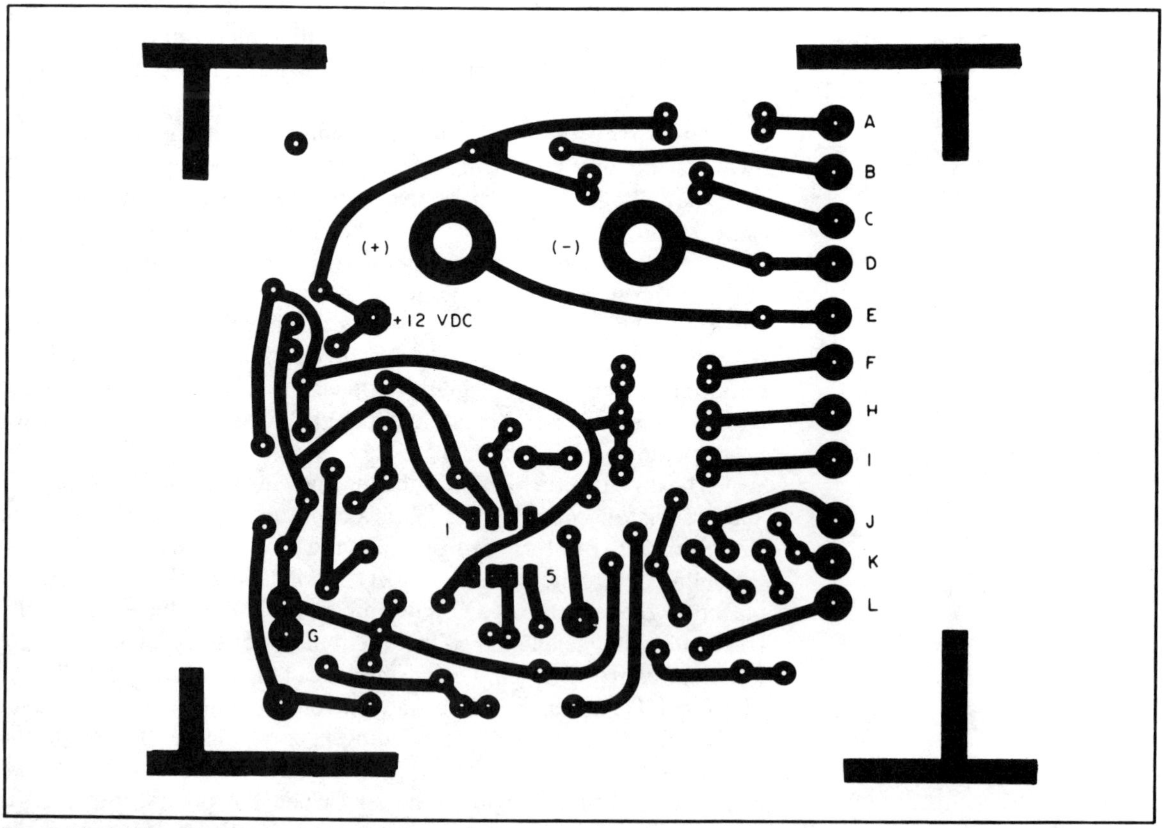

Fig. 5-18. This is a full-scale etching guide for the PC board of Dwell/Tachometer. Do not etch the board until you know the center-to-center distance of the studs.

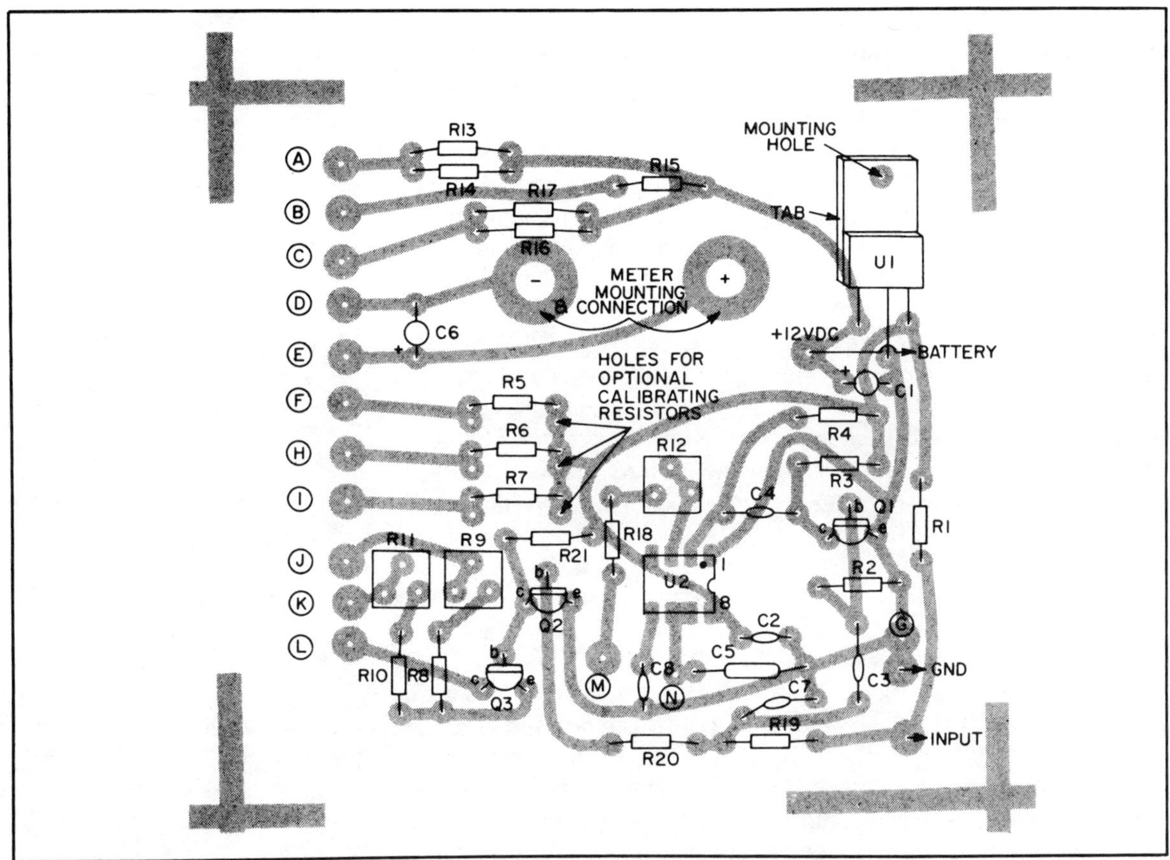

Fig. 5-19. The component layout diagram will guide you in the assembly of the board. Take care not to cover holes for calibrating resistors near R5, R6, and R7.

rpm of the engine. Q1 is used as a buffer transistor between the ignition system and the trigger input of a 555 timer IC which is connected as a one-shot multivibrator. Each time the ignition circuit produces a positive-gain pulse, U2 generates an 8-volt pulse of 2500 to 5000-microseconds duration, depending upon the number of cylinders of the engine under test. The output of U2 is fed to the meter through a calibrating potentiometer. C6 acts as a filter to smooth out the pulses to nearly pure dc and provides a steady meter reading, which is engine rpm.

The voltmeter section of the unit consists of R13, R14, R16, and R17. These components are used as multiplier resistors so that full scale meter current is generated when either 15 or 1.5 volts is fed to the power leads of the instrument. The function switch of the unit connects the proper resistors into the circuit as necessary for a full-scale voltmeter reading of either 15 or 1.5 volts.

U1 is a fixed, 8-volt regulator IC that provides the power to operate the dwell and tachometer sections of the unit. Because the circuit derives its power from the battery and alternator of the automobile under test, the 8-volt regulator ensures that the calibrated accuracy of the instrument is retained, regardless of varying voltages being generated by different charging systems.

Construction. Most of the circuitry is built on a printed circuit board, which mounts all components except the front panel switches and meter. The PC board is mounted to the rear of the meter by means of the two meter studs. This type of construction allows the entire circuit of the instrument to be contained in one module, and allows ease of assembly and service if ever necessary. The printed circuit layout, as seen from the copper side of the board, is shown in full size in Fig. 5-18, and the component layout as seen from the parts side of the board is shown in Fig. 5-19. If you are going to use a different meter than that specified in the parts list, be sure to take into account the center-to-center stud distance when laying out the printed circuit board.

Figure 5-20 shows a meter scale that can be used for the instrument. This scale has two ranges: 0 to 1500 rpm and 0 to 60 degrees dwell. To change the 0 to 1 mA scale on the meter, remove the plastic front of the meter and carefully remove the two small screws which hold the scale in place. You can then paste the scale of Fig. 5-20 over the back side of the meter scale and put it into place over the meter movement. Be careful not to disturb the meter needle during this operation, since it is very fragile. Figure 5-21 can be used as a front panel label which provides the FUNCTION and CYLINDER lettering.

The instrument is connected to the automobile ignition system with three wires, as shown in Fig. 5-22. Be sure to use different colors to help prevent misconnections when the instrument is placed in use. Rubber covered test lead wire is ideal for the purpose, and comes in several colors besides red and black. Alligator clips and boots can be placed on the ends of the wires for the connections to the automobile.

Connections between the printed circuit board and front panel switches are indicated on the schematic diagram and printed circuit layout by a group of 14 letters, A through N. Three additional wires are used for the three operating leads of the instrument. These connections are clearly marked on the parts location guide diagram.

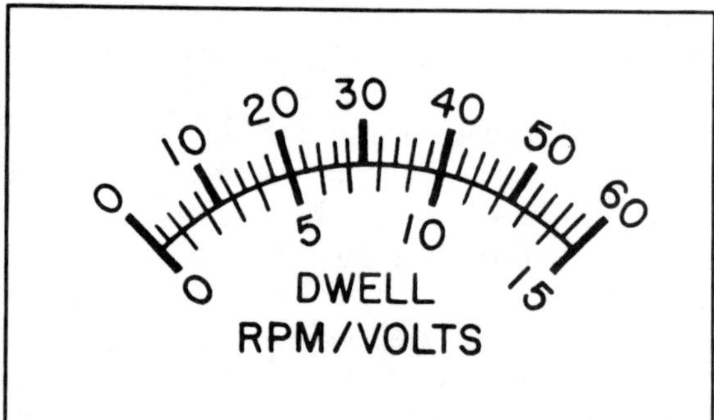

Fig. 5-20. This is a full-scale template for use on the meter face. It is designed to be used with the GC Electronics movement described in the part list, though it will likely fit other meter faces just as well.

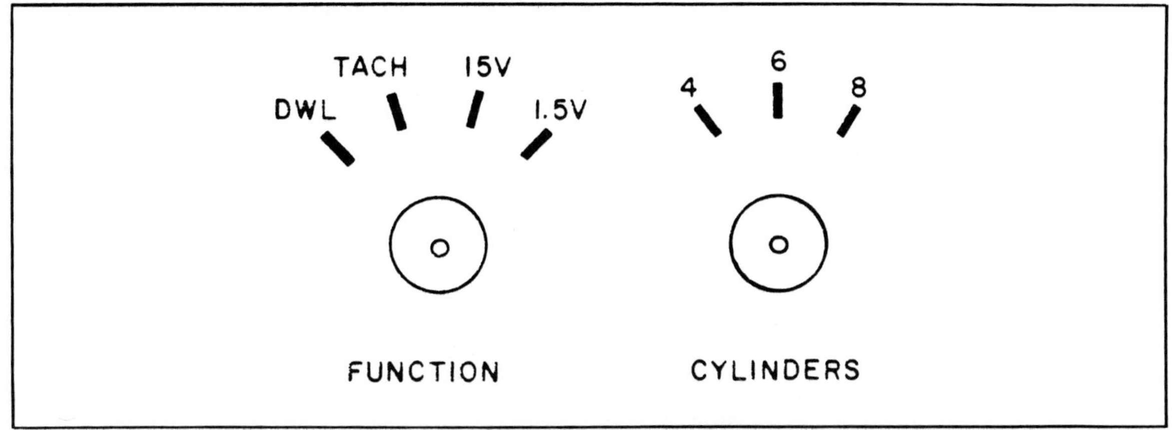

Fig. 5-21. This template, also full-scale, can be used on the cabinet front to illustrate the switch positions. If your configuration differs from this one, a good method of illustrating the front panel is to use dry-transfer lettering stencils.

After the unit is completely wired, double check to make sure that the transistors, integrated circuits, and electrolytic capacitors are mounted to the printed circuit board in the correct direction. These components are polarized and will be damaged if they are placed into the circuit improperly.

Checkout and Calibration. To check and calibrate the unit, you will need a variable dc power source of 0 to 15 volts, an accurate dc voltmeter, and an audio oscillator which can deliver at least 15 volts rms output. A Hewlett/Packard 200CD or equivalent is ideal.

Set the FUNCTION switch to 15 volts and connect the positive and negative leads of the instrument to the power supply. Connect the voltmeter across the output of the supply. Raise the voltage of the supply from zero to 15 volts while watching the instrument, which should agree with the dc voltmeter. If necessary, you can change the value of R14 to provide an accurate indication of 15 volts. Reduce the

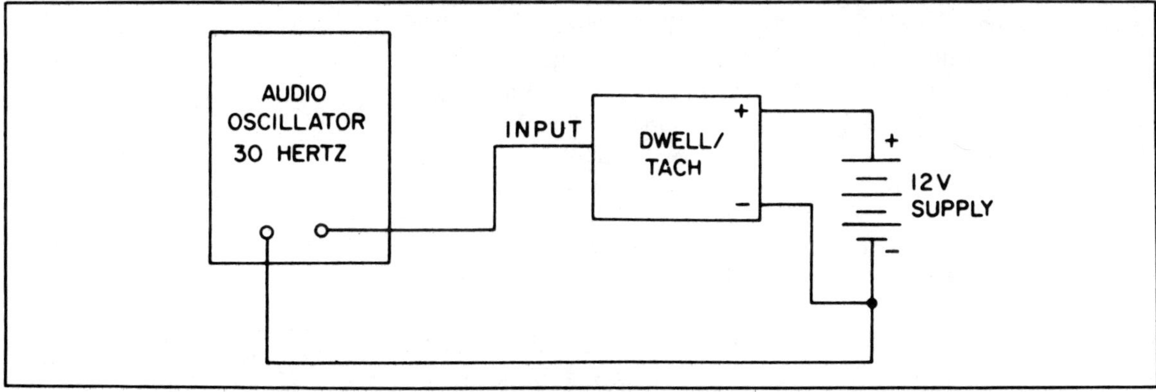

Fig. 5-22. This is the wiring setup utilized for final calibration. A well-regulated, 12-volt power supply is a necessity here. Alternatively, you can use the car battery if it is fully charged.

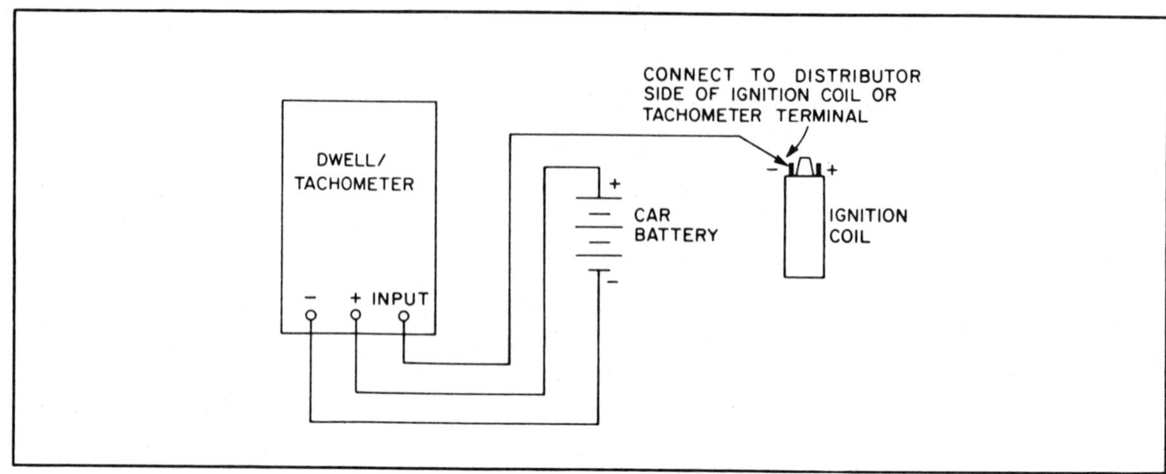

Fig. 5-23. Schematic representation of the manner in which the Dwell/Tachometer is connected to car's ignition system. No other connections are needed to use all of meter's functions.

output of the supply to 1.5 volts and set the function switch to 1.5 volts. If necessary, the value of R17 can be changed to provide an accurate indication of 1.5 volts.

The next check to be made is upon the dwell meter circuit. Set the FUNCTION switch to DWELL, and the power supply to 14-volts. Connect the sensing lead of the instrument to the negative side of the power supply. This should result in some positive meter reading. Set the CYLINDER switch to 8, and adjust R9 for a meter reading of 45 degrees on the 0 to 60 dwell scale. Set the CYLINDER switch to 6 and adjust R11 for a meter reading of 60 degrees. Check the meter reading with the CYLINDER switch set to 4. It should read 45 degrees. (This reading will be doubled during operation of the instrument, and is actually 90 degrees for 4 cylinder engines). Remove the sensing lead from the negative side of the power supply. The meter should read zero for all settings of the CYLINDER switch. This completes the dwell calibration.

To calibrate the tachometer section of the unit, connect the instrument, power supply,and audio oscillator as shown in Fig.5-22. Set the power supply to 14 volts output, and the audio oscillator to 30 Hz at 15 volts output or more. Set the FUNCTION switch of the instrument to TACH and the CYLINDER switch to 8. Adjust R12 for a meter reading of 450 on the 0 to 1500 scale of the meter. Check the reading of the meter with the cylinder switch set to 6 and 4. These readings should be 600 and 900 respectively. If necessary, you can parallel R5, R6, or R7 with resistors as required to attain proper calibration for all positions of the CYLINDER switch. The printed circuit layout has additional pads and holes for any extra resistors that may be necessary.

Operation. The Dwell/Tach is connected to the automobile system as shown in Fig. 5-23. Note that cars with factory installed

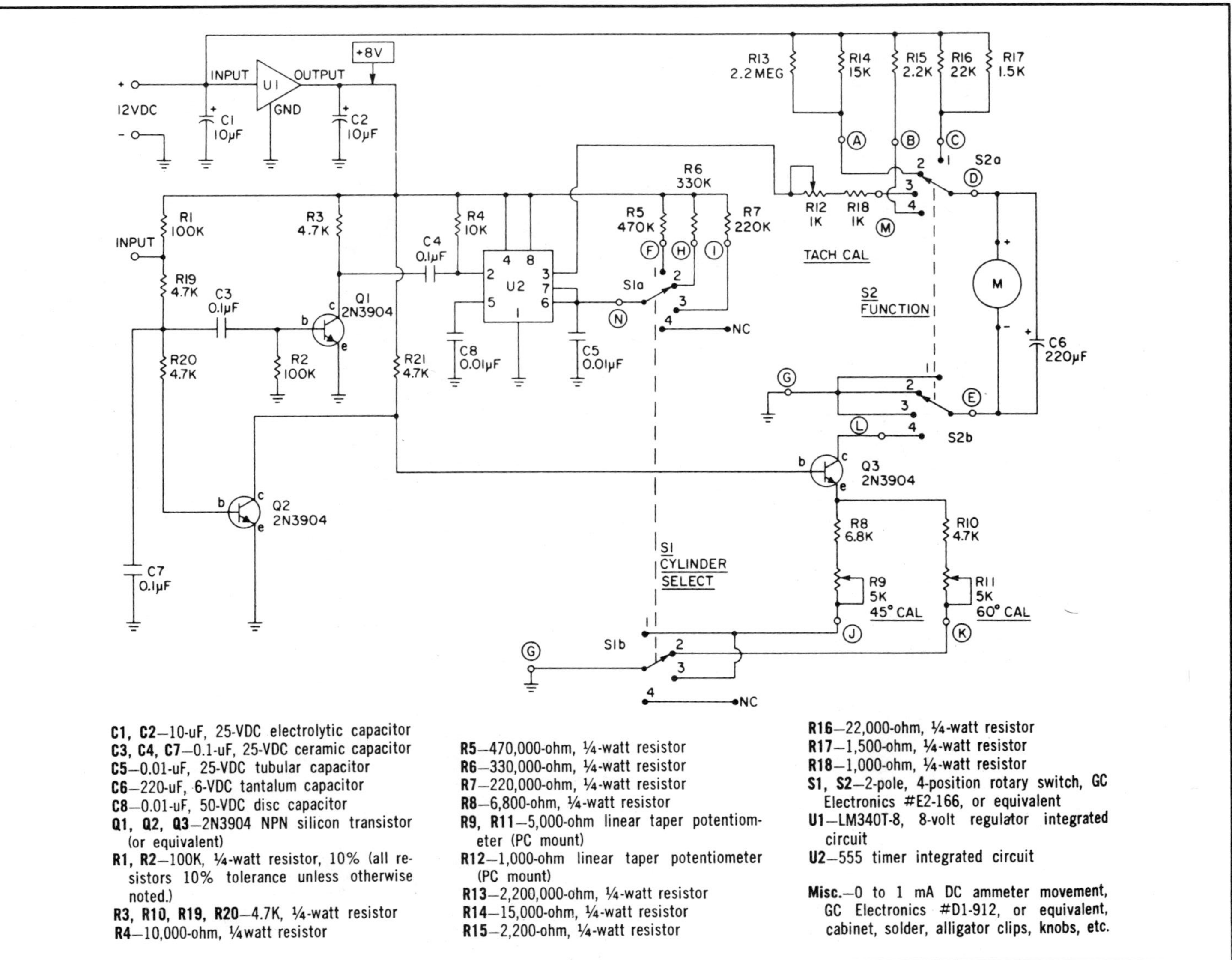

C1, C2—10-uF, 25-VDC electrolytic capacitor
C3, C4, C7—0.1-uF, 25-VDC ceramic capacitor
C5—0.01-uF, 25-VDC tubular capacitor
C6—220-uF, 6-VDC tantalum capacitor
C8—0.01-uF, 50-VDC disc capacitor
Q1, Q2, Q3—2N3904 NPN silicon transistor (or equivalent)
R1, R2—100K, ¼-watt resistor, 10% (all resistors 10% tolerance unless otherwise noted.)
R3, R10, R19, R20—4.7K, ¼-watt resistor
R4—10,000-ohm, ¼watt resistor
R5—470,000-ohm, ¼-watt resistor
R6—330,000-ohm, ¼-watt resistor
R7—220,000-ohm, ¼-watt resistor
R8—6,800-ohm, ¼-watt resistor
R9, R11—5,000-ohm linear taper potentiometer (PC mount)
R12—1,000-ohm linear taper potentiometer (PC mount)
R13—2,200,000-ohm, ¼-watt resistor
R14—15,000-ohm, ¼-watt resistor
R15—2,200-ohm, ¼-watt resistor
R16—22,000-ohm, ¼-watt resistor
R17—1,500-ohm, ¼-watt resistor
R18—1,000-ohm, ¼-watt resistor
S1, S2—2-pole, 4-position rotary switch, GC Electronics #E2-166, or equivalent
U1—LM340T-8, 8-volt regulator integrated circuit
U2—555 timer integrated circuit

Misc.—0 to 1 mA DC ammeter movement, GC Electronics #D1-912, or equivalent, cabinet, solder, alligator clips, knobs, etc.

Fig. 5-24. Dwell/Tachometer schematic and parts list.

electronic ignition systems will have a special terminal for the connection of the sensing lead of the instrument. Refer to the service manual for your car, or ask your dealer where this terminal is located. Once the instrument is connected to the automobile, the function switch can be set to DWELL, TACH, or 15 volts as necessary. Keep in mind that when measuring dwell on four-cylinder engines, you must double the meter reading. Be very careful not to switch the function to the 1.5 volt scale unless you have first checked the voltage of the circuit under test with the 15 volt scale to be sure that the voltage is less than 1.5 volts. This will avoid possible damage to your meter.

WIPER-TROL II

There's no doubt that many drivers want to build a project that gives the convenience of a single-flick automatic windshield wiper. If this system is not for your car, or if your car and you refuse to get together with a workable installation, you will know it *before* you sink valuable bucks and construction time into the project. We show you how to check your vehicle to be sure Wiper-Trol II will work—*before you build it.* Interested? Read on.

Basic Operation. In virtually all cars, turning on the wiper switch momentarily will cause the wipers to sweep once and then return to their park position. The operation of the wiper unit described here is simply equivalent to turning the wiper switch on then off. A 555-type timer controls the opening and closing of relay contacts connected in such a way as to simulate turning the dashboard switch on and off. The time interval between wipes can be varied, and the unit does not interfere with normal operation of the wipers.

Refer to Figs. 5-25 through 5-31.

Other wiper control units are available, but one has to buy them before finding out if they work on his car. Another problem with existing units is that some do not park the wipers after each sweep. Eventually the wipers can stop in the middle of the windshield. This happens when the rain lets up and there is more drag on the wipers, or when the car is stopped at a traffic light and the wiper motor slows down due to drain on the battery. In contrast, the wiper control presented here causes the blades to return to exactly their park position after each sweep.

There are two other useful features of this wiper unit. It has a button for one-shot operation of the wipers: during installation, it can be adjusted to give two sweeps instead of one, for each kick of the motor instead of one.

Before You Build. The object here is to determine whether Wiper-Trol II will work on your car before you build. Some dpdt switches (such as Radio Shack 275-1537) are used because they are more convenient than spdt. While car wiper circuits vary a great deal, most run four wires to the motor; therefore, the test described here will assume four wires. If your wiper motor has more wires, just use more switches. This will correspond to another relay in the control box. If your wiper motor and washer pump are housed together, use only the wires to the wiper motor.

There are two ways to determine how to connect the dpdt test switches into your car wiper circuit. One is to use a schematic for your

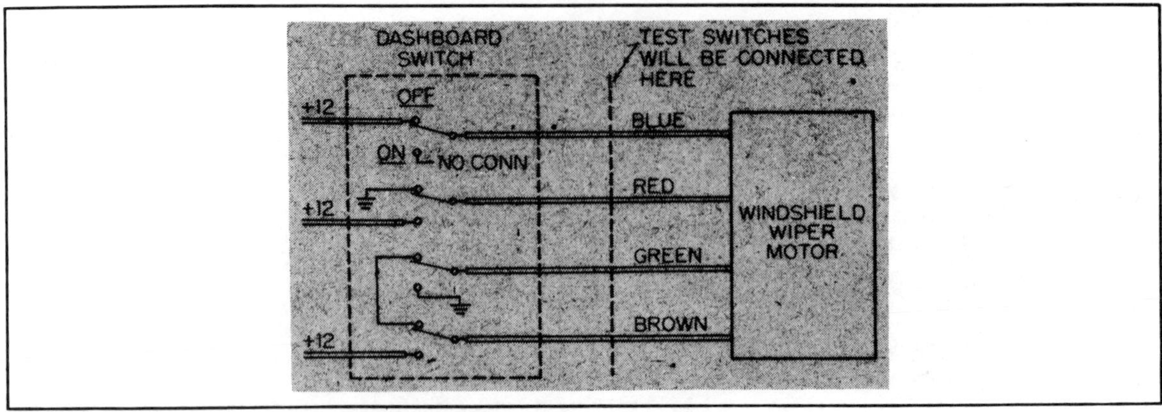

Fig. 5-25. This example of actual windshield wiper wiring is for a three-speed Chrysler Corporation system. It is shown in the low-speed position. While not entirely necessary in all cases, it is nonetheless a good idea to have a copy of your windshield wiper wiring diagram.

car that shows the inside workings of the wiper dashboard switch. The other is to determine the inside workings of the dashboard switch through a tracing procedure described later.

Let us first suppose you have a schematic. Figure 5-25 shows a typical wiring arrangement (in this case, a Chrysler wiper). For low-speed operation, the dash switch connects the red and brown wires to the car battery + terminal. Equally important, however, is the fact that the blue is connected to nothing for low speed, and the green is connected to ground. Your test switches would be connected as shown in Fig. 5-26. With the dash switch off, and the test switches at position A, the wipers should sweep. More important, however, with the test switches returned to B, the wipers should go to their park position because the dash switch is turned off. If the test works, the test switches can be replaced with the relay contacts of the control unit with confidence that the unit will work on your car.

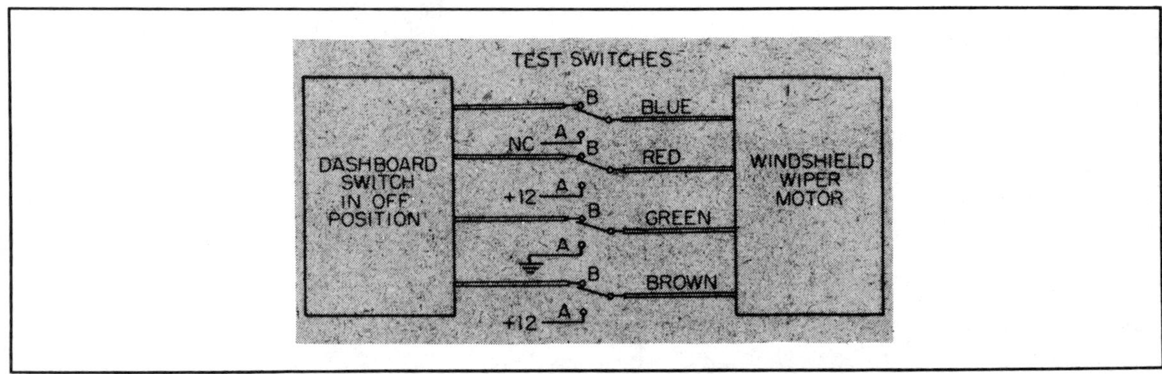

Fig.5-26. To fully ensure your success before you build this wiper delay device, connect inexpensive toggle switches to test the wiper-delay technique. You simulate the operation of Wiper-Trol II by throwing both switches at the same time to confirm your hook-up.

Finding Data. Wiring diagrams that show the operation of the dash switch can sometimes be found on the car manual. Some manuals come with the car; others can be obtained from car dealers or a library.

If a diagram for your dash switch cannot be found, a tracing scheme shown in the figure may be used to find out how to connect the test switches. The + battery lead is disconnected (for positive ground cars, disconnect to negative lead), and the ignition switch and wiper dash switch are both turned on. The wires to the motor are disconnected at the motor. Generally, there is a connector at the motor that can be just be unplugged. Now sketch a table like that shown, with the first column containing the color of each of the wires and the second column blank. The object now is to find what each wire coming from the switch side (denoted "S") is connected to, and write the data in the second column of the table. One ohmmeter lead is connected to an "S" wire, the other is connected to the + battery lead, then to the chassis, then to each of the other "S" wires. In this way, for example, one finds that the brown "S" wire is connected to the + battery lead, and the green "S" wire is connected to nothing. To conduct this test, one should have an ohmmeter that can distinguish between a direct connection, and a connection through a resistance of about 5 ohms. Some dash switches have resistors between their terminals, and can make it appear at first glance that one "S" wire is connected to two different points, when there may actually be a resistor between the "S" wire and one of the points.

After completing the table, the test switches can be tried, also as shown in Fig. 5-28. Once the switches have been connected, the car battery is reconnected and the ignition switch is placed in the on position, but the car engine should not be running. If placing the

Fig. 5-27. Perhaps the most interesting feature of the Wiper-Trol II is the single sweep (one-shot) mode. Should you accidentally fail to turn off the power, the wiper will single-sweep 10 minutes after the last push of the button.

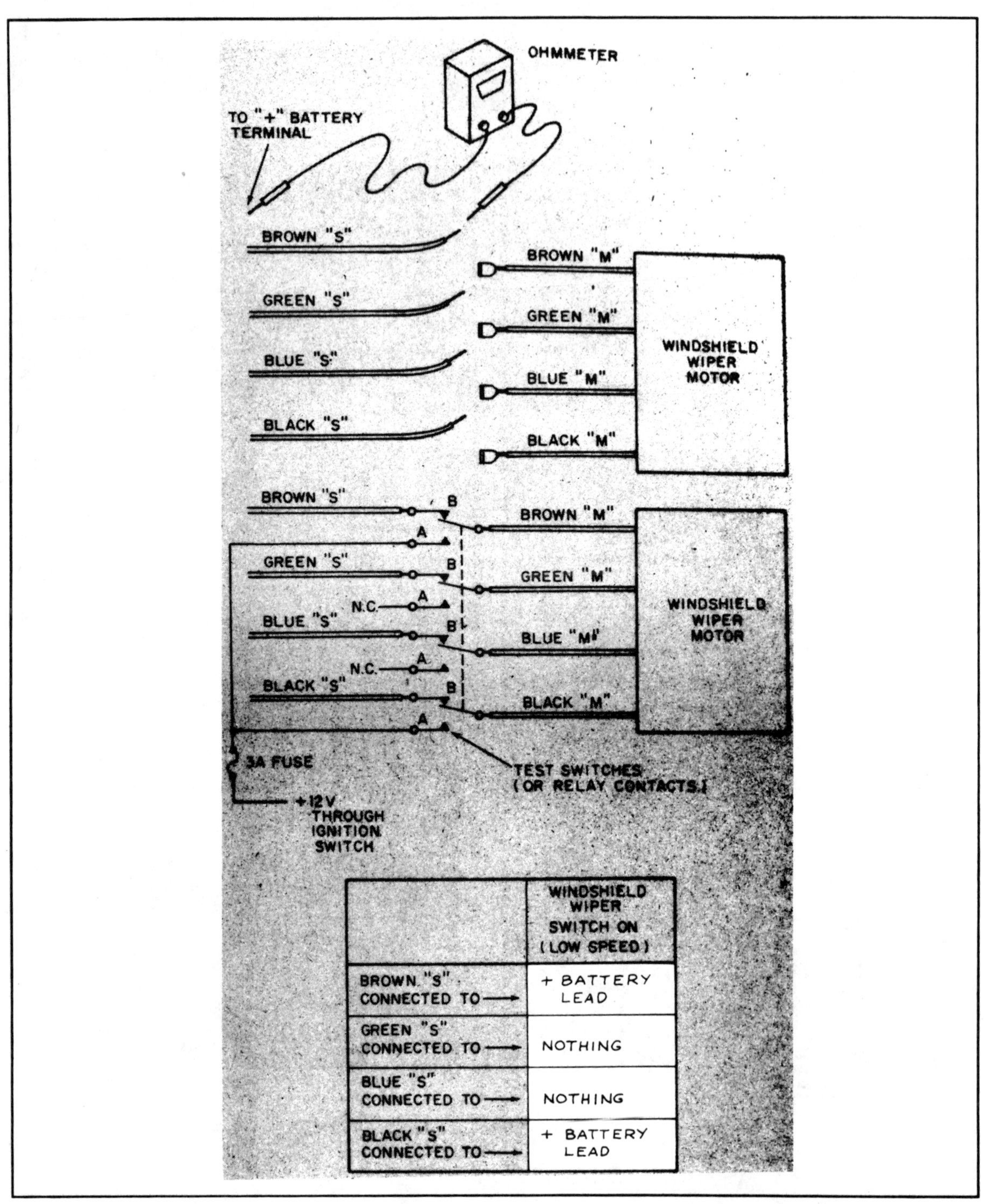

	WINDSHIELD WIPER SWITCH ON (LOW SPEED)
BROWN "S" CONNECTED TO →	+ BATTERY LEAD
GREEN "S" CONNECTED TO →	NOTHING
BLUE "S" CONNECTED TO →	NOTHING
BLACK "S" CONNECTED TO →	+ BATTERY LEAD

Fig. 5-28. An example of how you can determine if the Wiper-Trol II is right for your vehicle. Simply disconnect the battery lead, turn on the ignition and wiper control, and use an ohmmeter to furnish data to enter in the table.

Fig. 5-29. For the math-minded, relay-on *kick duration* time equals 0.7 (R4 + R5) C1 and is 1.3 to 4.8 seconds in duration. Interval between kicks is equal to 0.7 (R2 + R3 + R4 + R4 + R5) C1 and varies with R3 from three seconds to 50 seconds. Waveform shown is the output of 555-type timer Pin 3.

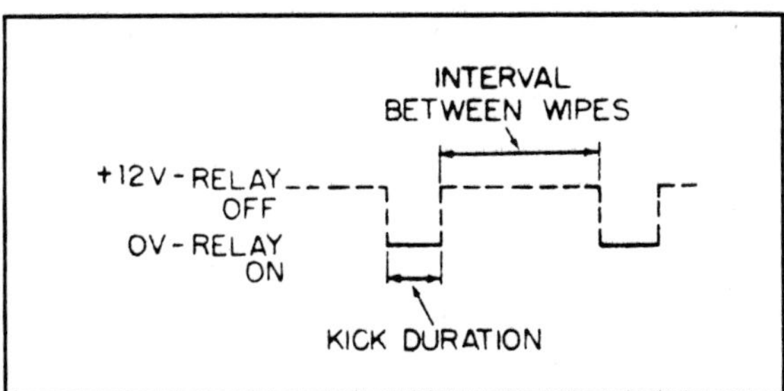

switches in the A position sweeps the wipers, and placing them in the B position causes the wipers to park, the control unit will work when the test switches are replaced with the relay contacts.

Other Hints. For those cars with relays built into the wiper circuitry, as on some GM cars, the wiper control unit can still work well, but the test procedure described above is best performed with a car wiring diagram in hand so that you are sure to trace connections *through the dash switch* and not through the car relay. Also, remember that an improper trace may cause some sparks to fly when the test switches are closed—so be prepared to open them quickly. A 3-amp fuse in the power lead to the relay contacts should prevent damage. The test switches should be operated simultaneously, and all leads to them should be double-checked before the experiment is tried. If the test switches do not operate the wipers, check the switches themselves, check your wiring and tracing, and, finally, try getting a description of the dashboard switch from the library, bookstore, or car service center. This may be the challenging do-it-yourself part.

As can be seen from the schematic, it requires only one IC and a handful of other components. The final assembly fits on a 3 × 3-inch perfboard and inside a small (3 × 2 × 4-inch) cabinet. Power to the IC

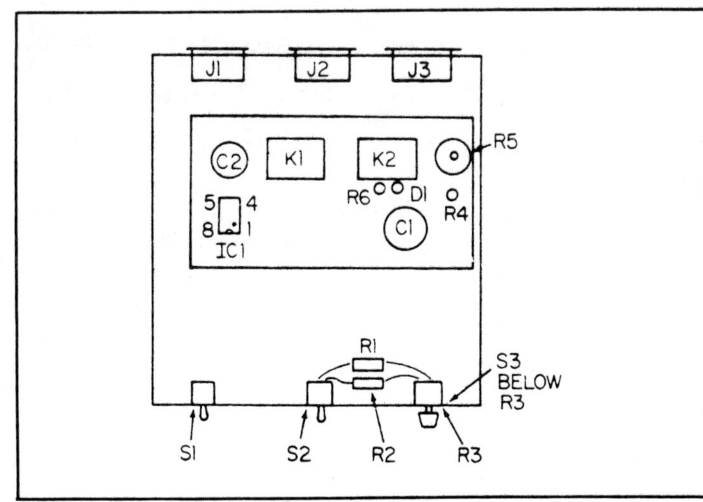

Fig. 5-30. Typical layout of parts on perfboard.

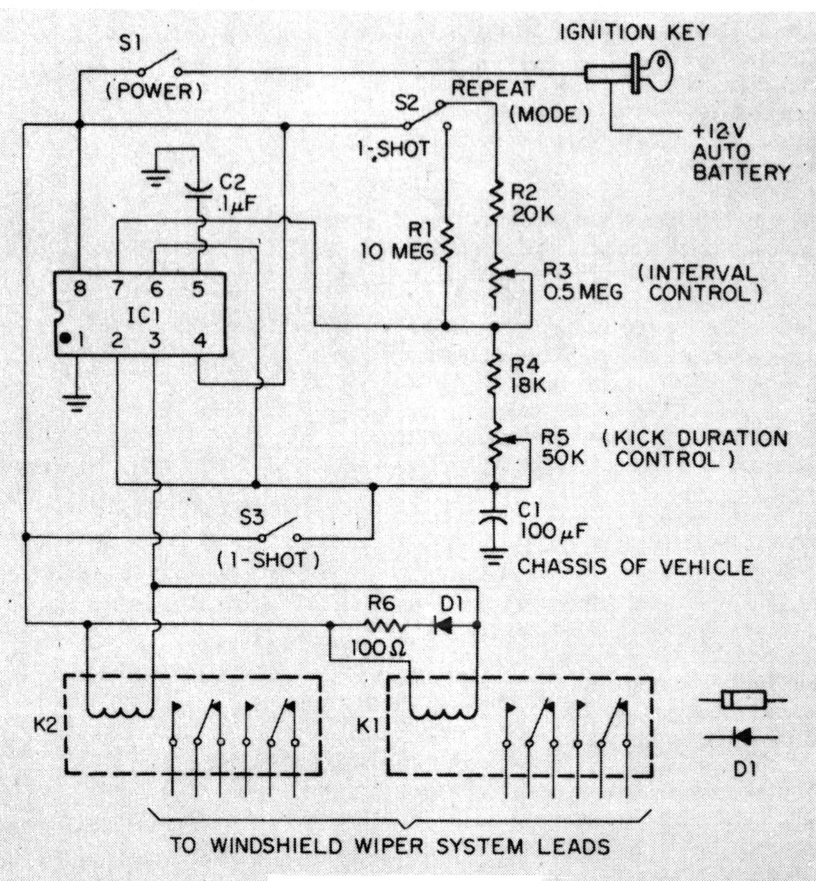

C1—100-uF, 35-VDC electrolytic capacitor (Radio Shack 272-1016 or equiv.)

C2—0.1-uF capacitor (Radio Shack 272-996 or equiv.)

D1—1-amp, 50-PIV silicon diode (Radio Shack 276-1101 or equiv.)

IC1—555-type timer IC (Radio Shack 276-1723 or equiv.)

K1, K2—dpdt relay, 3-amp contacts, 12-VDC coil (Radio Shack 275-206 or equiv.)

R1—10-meg, ½-watt resistor (Radio Shack 271-000 or equiv.)

R2—20,000-ohm, ½-watt resistor (Radio Shack 271-000 or equiv.)

R3—0.5-meg linear taper potentiometer (Radio Shack 271-210 or equiv.)

R4—18,000-ohm, ½-watt resistor (Radio Shack 271-000 or equiv.)

R5—50,000-ohm linear taper potentiometer, PC type (Radio Shack 271-219 or equiv.)

R6—100-ohm, ½-watt resistor (Radio Shack 271-000 or equiv.)

S1—spst subminiature toggle switch (Radio Shack 275-324 or equiv.)

S2—spdt subminiature toggle switch (Radio Shack 275-326 or equiv.)

S3—spst pushbutton switch (Radio Shack 275-609 or equiv.)

Misc.—perf board, hardware, case approx. 3 x 2 x 4-in. (Radio Shack 270-251 or equiv.), 3 amp fuse and dpdt switches for testing (see text), 4-pin chassis connector and mate (optional, see text) (Radio Shack 274-206 or equiv.), wire, solder, etc.

Fig. 5-31. Wiper-Trol II schematic.

should come through the ignition switch and S1, which is on the front of the control box. Switch S2 allows the unit to operate in a repeat mode, or a one-shot only mode. In the one-shot mode, the wipers can be kicked at will with a touch of pushbutton S3. With S2 in the one-shot position, R1 causes the wipers to sweep once every ten minutes as a reminder that power is on. In the repeat mode, the wipers are kicked by relays K1 and K2 at intervals determined by R3 which is mounted on the front panel of the control box. The repeat mode allows one-shot operation as well.

Set-Up. Resistor R5 is important because it allows you to adjust the duration of kick that is best suited to your car, but it is located inside the control box because the adjustment need be made only once. The figure shows that the kick duration is adjustable from 1.3 to 4.8 seconds, and the interval between "sweeps" is adjustable from 3 seconds to 50 seconds. Adjusting the kick duration to be a bit long (around 4 to 5 seconds) will cause the wiper to sweep twice before parking, thereby drying the windshield just a little bit better.

Relays K1 and K2 have *dpdt* contacts, but a *4pdt* relay may be used instead. The contacts must be of the break-before-make type, and should be rated for at least 3 amps, as should the wires connecting them to the wiper motor. Chassis connectors (female) are convenient for handling wires from the relay contacts. One arrangement is to run all normally closed contacts to connector 1, the center poles to connector 2, and the normally open contacts to connector 3. This places all wiper motor wires on one incoming male connector and all switch wires on another.

Another relay may be added in parallel with K1 and K2 if more contacts are needed; up to 150 mA may be drawn by the coils without harming the 555. Power connected to the contacts can come from the ignition switch to be sure wiper motor power is off when the car is unattended (when the ignition switch is off).

Troubleshooting. If the unit fails to properly operate the wiper motor even though testing with the *dpdt* switches was successful, there can be only a few simple reasons for the cause. Check to see that the 555 operates the relay coils, that the wiring to the wiper motor is correct, and that the relay contacts all open and close properly.

AUTOMOBILE IGNITION MAZE

Install a combination lock on you car's dashboard and a thief would have a better chance playing Russian Roulette. Switches S1 through S5 are spt rather than spst only to keep all external switch markings the same.

Tracing the circuit will show that only if switches S2 and S4 are down is the siren disabled. The siren sounds if any other switch is down or if S2 or S4 is up when the ignition is turned on. A simple wiring change lets you set any combination.

Refer to Fig. 5-32.

The switches can be "sporty" auto accessory switches sold individually or in switch banks such as G.C. 35-916. Provide labels such as "Carburetor Heater" or "Window Washer," and no one will know the car is wired for "sound."

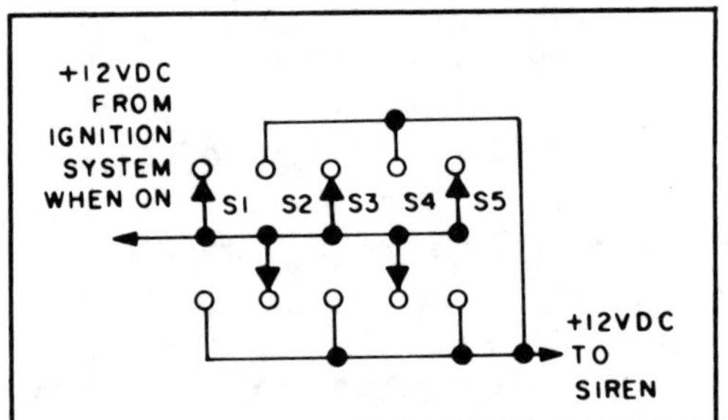

Fig. 5-32. Automobile Ignition Maze.

Chapter 6

Photography Projects

LONE RANGER LIGHT METER

Refer to Figs. 6-1 through 6-8.

Lone Ranger is a photographic light measuring instrument without the usual (needle-and-scale) mechanical meter. Instead, it uses light-emitting diodes (LEDs for short) to tell you what lens opening to use. In addition to cutting the cost by more than 50 percent, eliminating the meter has other advantages. The chance of damage from dropping is much less. People with no knowledge of photography can easily use this exposure indicator once taught the significance of the displays. Finally, because the readout is on LEDs, it's always easy to see, even in low light where an ordinary meter's needle might be hard to read accurately.

This comparator-LED light meter is ideal for the serious begin-

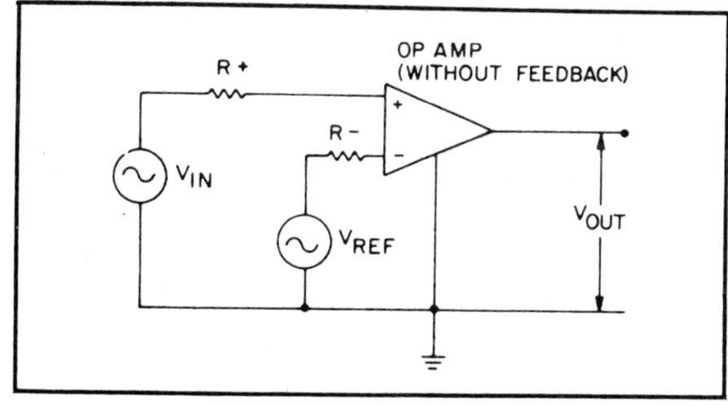

Fig. 6-1. Operational amplifier without feedback has extremely high gain. It can be used to compare an input signal voltage (unknown) with a known (reference) voltage, and indicate clearly (by its output's going to saturation, or by staying at its initial (very low) voltage that the unknown voltage is either below or above the reference voltage. This makes it a comparator.

ning or intermediate photographer because most people shoot with the same speed film most of the time. And if you do use two or three different speed films, it's easy to apply a conversion factor to the Lone Ranger's lens-opening scale.

It's a one-speed-range photographic light meter which tells you at what f-stop diaphragm opening to set your 35 mm or other precision camera lens. It provides readings for setting your camera lens opening between f-stops as large as 2.8 and as small as 32. These are based on one of the most popular black-and-white film for 35 mm use, Plus-X, a widely available fine-grain film.

Photo Basics. First before showing how the meter works, let's review some basic photography. The photographer is concerned with three numbers when making an exposure:

- ☐ the ASA rating (the speed) of his film
- ☐ the f-stop of the lens aperture
- ☐ the speed of the shutter.

Let's see how these factors interrelate. Suppose you take a correctly-exposed picture under light of intensity I, with f-stop n and exposure time equal to T. If the intensity suddenly jumps to $2I$, you must compensate by either reducing the aperture (multiplying the f-stop by 1.4) or by reducing the exposure time by half—$T/2$. And if the light intensity is reduced by half you would compensate either by making the f-stop 1.4 times larger, or by increasing the exposure time to $2T$. This assumes, naturally, that the film's speed (ASA) remains constant.

Now suppose that a correctly-exposed photograph is made under light of intensity I, with f-stop $=n$, and exposure time $=T$. To take the same picture with a film whose ASA rating is twice that of the original, you'd compensate by making the f-stop $= 1.3n$ or by making the exposure time $=T/2$. To take the same picture with a film whose ASA rating is half that of the original film, make the f-stop $=n/1.4$, or make exposure time $=2T$. Now let's look at an electronic circuit to measure the ambient light.

Use a High-Gain Amplifier. Suppose we take a high-gain differential amp and place a known voltage on one input, an unknown on the other. Since we're using the amp open-loop (without the usual feedback), only a small voltage difference at the two inputs is required to send the output either to saturation, or to cut-off. Specifically, if the voltage at the non-inverting (+) input is a few millivolts greater than that on the inverting (−) input, the output will go high. Likewise, if the voltage on the inverting input is the greater, the output will go low.

There are limitations to the size of the voltages which may be compared. For the LM339, input voltages should be less than supply voltage (V)—1.5 V. Furthermore, these input voltages should be much greater in magnitude than a few millivolts, to swamp out measurement errors due to the inherently imperfect nature of the comparator itself.

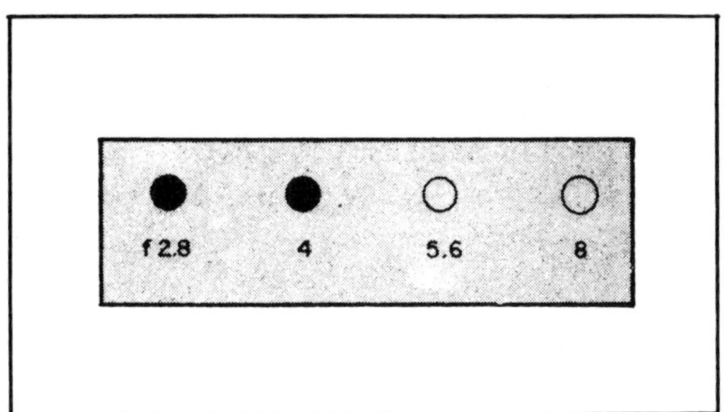

Fig. 6-2. This is the way the LED readout of the Lone Ranger would look if the amount of light being measured was enough for a camera opening of f-stop 5.6. The two LEDs at the left are dark, and the two on the right are lighted.

Between these extremes a comparator can give a very accurate answer to the question, "Is the unknown voltage above or below the reference voltage?"

The LM339 incorporates four comparators on a single chip. If one input of each comparator reads some common, unknown, voltage, while the other four inputs connect to different reference levels, then the size of the unknown voltage can be estimated by observing the output states of the comparators.

Figure 6-7 shows the LM339 as the heart of a light meter. All the inverting inputs go to the junction of PC1 and R1, and thus sense a voltage whose magnitude increases as the intensity of the light being measured increases. C2 bypasses any interference caused by floures-cent lighting in the vicinity. The non-inverting input of each comparator goes to a reference voltage, with section A connected to the lowest reference voltage and section D to the highest. Consequently, in very dim light all four comparator outputs will be at cutoff, hence all four LEDs will be extinguished.

As the light intensity increases, section A will be the first to change state (rise toward saturation) and thus cause LED1 to light. At

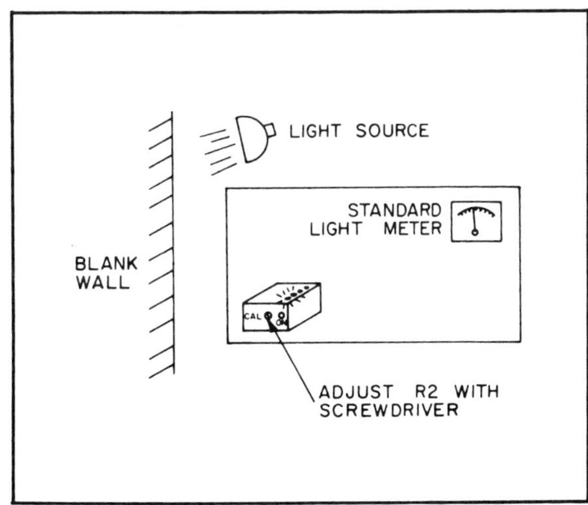

Fig. 6-3. Setup used to calibrate the Lone Ranger light meter. You'll need to borrow an old-fashioned (analog) light meter for this procedure.

higher intensities LED1 and LED2 will both be turned *On*. The reference voltages I used were chosen to correspond to differences in lens aperture of one f/stop. Thus, a display like the one shown here would indicate that the correct photographic exposure is between f/4 and f/5.6.

Extending the Meter's Range. Notice that in contrast to the continuous readout of an analog meter, this comparator system of voltage measurement indicates proper exposure as being between two levels. In order to get better resolution (more detailed information as to lens opening) we would need more comparators. We would also need more comparators if a larger measurement range is desired. To accomplish such a range expansion we could add another LM339—inputs 4, 6, 8, and 10 would go to the junction of PC1 and R1, while pins 5, 7, 9, and 11 would go to new (added) reference voltages. However, there is a cheaper method of range expansion. We simply install a variable aperture in front of the photocell. In this way, the measurement range of the photometer is doubled to 8 stops, by using two apertures whose areas are in the ratio of 16:1.

The total measuring range of this instrument thus spans from f/2.8 to f/32 with ASA 125 film (such as Plus-X) at a shutter speed of 1/30th second. Later on we'll discuss the simple mathematical conversion necessary to allow use of the light meter with different film speeds and different exposure times.

Building Lone Ranger. Actual construction of the Lone Ranger meter is not critical but will require some care because of its small size. A printed circuit board was used in the Lone Ranger prototype, and although it is not necessary that you use the printed circuit, it

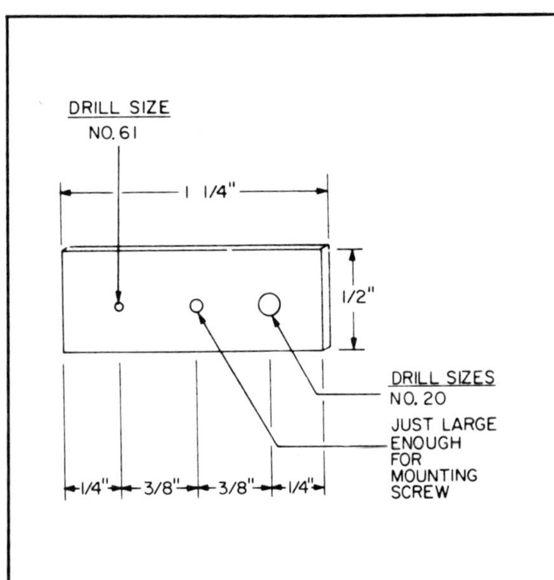

Fig. 6-4. Dimensions of the range-extender for Lone Ranger. It's a simple piece of aluminum with two different-size holes in it. The middle hole is for mounting the strip to the front of Lone Ranger.

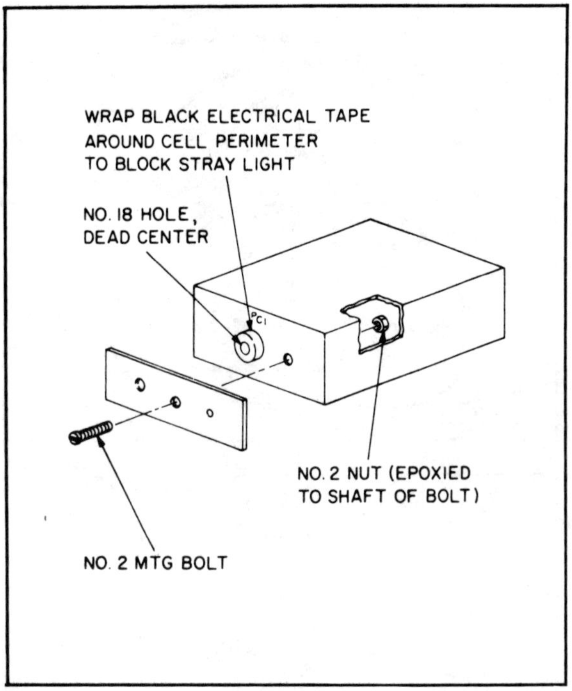

Fig. 6-5. Here are the details for mounting the range extender to the front of the case and for taping around the photocell to keep it from receiving stray light, which can cause misreading of the ambient light.

would be wise to copy the same general layout as the prototype. My Lone Ranger is housed in a 3¼ × 2⅛ × 1⅛-inch plastic minibox. If you use the same box, note that the mounting post in the upper-right-hand corner must be removed to make room for S1. A soldering gun with a cutting tip was used to slice out the mounting post, leaving three posts to hold down the metal cover of the box. If you are inexperienced in small-scale construction, by all means use a larger box. Regardless of the box size used, however, the following construction details given will still apply.

When the board has been completed, mount the IC socket, trimmer R2, and all resistors and capacitors. Next, solder the negative lead from the battery clip to its hole near pin 12 of the IC socket. Solder a 2-inch length of flexible wire to the hole indicated in the upper-right-hand corner of the board. This wire will later be connected to S1. Now mount the photocell so that its light-sensitive face is perpendicular to the board and facing toward its upper border. Finally, mount the four LEDs into the circuit board, but be sure to observe proper orientation. The tops of the LEDs should all extend the same distance above the board—about ⅞ inch if you have a cabinet of the same depth. Now plug IC1 into its socket and set the board aside temporarily.

The range selector is just a simple aluminum plate (about 18 gauge) with the dimensions shown in the diagram. Note that two holes, one No. 20 and one No. 61, must be *carefully* drilled. Further note that the plate must be absolutely flat. Don't cut it out with tin snips. Use a

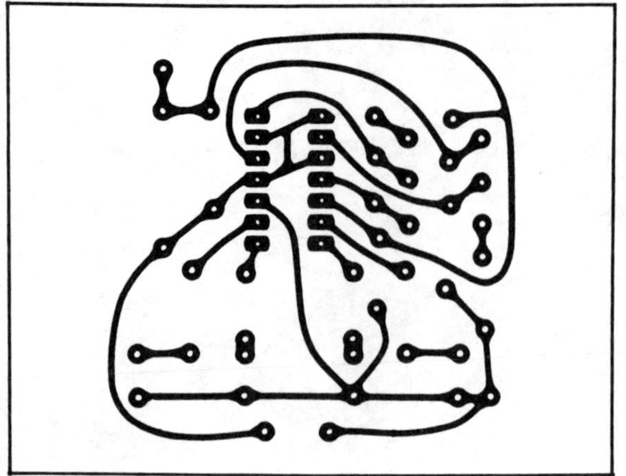

Fig. 6-6. Full-size template for the printed circuit board. See the text for suggestions on making printed circuit boards if this is your first.

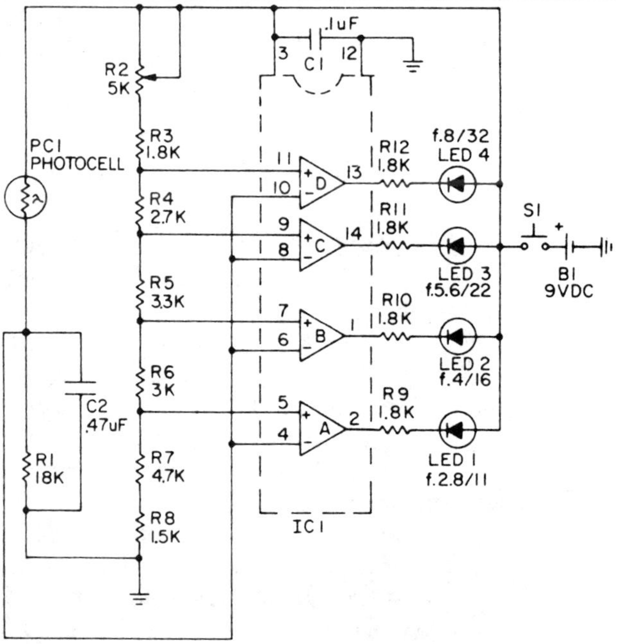

C1—.1-uF capacitor
C2—.47-uF capacitor
IC1—Quad comparator integrated circuit LM 339
LED1, 2, 3, 4—Light-emitting diodes
PC1—Cadmium sulfide photocell
R1—18,000-ohm, ½-watt resistor
R2—5,000-ohm potentiometer, printed circuit board-mounting
R3, 9, 10, 11, 12—1800-ohm, ½-watt resistor
R4—2700-ohm, ½-watt resistor
R5—3300-ohm, ½-watt resistor
R6—3000-ohm, ½-watt resistor
R7—4700-ohm, ½-watt resistor
R8—1500-ohm, ½-watt resistor
S1—SPST momentary on switch
Misc.—Minibox 3¼-in. x 2⅛-in. x 1⅝-in. (or larger), socket for IC1, 9-VDC transistor radio battery, clip for battery, wire solder, printed circuit board kit, etc.

Fig. 6-7. Lone Ranger schematic and parts list. A light meter without the meter. The circuit enables you to build a meterless light meter to solve your f-stop woes. The Lone Ranger uses light-emitting diodes to tell its story, rather than all too breakable meter. It's absolutely great for the beginning and intermediate shutterbug to use for great picture results.

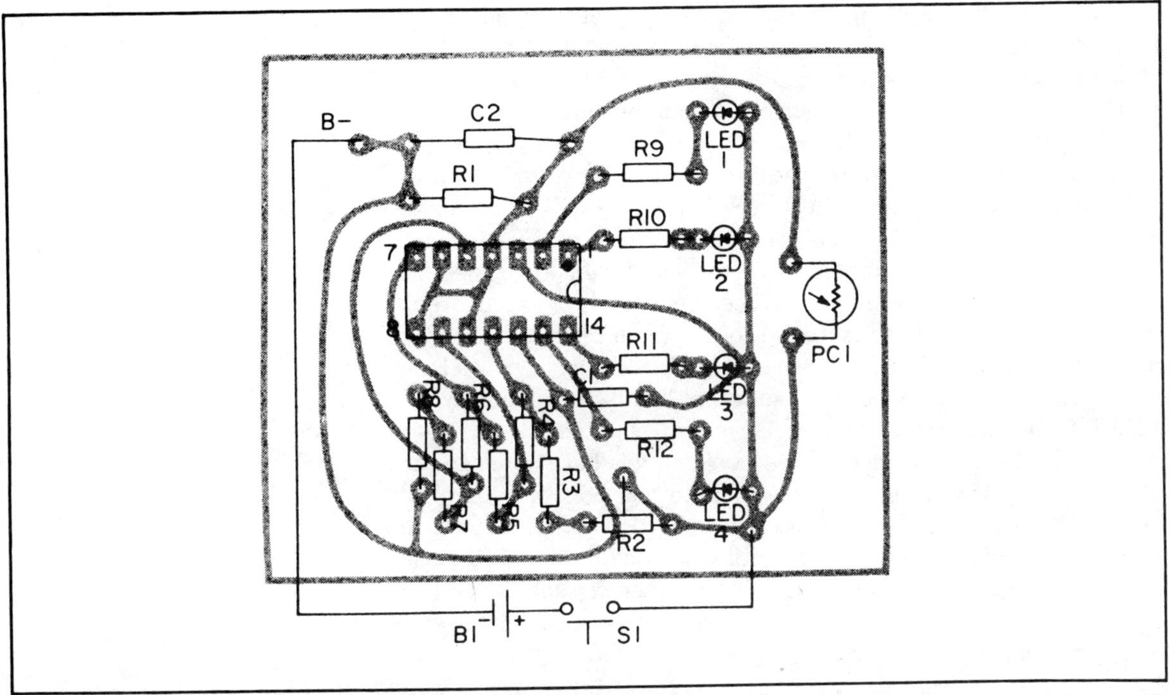

Fig. 6-8. Placement of parts on Lone Ranger's printed circuit board. If you use perfboard instead, you can put the parts wherever you want, but you'll find this general arrangement most convenient.

nibbling tool or hacksaw, which will cut the aluminum without distorting it. Now use a file to round off all the edges, and then buff it with steel wool. This will make the range selector rotate readily when you're out shooting.

More On Construction. The drawing of the cabinet shows how to mount the photocell relative to the range selector. When the proper holes have been drilled, mount the range selector with No. 2 hardware and tighten until the fit is just snug. Use a drop of epoxy to lock the nut to the shaft of the bolt, and let the cement dry.

S1 may be placed wherever it is convenient. Be sure to drill a hole to allow calibration-adjustment of trimmer R2 from the outside. Now locate and drill four holes in the cover to allow the LEDs to be visible. The exact location of these holes will depend upon the dimensions of your case and the dimensions of your board. Simply insert the board into the bottom of the box and measure how far from the sides each LED's center is located. Transfer these dimensions to the cover and drill four No. 22 holes.

Mount the board in the cabinet so that the photocell lies directly behind and flush against its mounting hole. If you used the same size box as I did the ¼-inch spacers will be needed between the board and the bottom of the case to allow the LEDs to protrude slightly through the thin metal cover. After the board has been securely mounted, take a ¼-inch wide, 1½-inch-long strip of black electrical tape and wrap it around the perimeter of the board and the case is covered. Solder the positive lead from the battery clip to one side of S1. To the other terminal of S1 solder the short lead from the circuit board. Finish off by mounting the cover and applying press-on decal labels as desired.

Calibrating Lone Ranger. Set the range selector to the low-light measurement position (the larger hole), then point the meter towards a bright light bulb and depress S1. One or more LEDs should light, depending upon the brightness of the source. If not, go back and check whether any components have been improperly oriented. When all is working well, only the calibration of the meter remains. Borrow a good light meter for this task. Choose a large, preferably blank wall and evenly illuminate it (avoid using fluorescent light sources, however). Adjust the light source and the distance until your reference meter indicates f/8 at ASA 125 and 1/30 sec. When you have obtained the correct reading on your reference meter, hold your Lone Ranger in the same spot and point it in exactly the same direction that the reference meter has been facing. Press S1 and adjust R2 so that LED4 (the one farthest) extinguishes. Now turn R2 back the other way until LED4 just comes back *on.* The meter is now calibrated. To use the meter with different film and shutter speeds, consult the following table.

Film Speed ASA/ISO	Exposure Correction
400	+2
250	+1
125	0
65	−1

Shutter Time	Exposure Correction
250	1/8
125	1/15
30	1/30
15	1/60
—	1/125

ASA = 123
+—go to higher f-stop
−—go to lower f-stop

Additional Circuit Uses. You may have noticed that the comparator circuit presented here has great potential. A thermistor might be submitted for photocell PC1 and the circuit becomes an electronic thermometer. Or mount a potentiometer so that its control shaft spins as another shaft rotates. The LED display would then indicate angular position, perhaps for an antenna rotor. The information here plus your own imagination should produce many new devices.

DARKROOM COLOR ANALYZER

One of the shutterbug's most satisfying accomplishments is producing his own color prints. For years the time spent on and the cost of making color prints were discouraging, but with modern color chemistry you can turn out quality color prints in less time than for black and white (about three minutes), and the prints will be far superior to anything you're likely to get from a color lab.

One thing that takes the drudgery out of color work—besides the chemistry—is a color analyzer, a device that gives you the correct filter pack and exposure time at the very first crack. More often, the very first print made with the analyzer will be good. At most, it will take perhaps 0.10 or 0.20 change of filtration for a superb print. This is a lot less expensive and time-consuming than making test print after test print. In fact, it's really the color analyzer that puts the fun into making your own color prints!

Refer to Figs. 6-9 through 6-14.

Color Analyzers Are Not Cheap. A decent one costs well over

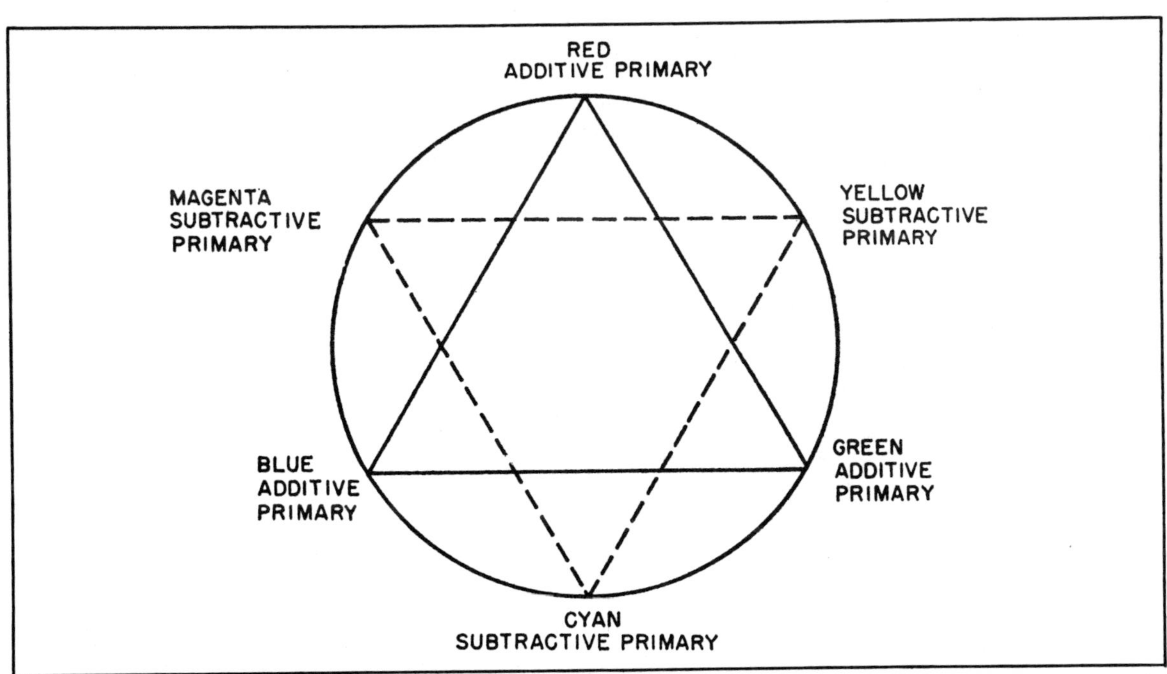

Fig. 6-9. Any one of the primary colors on this circle is composed of its immediately adjacent colors in equal amounts. It is the balancing of additive primary colors of photographic light sources and subtractive-type color filters that provides color print photography.

$100, and a good one runs well over $200. But if you've got even a half-filled junk box you can make your own color analyzer for just the junk parts and perhaps $10 or $15 worth of new components.

A color analyzer is basically a miniature computer. You make a near perfect print the hard way—by trial and error—and then calibrate the analyzer to your filter pack and exposure time. As long as you use the same box of paper and similar negatives, all you need to do to make a good color print is focus the negative, adjust the filter pack and exposure so the analyzer reads "zero," and hit the enlarger's timer switch. Even if you switch to a completely different type of negative, the analyzer will put you well inside the ballpark, so your second print is a winner. (And even if the filtration is off, the exposure will probably be right on the nose.)

Construction. The color analyzer shown was specifically designed for the readers of this magazine—essentially an electronic hobbyist with an interest in photography. All components are readily available in local parts stores or as junk-box parts. Several protection devices have been designed into the circuit so accidental shorts won't produce a catastrophe. The printed circuit board template has foils for both incandescent and neon meter lamps, as well as extra terminals so you can use either a socket and plug or hard wiring for the color comparator and exposure sensor. In short, you can make a lot of changes to suit your individual needs.

The template for IC1 uses a half-minidip. Signetics V type package lead arrangement. However, you can also use an IC with a round (TO-5) configuration. If anything is wrong with the IC you can get the TO-5 out easily. The half-minidip removal might result in destruction of the PC board. We'll explain how to install the TO-5 IC on the PC board later.

You can either buy or make the printed circuit board (see parts list). Either way, the first step is to prepare the printed circuit board. If you do it yourself, make it any way you like, using free-hand or template resist. Nothing is critical, but be certain there are no copper shorts between the terminals for IC1. Use a number 56 bit for all holes. Then use a larger bit for transformer T1's mounting screws (No. 4 or No. 6 screws), a ¼-in. bit for resistor R6, and a No. 30 to 40 bit for the line cord connections (any bit that will allow the linecord wires to pass through the board).

Assemble the power supply and check it out before any other components are installed. Install transformer T1 first. Any 24-volt or 25.2-volt center-tapped transformer that will fit on the board will be fine. Get something small, like 100 milliamperes.

Bridge rectifier BR1 is the low-cost surplus found in many distributors. This type has the positive and negative outputs at opposite ends of a diamond. The ac connections are the remaining opposite ends. Note that BR1 is installed in such a manner that its negative

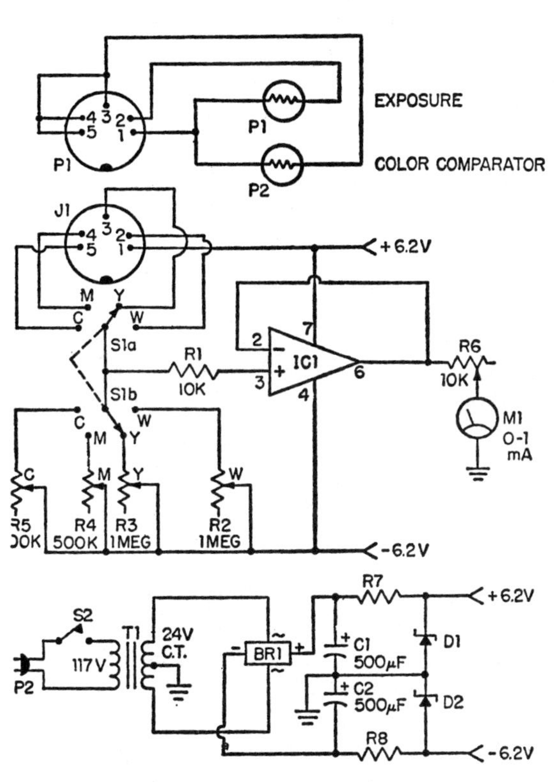

BR1—Bridge rectifier, silicon, 50 PIV, 0.5 amp or higher (Radio Shack 276-1151 or equiv.)
C1, C2—500 μF capacitor, 10 VDC or better (Radio Shack 272-957 or equiv.)
D1, D2—Zener diode, 6.2 volts, 1 watt (Radio Shack 276-561 or equiv.)
IC1—Type 741C operational amplifier, see text (Radio Shack 276-010 or equiv.)
J1—DIN type 5-pin socket (optional, see text) (Radio Shack 274-005 or equiv.)
M1—0 to 1 mA DC meter, see text (Radio Shack 22-052 or equiv.)
P1—DIN type 5-pin plug (optional, see text) (Radio Shack 274-003 or equiv.)
P1, P2—Photocell, Clairex CL5M5L, **do not substitute**
R1—10,000-ohm, ½-watt resistor (Radio Shack 271-000 or equiv.)
R2, R3—1-megohm potentiometer, see text (Radio Shack 271-211 or equiv.)
R4—500,000-ohm potentiometer, see text (Radio Shack 271-210 or equiv.)
R5—100,000-ohm potentiometer, see text (Radio Shack 271-092 or equiv.)
R6—10,000-ohm trimmer potentiometer (Mallory MTC-14L4 for exact fit on PC board; Radio Shack 271-218 or equiv. for point to point wiring)
R7, R8—820-ohm, ½-watt resistor (Radio Shack 271-000 or equiv.)
R9—100,000-ohm, ½-watt resistor (Radio Shack 271-000 or equiv.)
S1—Rotary switch, 2-pole, 4-position (Allied Electronics 747-2003; adjust stops for 4 positions)
S2—Switch, SPST (Radio Shack 275-1551 or equiv.)
T1—Transformer, 117 volt primary to 24 or 26.6 volt secondary, see text (for point-to-point wiring, Radio Shack 273-1512 or equiv.) Note—you can also use two less expensive 12 volt transformers with secondary windings connected in series-aiding, if you have the space.
Misc.—Cabinet, pilot lamp for meter, 2-in. or 3-in. size Kodak Wratten filters #70, #98, and #99 (available from photo supply dealers), calibrated knobs, wire, solder, hardware, etc.

Fig. 6-10. Darkroom Color Analyzer schematic and parts list.

output is farthest from transformer T1 while the positive output is nearest to T1. Make certain your bridge rectifier has the same lead configuration, if it is different, modify the printed circuit template to conform to the rectifier you're using. Get it right the first time.

Finally, install C1 and C2, R7 and R8, and zener diodes D1 and D2. Take care so the capacitors and zener diodes are installed with the polarity correct. If the capacitors have their negative leads marked with an arrow or line, these markings face the *opposite edges* of the PC board (negative to the outside). The zener diodes are installed so that their cathodes (the banded ends) face each other towards the center of the board.

Initial PC Checkout. When the power supply is completed, temporarily connect a linecord. Connect the negative lead of a meter rated 10 volts dc or higher to the foil between T1's mounting screws (that's ground). Connect the meter's positive lead to the junction of R7 and D1, which is in the center of the board; the meter should indicate approximately—6.2 volts dc. Then connect the positive meter lead to the R8 and D2 junction, which is near the edge of the board. You should get approximately −6.2 volts dc. If the voltages are far apart in value, or if the polarity is wrong, make certain you find the mistake *before* installing IC1.

Disconnect the linecord and complete the PC assembly. If you use a 24-volt or 28-volt pilot lamp to illuminate the meter you connect to the holes adjacent to T1's secondary (24 V) leads, if you plan to use a neon illuminator, install a 100,000-ohm resistor (R9) on the PC board and connect the lamp to the holes marked "neon." The lamp must have

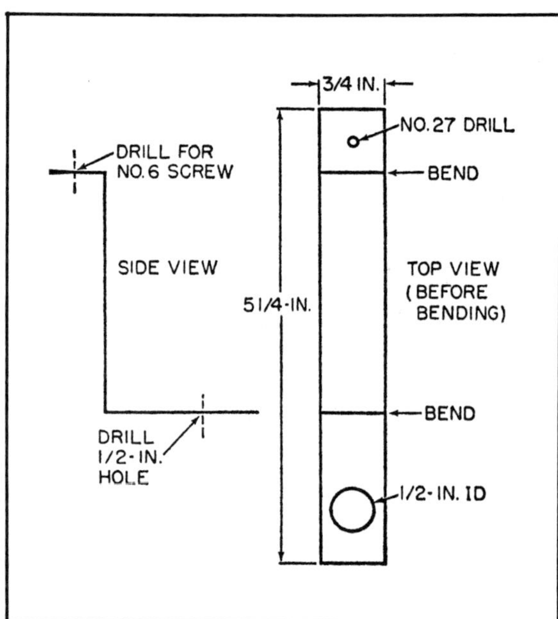

Fig. 6-11. The color comparator photocell Z-bracket is installed under a light integrator. If your enlarger has a filter holder under the lens, attach the Z-bracket to the holder.

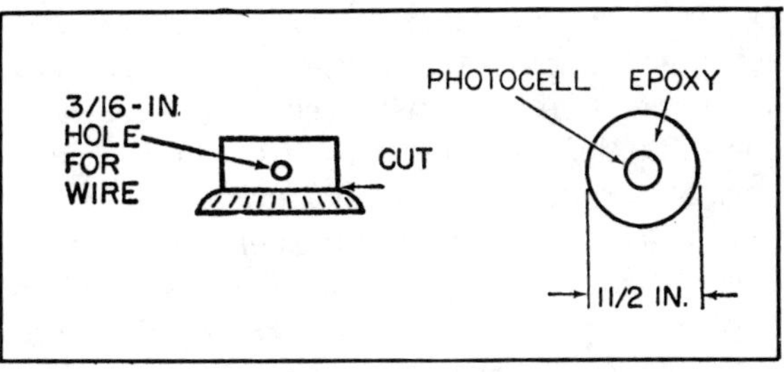

Fig. 6-12. The exposure sensor photocell is mounted in anything that will keep it in place on the easel. This example was epoxy cemented into a large control knob after the outside dial section was ground off. In typical operation, the sensor is placed under the lens with the light integrator or filters.

as little illumination as possible. Incandescent 24-volt or 28-volt lamps must be the miniature or "grain of wheat" type rated approximately 30 to 60 mA; the lamps come with attached leads. Do not use pilot lamps of the 100 to 500 mA variety. The excessive light will confuse the analyzer.

To install IC1 when it is the metal can TO5 type, fan out the number 1 to 4 leads and number 5 to 8 leads so they form two straight lines. Note that the lead opposite the tab on a TO5 package is No. 8. Insert the leads into the board leaving about ¼ inch between the IC and the board. The IC is correctly installed if the tab faces *away* from the transformer towards the nearest edge of the PC board. Solder IC1 and cut off the excess lead length.

The edge of the PC board nearest IC1 has four sets of paired foil terminals. These are provided as mounting terminals if you connect the photocell comparator and sensor without the use of a plug and jack. However, we strongly suggest the use of the specified DIN type connectors as they allow for easy repairs if the connecting wires break. (The connectors aren't *that* costly.)

Potentiometers R2 through R5 can be linear or audio taper, though audio taper gives a slightly smoother adjustment—use whatever you have in stock.

Meter M1 should be 0-1 mA with a zero-center scale. But these are expensive, so you can substitute any standard 1 mA meter you want. You will simply calibrate the instrument for zero-center.

If you use a neon pilot lamp mount it directly above the meter and shield the forward brilliance with a piece of black tape; the lamp should radiate straight down onto the meter scale. If you use the meter in the parts list, remove the front cover by pulling it forward. Then remove the meter scale. Place a black dot approximately 3/16-inch wide at the center of the scale. If you want, you can also modify the meter for the incandescent lamp. Drill a ¼-inch hole in the lower right of the meter *from the rear*. Position the meter in the cabinet and mark the location of the meter hole on the panel. Remove the meter and drill a ⅜-inch hole in the panel. When the meter is installed you can pass a "grain of wheat" lamp through the panel into the meter. Reassemble the meter and complete assembly.

The Comparator. The photocells used for the comparator and exposure sensor, P1 and P2, must be Clairex type CL5M5L. Make no substitutions. From a piece of scrap aluminum ¾ to 1-inch wide fashion a Z-bracket to the dimensions shown. Drill a ½-inch hole close to the end of the longer Z-leg. Fasten the other end of the Z-leg to your enlarger's under-lens filter holder. If your enlarger does not have a filter holder, or if it has a permanent swing-away red filter under the lens, mount a Paterson swing-away light integrator (available from local photo shops) under the lens. Fasten the short leg of the Z-bracket to the integrator—which has pre-drilled holes—in such a manner that the ½-inch hole is on the optical center of the lens. Then cement photocell P2 in the hole and attach the connecting wires; these can be extra-thin zip cord such as used for short-length speaker connections. Photocell P1, which measures the exposure light, can be mounted in anything heavy enough to hold it in place on the easel.

When the complete analyzer is assembled, attach oversize calibrated knobs such as the Calectro E2-715 to R2 through R5. The knob calibrations are important so they should run out to the very edge of the knob skirt. If the calibrations don't run to the edge you won't be able to preset the controls with any reasonable degree of accuracy. Place a fine line or other indicator directly above each knob.

Checkout. Connect the photocells to the control unit and apply power. Don't worry if the meter pins at either end of the scale. Set switch S1 to the extreme clockwise position and adjust R2 through R5 until you find the control that changes the meter reading. Mark the switch and the control "C" for cyan. (We suggest you paint the cyan

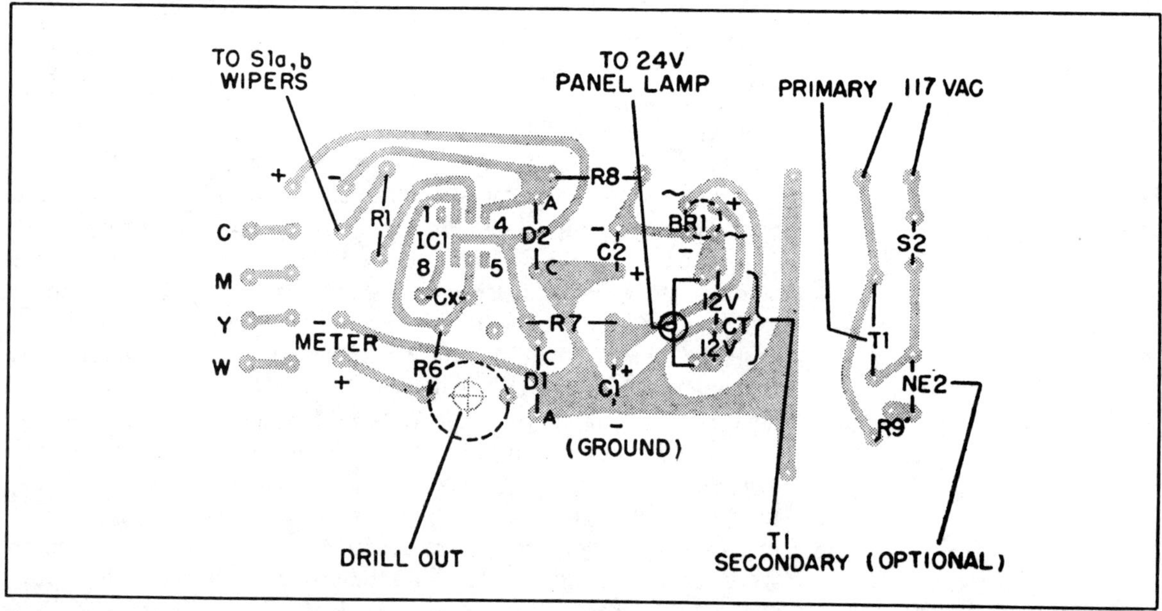

Fig. 6-13. There are a few parts on the PC board, and nothing is critical. Modify the board if you wish. Trimmer potentiometer R6 should be a flat mount, so it can be adjusted through a hole in the cabinet.

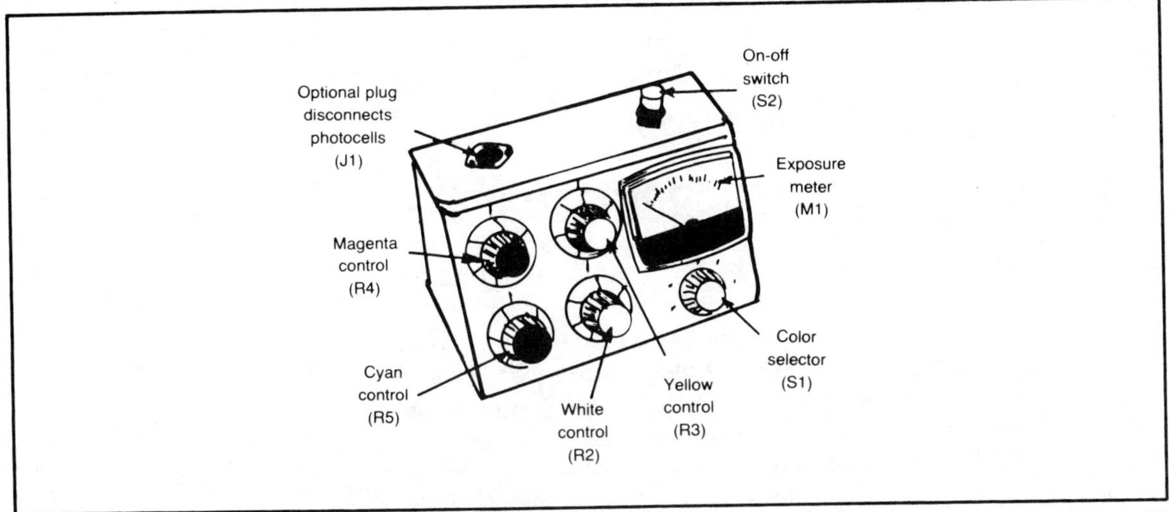

Fig. 6-14. To avoid upsetting a control setting while groping for the on-off switch in the darkroom, mount switch S2 as far as possible from the controls.

knob insert a blue-green. Also paint the other knobs the appropriate color.) Advance S1 one position clockwise, find the correct knob and label both "M" for magenta. Advance the switch another position clockwise, find the knob and label both "Y" for yellow. The last switch position and knob is labeled "W" for white (white light exposure). Make certain the C, M, and Y controls are reading P2, the color comparator mounted under the lens.

Set S1 to any position, turn on bright room lights, and adjust the associated color control until the meter pins, or approaches full scale deflection. Make certain the control is adjusted for the maximum meter reading. Adjust trimmer control R6 so the meter pointer just pins (don't be afraid to pin the meter). Depending on the amount of light the meter pointer will pin right (for bright light) and left (for dark or very low light). This is normal, there will be no damage to the circuit or the meter. (Note: If you use a zero-center meter the pointer will barely pin on both sides.)

Install the Z-bracket under the lens. If your enlarger uses a filter holder under the lens insert a diffusion screen or glass, or a Beseler Light Integrator or similar ground glass in the filter holder. You are now ready to make color prints.

SUPERMATIC T/S SYNCRONIZER

No doubt about it . . . there's an art to making home slide shows interesting! Of course, a little science can also help, which explains why we've designed this unique two-way slide synchronizer circuit. Like any other synchronizer, it links your tape recorder with your slide projector, automatically advancing the slides in step with a taped narration. But unlike any other unit, it will also reverse the slides. The only requirement is that your projector be equipped with a two-way (forward and backward) hand-held remote control unit.

Refer to Figs. 6-15 and 6-16.

Backward slide selection sounds nutty until you think about it. When you do, you'll have no trouble imagining the various visual special effects you can achieve by cycling your projector backwards as well as forwards.

The synchronizer circuit will work with any stereo tape recorder. One channel contains the taped narration; the other channel contains

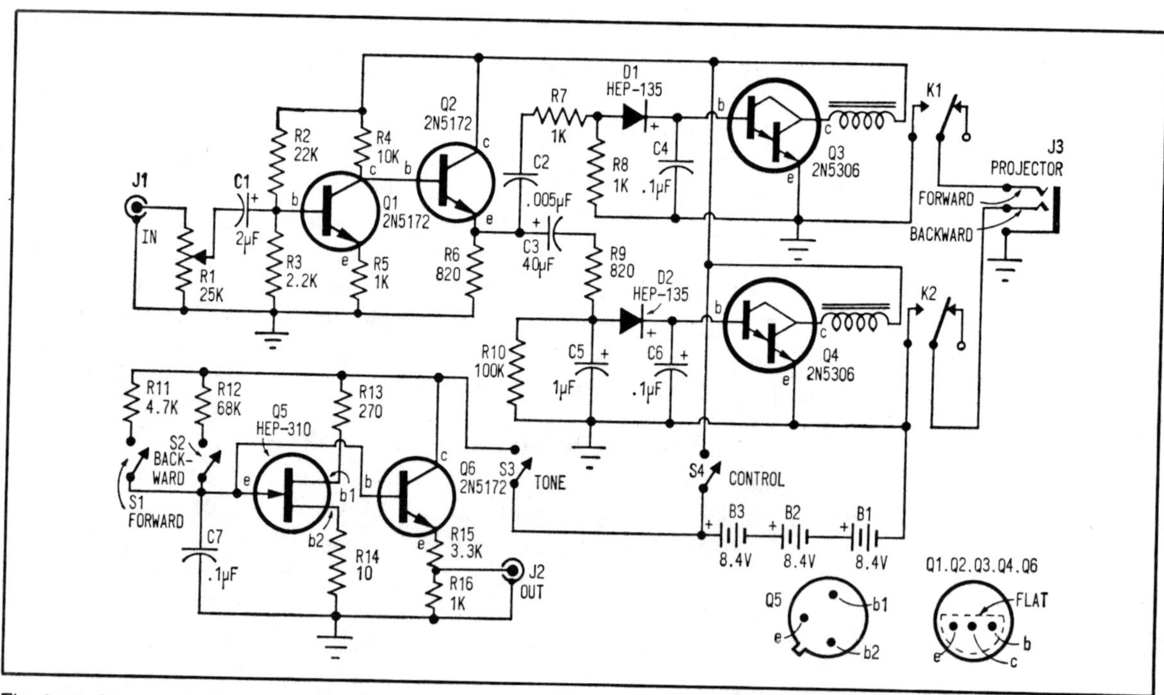

Fig. 6-15. Supermatic T/S Synchronizer schematic. Though Synchronizer will work with any stereo tape recorder, it's ideally suited for those with preamp outputs. Entire unit is housed in 3 × 5 × 7-inch aluminum chassis box, with the majority of components mounted on single perfboard. Pushbutton switches S1 and S2 in the tone generator are mounted on top of the case.

the electronic control signals that automatically cycle the projector via the synchronizer circuit.

How It Works. The circuit consists of two independent tone-operated relay circuits. One (built around transistor Q3) responds to a tone of 1000 Hz (this cycles the projector one slide forward); the other (built around transistor Q4) responds to a tone of 100 Hz (this cycles the projector one slide backwards). Each of these circuits is made up of a frequency-selective filter network and a relay driver/amplifier transistor (actually a dual transistor mounted inside a single case). Both circuits are driven by a two-transistor tone amplifier stage formed by transistors Q1 and Q2.

In operation, the circuit monitors the output of the "control" channel—it can be either left or right—of the recorder. When it "hears" either a high-frequency or a low-frequency tone, the appropriate relay closes and activates the changer mechanism in the projector.

Where do the tones come from? You place them on the control channel at the same time you record the narration on the other channel.

Fig. 6-16. Supermatic T/S Synchronizer parts list.

B1, B2, B3—8.4-V mercury battery (Mallory TR146X, Burgess H-146 or equiv.)
C1—2-uF, 15-VDC electrolytic capacitor
C2—.005-uF, 15-VDC disc capacitor
C3—40-uF, 15-VDC electrolytic capacitor
C4, C6, C7—.1-uF, 100-VDC paper capacitor
C5—1-uF, 15-VDC electrolytic capacitor
D1, D2—Silicon diode (Motorola HEP-135 or equiv.)
J1, J2—Phono jack
J3—Two-circuit audio jack
K1, K2—Spdt general-purpose relay; 2500-ohm coil, 7 mA pull-in current (Ohmite GPRX-82T or equiv.)
Q1, Q2, Q6—Silicon transistor (GE 2N5172)
Q3, Q4—Darlington transistor (GE 2N5306)
Q5—Unijunction transistor (Motorola HEP-310)
R1—25,000-ohm, linear-taper potentiometer
R2—22,000-ohm, ½-watt resistor
R3—2200-ohm, ½-watt resistor
R4—10,000-ohm, ½-watt resistor
R5, R7, R8, R16—1000-ohm, ½-watt resistor
R6, R9—820-ohm, ½-watt resistor
R10—100,000-ohm, ½-watt resistor
R11—4700-ohm, ½-watt resistor
R12—68,000-ohm, ½-watt resistor
R13—270-ohm, ½-watt resistor
R14—10-ohm, ½-watt resistor
R15—3300-ohm, ½-watt resistor
S1, S2—Spst pushbutton switch
S3, S4—Spst toggle switch
Misc.—3 x 5 x 7-in. aluminum chassis box, perf board, push-in terminals, battery connectors, two-circuit plug for J3, angle bracket, wire, solder, hardware, etc.

To do so, you simply activate the two-frequency tone generator built into the circuit. This generator consists of unijunction transistor Q5 (which is wired in a relaxation oscillator circuit) and output transistor Q6. By pushing either one of the two pushbutton switches (S1 or S2), you generate either the high- or low-frequency tone. The generator has its own power switch (S3), since it is not used when the control circuit is working, and vice versa.

The circuit will work best with a tape recorder that has a preamplifier output jack. This way, the preamp output signal of the control channel can be fed directly to the synchronizer circuit. Though the circuit will work with the input connected (via a shielded lead) across the speaker terminals of a tape recorder's power amplifier, the volume control usually must be turned up rather high in such cases. Further, the control signals themselves will be audible. Therefore, if possible, it's best to disconnect the speaker and temporarily replace it with a dummy load resistor having approximately the same value as the speaker's nominal impedance (generally 4 ohms). Do *not* use this resistor if the circuit is connected to the preamp output.

Building It. The entire circuit can be housed in a 3 × 5 × 7-inch aluminum chassis box with room to spare. All components (except the panel-mounted controls, relays, and jacks) are wired on a piece of perforated phenolic chassis board, using push-in terminals as soldering points.

Work carefully, and be sure to observe polarity when you mount the diodes and the electrolytic capacitors. And solder quickly, with a miniature-tipped iron when you mount the semiconductor components—these are easily damaged by excessive heat.

We chose a battery supply rather than ac-operation for two reasons:

☐ The relatively low current drain makes battery power economical. A set of batteries should last through well over a year of slide shows.

☐ The use of batteries simplifies hum pick-up problems and eliminates complex grounding requirements.

Note that you must use the 8.4-V mercury batteries called for in the parts list.

If you wish, you can mount the batteries in individual holders. Considering their long service life, though, it's just as effective to tape the three batteries together into a single battery pack and cement it in place with a dab of contact cement. Use snap-on connectors to wire the batteries to the circuit. At replacement time, simply break loose the cement bead and install a new set.

The relay contact wiring shown in the diagram will control Kodak Carousel projectors as well as others using a three-wire control system. Essentially, the mechanism cycles forward or backward when either the forward or backward (i.e., reverse) control wire is connected—for a brief time—to a common control wire.

Easiest way to connect the device to your projector is to buy an extra hand-held control unit and cut off the hand-switch assembly. Use an ohmmeter or continuity checker to determine which wire in the cable controls forward motion, which controls backwards motion, and which is the common. Do this by connecting the ohmmeter across different pairs of wires leading into the switch unit as you press the buttons.

Next, connect the cut end of the cable to a three-conductor (two-circuit) audio plug so the appropriate cables are routed to the appropriate relay contact terminals as indicated on the diagram.

Using It. To record the control signals, connect the device's output jack to either your recorder's mike or line input, for the channel you've chosen as control channel. Set the recorder's input gain control for this channel high enough so that the tones just overload the recorder (the distortion lamp comes on, or the vu meter reads in the yellow-red region, when you press either S3 or S4). Be sure to use shielded cable between recorder and synchronizer.

As you record the narration on the other channel, press either S1 or S2 to place a forward or reverse command on the tape, as desired. Hold the switches down for slightly longer than the time you would hold down the buttons on the hand-held control unit if you were working the projector yourself.

To control the projector, connect the synchronizer's jack to the control channel's output (as already described), and plug the projector's control cable into the three-conductor panel-mounted jack, J3.

Incidentally, input control R1 is provided for use with tape machines that don't have output level controls. If your machine has one, simply set R1 for maximum resistance (minimum attenuation) and bring up the output level until the circuit activates the projector reliably. If the machine doesn't have an output control, set R1 to minimum resistance and back off its setting until the synchronizer works properly.

PHOTO TIMER

Featuring illuminated digit-set dials, automatic reset, and safelight control, the Photo Timer eliminates error-prone juggling of room light, safelight, and timer switches and dials. You can set the timer in complete darkness and you can be sure the safelight was off when you used your enlarger printmeter. The large easily-read dial indications make the time a joy to use. The timer also includes push-to-start and push-to-stop buttons.

Using the 555 precision IC timer, the timer circuit is not affected by line voltage changes. Timing is adjustable from 1 to 119 seconds in one-second steps. Accuracy and repeatability depend only on the accuracy of the timing resistors and quality of the timing capacitor. The Photo Timer is easily constructed at low cost.

Circuit Operation. The schematic diagram shows a 555 precision timer connected as a one-shot timer with automatic reset. The timing interval is determined by timing capacitor C1 and by timing resistors selected by switches S1 and S2. Assuming pin 5 of IC1 is disconnected from calibration pot R9, the time interval T (seconds) equals 1.1 times R (megohms) times C (microfarads). Timer-output at pin 3 controls both normally-off load relay K1 and normally-on load R6. If one load is deenergized, the other is energized and vice-versa.

Refer to Figs. 6-17 and 6-18.

With C1 initially held discharged by IC1, timing commences when start button S4 is depressed causing a triggering pulse at trigger pin 2. The relay closes instantly and C1 begins to charge through the timing resistor. When the voltage of C1 rises to two-thirds of the dc supply voltage, IC circuits are activated causing the relay to open and C1 to discharge completing the cycle with automatic reset. A timing cycle in progress may be terminated by depressing stop button S3.

Calibration pot R9 varies the timing control voltage at pin 5 accounting for tolerances of timing capacitor C1. Provided with both normally-on and normally-off loads, the IC circuit draws a fixed load current from the power supply. Resistor R7 sets the dc supply voltage to about thirteen volts. Voltage clamp zener diode D2 limits the supply voltage to safe values if the supply voltage should rise. Timing is not affected by changes in supply voltage. Rectifier diode D1 eliminates voltage spikes at K1 which would re-cycle the timer.

Construction. Build the Photo Timer in a 9 × 5 × 3 inch metal cabinet. Begin construction by cutting out two 2½ inch dial discs from 1/16-inch thick red or white translucent plastic. The discs are easily cut using a holesaw. Chuck the discs in a mandrel and true up the

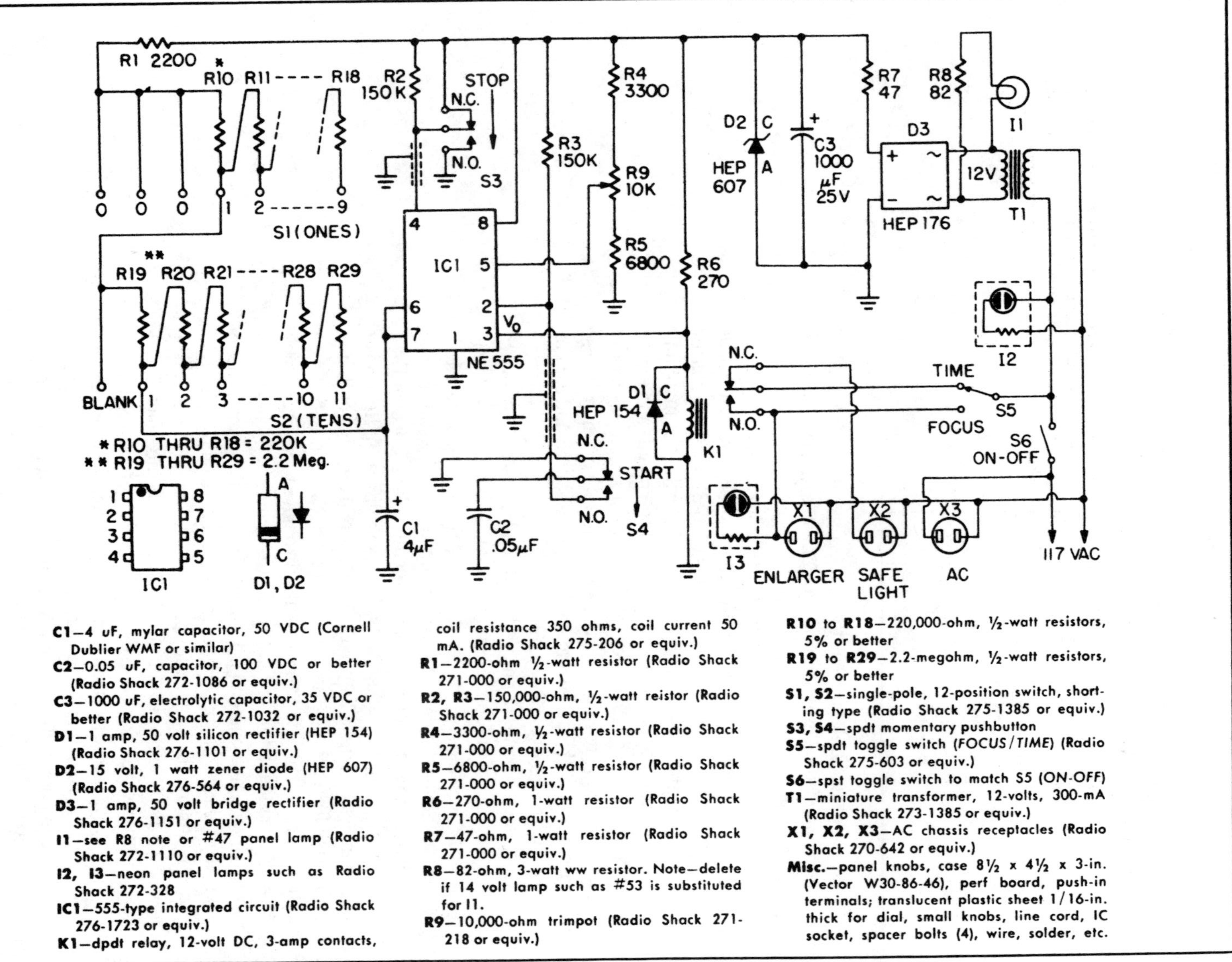

C1—4 uF, mylar capacitor, 50 VDC (Cornell Dublier WMF or similar)
C2—0.05 uF, capacitor, 100 VDC or better (Radio Shack 272-1086 or equiv.)
C3—1000 uF, electrolytic capacitor, 35 VDC or better (Radio Shack 272-1032 or equiv.)
D1—1 amp, 50 volt silicon rectifier (HEP 154) (Radio Shack 276-1101 or equiv.)
D2—15 volt, 1 watt zener diode (HEP 607) (Radio Shack 276-564 or equiv.)
D3—1 amp, 50 volt bridge rectifier (Radio Shack 276-1151 or equiv.)
I1—see R8 note or #47 panel lamp (Radio Shack 272-1110 or equiv.)
I2, I3—neon panel lamps such as Radio Shack 272-328
IC1—555-type integrated circuit (Radio Shack 276-1723 or equiv.)
K1—dpdt relay, 12-volt DC, 3-amp contacts, coil resistance 350 ohms, coil current 50 mA. (Radio Shack 275-206 or equiv.)
R1—2200-ohm ½-watt resistor (Radio Shack 271-000 or equiv.)
R2, R3—150,000-ohm, ½-watt reistor (Radio Shack 271-000 or equiv.)
R4—3300-ohm, ½-watt resistor (Radio Shack 271-000 or equiv.)
R5—6800-ohm, ½-watt resistor (Radio Shack 271-000 or equiv.)
R6—270-ohm, 1-watt resistor (Radio Shack 271-000 or equiv.)
R7—47-ohm, 1-watt resistor (Radio Shack 271-000 or equiv.)
R8—82-ohm, 3-watt ww resistor. Note—delete if 14 volt lamp such as #53 is substituted for I1.
R9—10,000-ohm trimpot (Radio Shack 271-218 or equiv.)
R10 to **R18**—220,000-ohm, ½-watt resistors, 5% or better
R19 to **R29**—2.2-megohm, ½-watt resistors, 5% or better
S1, S2—single-pole, 12-position switch, shorting type (Radio Shack 275-1385 or equiv.)
S3, S4—spdt momentary pushbutton
S5—spdt toggle switch (*FOCUS/TIME*) (Radio Shack 275-603 or equiv.)
S6—spst toggle switch to match S5 (ON-OFF)
T1—miniature transformer, 12-volts, 300-mA (Radio Shack 273-1385 or equiv.)
X1, X2, X3—AC chassis receptacles (Radio Shack 270-642 or equiv.)
Misc.—panel knobs, case 8½ x 4½ x 3-in. (Vector W30-86-46), perf board, push-in terminals; translucent plastic sheet 1/16-in. thick for dial, small knobs, line cord, IC socket, spacer bolts (4), wire, solder, etc.

Fig. 6-17. Photo Timer schematic and parts list.

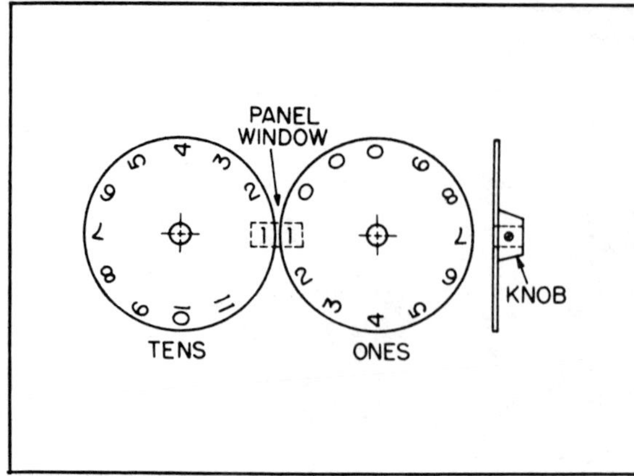

Fig. 6-18. Digit-set discs can be made at home from a plastic sheet. For a neat job, use press-type numerals after cutting out the discs, if you draw a temporary base line for each double digit and use it to align numbers.

edges. Ream the center hole to clear the shafts of switches S1 and S2. Drill through a pair of small panel knobs and cement a knob to each disc using epoxy cement.

Drill suitably spaced (one disc diameter) slightly undersize holes in the panel and ream for a close fit for the shafts of the switches. Cut the perforated board to size and drill four holes for 6-32 × 2-inch spacer bolts which support the board behind the panel. Drill four matching holes in the panel, bolt the board directly against the panel, and locate and machine holes in the board to accept the switches. Cut out a ⅜ × ¾-inch window in the panel midway between the switch shaft holes. Complete machining of the panel and apply panel labelling and a clear protective coating.

Install switches S1 and S2 on the board, dial discs on the switch shafts, and trial mount the assembly on the panel. The dial discs should rotate with little wobble and no contact with the panel. If needed, enlarge a switch hole on the board to correct any disc tilt by shifting the switch slightly. Remove the circuit board assembly from the panel and affix the discs at the top end of the shafts. This simplifies application of dry transfer numerals at the edges of the disc while using the switches to index the disc for each position. Label the "ones" dial with three zeros and 1 through 9. Label the "tens" dial 1 through 11 leaving a blank space. You can remount the assembly on the panel and check and correct any badly aligned numerals. Using 1/16 inch aluminum, make the compartment partition supporting transformer T1 and relay K1. The partition is secured by two of the spacer bolts. Cut out a portion of the flange of the partition to avoid interference with the "ones" dial disc. Make a bracket to accept the socket of K1 and affix to the partition. Wire the ac sockets, neon panel lamps (supplied with external voltage dropping resistors), and toggle switches before installing T1 and K1. Wire the normally open poles of the DPDT relay in parallel to double the current rating.

Install a large rubber grommet on the circuit board directly behind the panel window to accept panel lamp I1. Tint the lamp with red transparent lacquer. Complete wiring of the board using a socket for IC1. Carefully observe polarities of D1, D2, and C3. Use shielded wire for connections to pushbuttons S3 and S4. Install resistors R10 through R29 directly on the switches. It's usually a simple operation to defeat the switch detent stops on S1 and S2 allowing continuous rotation of the dials. Set the switches to pick up R10 and R19 and position and secure the dial discs for 11 seconds readout.

Capacitor C1 should be a mylar, polycarbonate, or polystyrene low-leakage, low-loss type. C1 was made up by connecting two 2 μF capacitors in parallel but you can use a single 4 μF capacitor. You can use a 5 μF capacitor by changing R10 through R18 to 180,000 ohms and R19 through R29 to 1.8 megohms.

Checkout And Calibration. Using a VOM, verify the presence of approximately thirteen volts dc across C1, about 50 mA current in R7, and about five volts ac across lamp I1. If you have substituted for T1, it may be necessary to resize R7 and R8 accordingly. To calibrate the Photo Timer, plug a sweep second electric clock into socket X1. Turn S6 on and set S5 to time. Set the dials for fifteen seconds. Depress start button S4 and observe elapsed time on the clock. By trial settings, set R9 so that the clock runs for fifteen seconds. Next, set the dials for 119 seconds and observe elapsed time. If you have used high-quality capacitors for C1 the interval should check close to 119 seconds with some allowance for inaccuracy of timing resistors.

Put It To Work. Plug the enlarger into socket X1 and safelight into socket X2. Plug the enlarger exposure meter into socket X3. Set S5 to Focus when focusing or using the exposure meter. The safelight will now be off as is required for use with any enlarger exposure meter. To expose the print to the set time interval, switch S5 to Time and depress the start button S4. During exposure, the safelight will be off but will return automatically upon completion of the exposure. Panel lamp I3 will be on during the exposure interval. If you have inadvertently overlooked setting of the timer or lens opening and have initiated the exposure, you can terminate the exposure with return of safelights by depressing stop button S3.

THUNDERBOLT FOR STOP-ACTION PHOTOS

How would you like to capture the sphere-capped minaret of a drop of water at the precise moment it strikes the surface of a pool, or a bursting balloon with the piercing dart still in mid-air? All you need is this easily-constructed, sound-activated, electronic flash—Thunderbolt.

Sound-activated switches have been around a long time. The first one I built many, many years ago weighed 25 pounds and would have cost nearly $100 if I hadn't cannibalized some old radios for the parts and *tubes*. When I built Thunderbolt a few months ago, it cost just a few dollars and weighed in at about eight ounces. What made the difference? Solid-state components, including a silicon-controlled rectifier, make it lighter and cheaper—and it works much better and faster.

How It Works. Sound picked up by a microphone is boosted by an amplifier which feeds the signal in the form of a rectified pulse (via R3 and D1) to the gate and cathode of the SCR. The SCR is internally like three diodes connected (alternately) in series—positive-negative-positive—so it acts like a conventional rectifier in the reverse direction. Thus, the SCR's forward conduction is controlled by operating the "switch," or *gate*. Since the sound we are picking up is a single, sudden sound of short duration, it acts like a pulse, when magnified by the amplifier, and it causes the SCR to conduct. An electronic flash unit connected across its anode and cathode detects this conduction as a direct short so it flashes.

Refer to Figs. 6-19 and 6-20.

In practice, you will find a wide latitude of application techniques possible. You can control the microphone's sensitivity so it will respond only to certain higher level sounds, if the ambient noise level is high. Additionally, you can select the time at which an event is photographed by varying the distance between the event and the mike.

Various Applications. Let's say, for example, the event to be photographed is a coin dropping into water. By placing the mike very close to the container of water, and by turning up Thunderbolt's sensitivity control, you can freeze the coin as it first touches the water. On the next shot, repeat the event, but place the mike farther away from the point of impact. The sound must now travel farther to reach the mike and the flash will go off at a later stage in the splash sequence.

By repeating this process, you can get a series of shots to cover the entire sequence from the coin first touching the water, to the final catapulted droplet falling back into the water. It could be a club flattening of golf ball, a dart bursting a balloon, a hammer shattering a light bulb, or a (patient) athlete diving into a swimming pool. Any event which produces a sound, faint or deafening, can be recorded on film at the decisive moment chosen by you.

The great advantage of Thunderbolt is that it is totally electronic, as opposed to the electromechanical heavy-weights of a few years ago. The older devices depended on mechanical relays and electromagnets to close a switch. This mechanical energy transfer added milliseconds to reaction time. Even that is a significant interval when you are planning to break up into sequences such events as bursting fire-crackers and shattering lightbulbs. Once the sound gets to Thunder-bolt's mike the reaction time approaches the speed of light. That's about as fast as you're going to get—in *this* world.

Putting Thunderbolt together is easy, because the most compli-cated part—the amplifier—is a module, ready to wire into a circuit with just a few simple connections and a handful of other parts.

Building It. Begin by selecting an amplifier. Almost any inex-pensive module will do as long as it has an output transformer. Note that most modern transistor amps don't have an output transformer. Radio Shack sells suitable amp modules. Any amp capable of delivering a couple hundred milliwatts is sufficient. I scavenged the amp for my Thunderbolt from an old, discarded portable tape player. You can find many of these old reel-to-reel relics in second-hand stores for a dollar or two. All you need do with these old units is carefully trace and identify the input and output leads and the battery supply leads. If you get a unit that's fairly intact, it may even have volume and tone control pots which are of the correct value for your Thunderbolt.

The cabinet shown in the parts list will accommodate almost any transistor amp you select. You could even get ambitious and build a simple transistor amp. Most any old tube amp will also work fine, though it'll be quite bulky.

Just which mike jack, you use will depend on the plug on the microphone you use. It may be a standard phone jack, or the miniature type—whatever, as long as it matches your mike plug. When you have all the parts in hand, arrange them on the cabinet's front panel and mark the panel for the mounting holes to be drilled. Parts placement is not critical, but the leads to R1 and R2 should be kept short. If you locate S1 close to the *sensitivity* control, R2, then you can use point-to-point wiring for the SCR, D1 and R3. They are rigid enough to be self-supporting if the leads are kept short; otherwise a tie-point terminal can be used. Follow the schematic and wiring illustrations carefully and you'll have no trouble. You must use shielded (coax audio cable) for the input connections from R1 to your amp.

Hookup To Flashgun. After making all the connections, double check your work. Be sure you have connected the SCR's three leads correctly and check the polarity of D1. When you are sure everything is in order, you'll need to make a connector cord for your flash unit. Insert

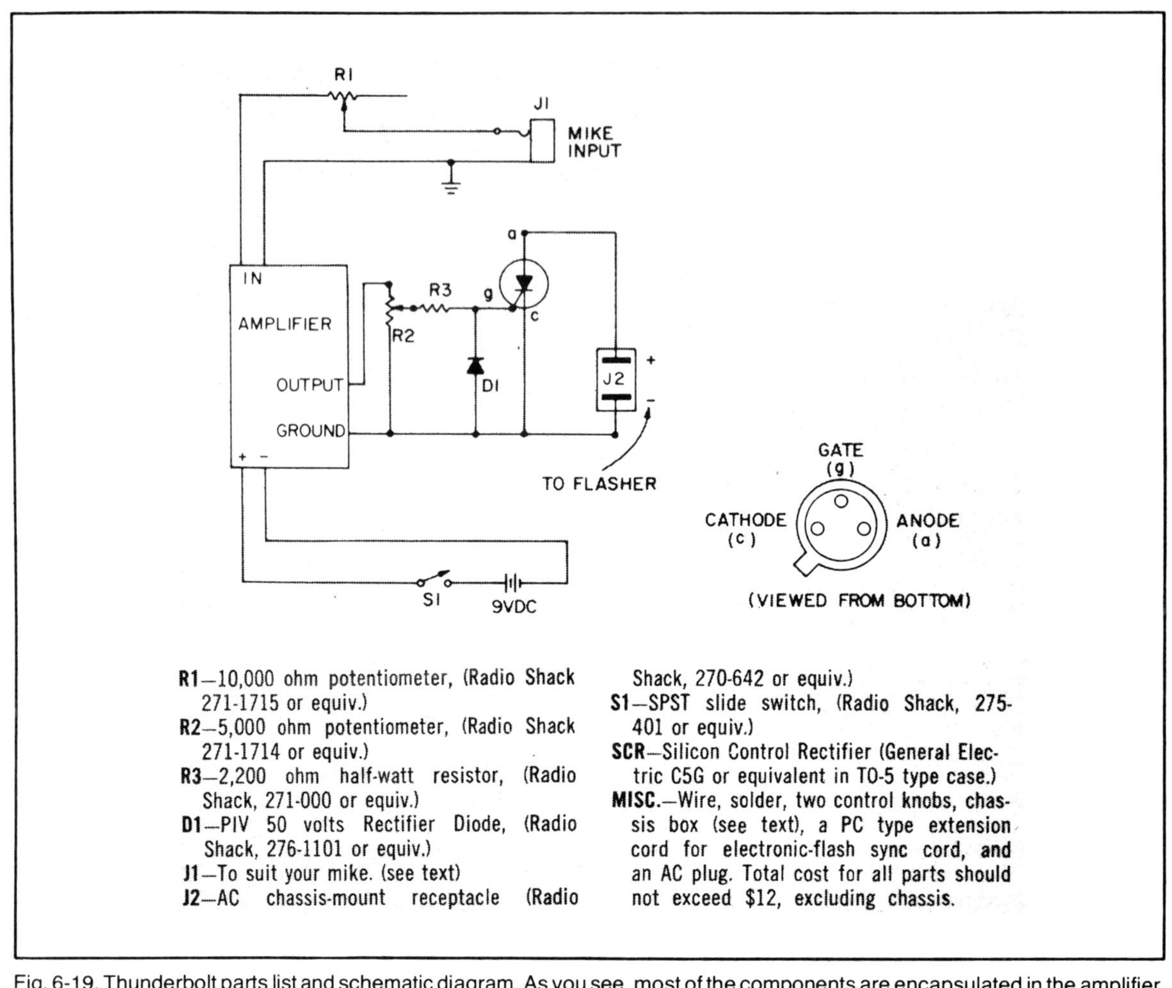

R1—10,000 ohm potentiometer, (Radio Shack 271-1715 or equiv.)
R2—5,000 ohm potentiometer, (Radio Shack 271-1714 or equiv.)
R3—2,200 ohm half-watt resistor, (Radio Shack, 271-000 or equiv.)
D1—PIV 50 volts Rectifier Diode, (Radio Shack, 276-1101 or equiv.)
J1—To suit your mike. (see text)
J2—AC chassis-mount receptacle (Radio Shack, 270-642 or equiv.)
S1—SPST slide switch, (Radio Shack, 275-401 or equiv.)
SCR—Silicon Control Rectifier (General Electric C5G or equivalent in TO-5 type case.)
MISC.—Wire, solder, two control knobs, chassis box (see text), a PC type extension cord for electronic-flash sync cord, and an AC plug. Total cost for all parts should not exceed $12, excluding chassis.

Fig. 6-19. Thunderbolt parts list and schematic diagram. As you see, most of the components are encapsulated in the amplifier, which is described fully in the text. Virtually any amplifier will do, as long as it produces a couple hundred milliwatts. The diagram below shows the connections to SCR1.

the PC plug of your flash extension cord into the flash unit's sync cord. Both ends of some brands of extension cords look almost alike and you don't want to cut off the wrong end. With one end plugged into your flash unit to make sure, cut off the other end close to the plug. Strip off the insulation and carefully separate the braided shielding from the inner conductor of the coax cord. There is little or no standardization in the photo industry, so you can't be sure that the inner conductor of any given sync cord is connected to a positive voltage when plugged into a flash unit. Some units have a positive ground and some have a negative ground. In order to make sure your Thunderbolt will work with any flash unit, you need a plug which can be reversed for any polarity match. You may have more than one flash unit and they may not be the same, hence the adapter cord.

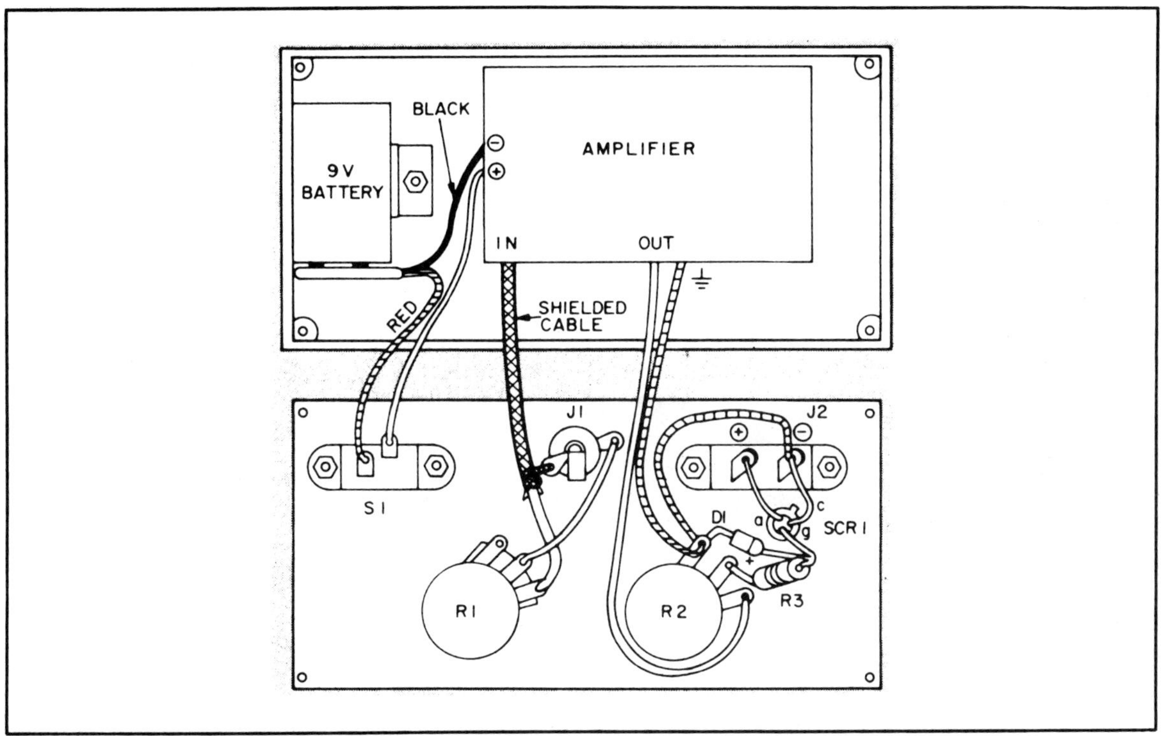

Fig. 6-20. Diagram of the parts layout for Thunderbolt. The components are pretty evenly distributed between the front panel and the box, and they are connected by a few hookup wires and some shielded cable. Make sure that the wires and the cable are long enough and flexible so that there won't be any trouble as you close the box after construction.

Plug one end of your modified cord into Thunderbolt and the other into the flash unit's sync cord. Set the *sensitivity* control, R2, at the center of its rotation and *input* control, R1, fully counterclockwise. Plug in a microphone and apply power to both the switch unit and the flash unit. The flash may fire once or twice by itself before it settles down. If the flash unit keeps firing as fast as it recycles, reverse the plug in J2 to get the correct polarity match.

With the polarity established, whistle or hum into the mike as you slowly turn R1 clockwise. This flash should go off. From this point, it's a matter of see-sawing controls R1 and R2 back and forth until you get the hang of your mike's sensitivity. The best way to discover what your Thunderbolt can do is to use it in a closely controlled test set-up. This procedure is easier with an assistant, so recruit a friend.

Against a dark background, set up a clear glass, or bowl, of water. Place the mike as near to the bowl as possible without it getting into the picture area. Position the flash on a tripod and aim it at the bowl. The camera, also tripod mounted, should be aimed at the bowl at a 45 degree angle to the flash. Focus on the surface of the water and compute your f-stop as you normally would for a flash shot using the flash's guide number divided by its distance to the subject. (For instance, if the manufacturer's recommended guide number for your flash is 45 when used with ASA 25 film and your flash is placed five feet from the subject, divide 45 by 5. Since the answer is 9, choose the closest f-stop, which is f-8).

Set the camera's speed control on "B" as you would for a time exposure. Attach a locking type shutter release cable to the camera and position your assistant close to the bowl, but out of the camera's field of view.

Turn off all the lights in the room and open the lens with the shutter release cable, but *do not* remove the lens cap yet. With your assistant poised over the bowl, ready to drop a coin into the water, turn on the flash unit and the switch unit. The flash may go off, triggered by the sound of its own switch, which is why you've left the lens cap on. Wait for the flash to recycle, then snap your fingers. It should go off again. When it recycles, remove the lens cap and give your assistant a visual signal to drop the coin. As the coin hits the water, the flash will go off and you can close the shutter and replace the lens cap.

On the next shot, move the mike a foot or two farther away and repeat the process. On successive shots, move the mike exactly the same distance farther away. You should have a complete sequence after about six to eight shots.

The film should be a very slow film—that is one with a very low ASA number, such as Plus-X by Kodak, which has an ASA of 125. If you have a setup that requires you to have more room illumination in which to work, use an Ortho-type film which is insensitive to red light. With this film, you can use a fairly bright red light in the room without affecting the film image while the lens is open.

Once you've done a series such as the water bowl and coin, you will know what Thunderbolt can do for you and how to predetermine its sensitivity to a given sound. When you have all its parameters for operation understood and set up, you can start thinking of things to do with Thunderbolt. Its applications are virtually limitless, since the principle of stopping sound-producing motion is an especially fascinating one. You can use it indoors in ordinary ways, such as the coin and bowl technique, or you could even use it for crime detection, by fixing it at night on a window or door you expect an intruder to come through. Any sound he makes will take his picture. Good luck!

DARKROOM PRINTING METER

Try to grind out wallet-size prints or enlargements from a full 36-exposure roll in only one evening and you'll know just how frustrating life can be. Every change in magnification and negative density means a different exposure. And if you use test strips or exposure guides to hit the correct exposure you're making at least two prints for every one you need.

Refer to Figs. 6-21 through 6-24.

The way to take all this drudgery out of your darkroom work is to use an electronic printing meter, a device that takes only seconds to indicate the correct exposure, regardless of whether the enlarger is at the top or bottom of the rack, or whether the exposure and negative development is over or under.

A quick example will illustrate how easy it is to make prints with a printing meter. Let's assume you have just chocked the negative in the enlarger and have cropped the picture exactly the way you want it. Now you take the probe from a printing meter—which you have previously calibrated for a 10 or 20-second exposure—place it on the easel at the point of maximum light transmission through the negative (the black reference in the print—deepest shadow) and adjust the lens diaphragm until the printing meter's pointer indicates some reference value you have previously selected.

That's the whole bit. Expose the paper for your normal 10 or 20-second exposure and the first print will be a good print—maybe even a great print. If you're grinding out wallet-size jobs for the whole family, each print from each frame will have the same excellent quality.

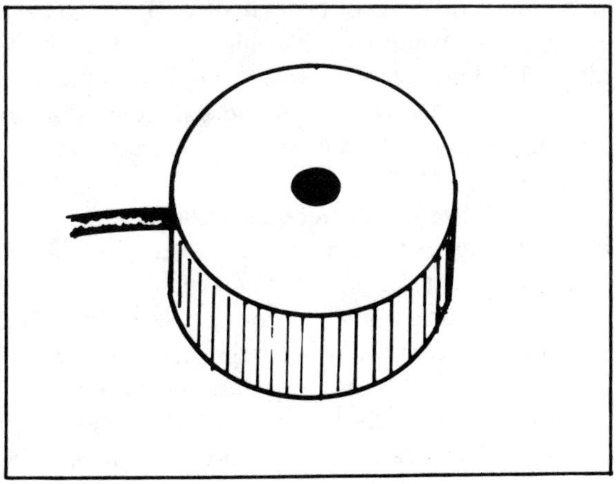

Fig. 6-21. The sensor is really a large tuning knob with photoresistor PR1 embedded in epoxy, plastic, or RTV rubber adhesive.

A Hint. The key to successful use of a printing meter lies in the fact that, except for some particularly artistic work, any print will look decent to excellent if there is some deep black, even if it's just a spot of black; for the black to highlight or border-white contrast gives the visual appearance of a full contrast range, even if the greys are merged. For those who do portraiture, a printing meter can be user-calibrated for "flesh tones."

The printing meter shown in the photographs has been especially designed for construction and use by the typical photographer/ electronics hobbyist. It features a calibration—called "speed"—adjustment to accommodate slow to fast enlarging papers (such as Polycontrast and Kodabromide) and readily available parts, many of which will be found in the typical experimenter's junk box. The layout is not critical—any cabinet can be substituted; there are no critical shielded circuits (not even shielded wire is used); and except for the photoresistor sensor, just about any component quality will do. There is absolutely no sense in building the project with the best components money can buy because the best components won't affect the final performance one iota.

Construction. The unit shown is assembled in a 5¼×3×5⅞-inch metal utility cabinet. Connecting jack J1 is optional as the photoresistor sensor, PR1, can be hard-wired into the circuit. If you use a jack, note that it must be the three-terminal type such as is used for stereo connections; the ground connection is not used since neither PR1 lead is grounded. Do not use an ordinary phone or phono jack as they will ground one of the PR1 leads. Plug P1 must similarly be a matching three-terminal stereo type. Either miniature or full-size jacks and plugs can be used.

Power switch S1 can be anything you care to use—lever, slide, or toggle. Use the least expensive slide switch if you're trying to keep the cost down.

The meter, M1, is an illuminated 0-1 mA S-meter. This meter was selected because it has built in pilot lamps with 6- and 12-volt connections. When 12-volt-connected to T1, which is 6 volts, the pilot lamps are dim enough not to affect the sensor and bright enough so that you can see the pointer in the darkroom. Meter M1 mounts in a 1½-inch hole, which can be cut with a standard chassis punch (if you have the punch).

Sort Them. The meter scales are jammed with numerals that can be confusing in the darkroom so the best bet is to paint out the unwanted "calibrations" using Liquid Paper, or Liquid KO-REC-TYPE, products used to correct typewriter errors (available in stationery stores). First, snap the plastic cover off the meter. It might feel secure but it's not. Grasp the top of the cover and force the cover outward and down, taking care that when it snaps free the pointer isn't damaged. Next, remove the scale by taking out the two small screws and sliding the scale out from under the pointer. Do not attempt to paint the scale while it is mounted in the meter as a single drop of the fast-setting correction fluid can ruin the meter if it gets into the pivot bearing. When reinstalling the scale, hold the screws with a tweezer or long-nose pliers until you "catch" the first few threads. When the scale is secure, snap the meter's cover into position. (On the unit shown all scales and markings other than 0-to-1 have been painted out, as the 0-to-1 scale is the most convenient to see under dim lighting.)

Note that meter M1, power switch S1, and jack J1 have been positioned on the front panel so as to provide the maximum room for the speed control's calibrated knob. Use the largest possible knob as the greater the calibrations the easier it is to reset the control to a desired paper speed.

Power transformer T1 can be any 6.3-volt filament transformer rated 50 mA or higher. (A 6-volt transformer scrounged from a portable cassette recorder will work just fine.)

Power Filter. If the line voltage in your home is known to be reasonably constant, assemble the unit as shown in the schematic. If your local utility likes to bounce the line voltage to vary (indicated by dimming lights), install zener diode D5 across points A and B. The zener will provide a regulated 6 volts, with the slightly lower circuit voltage (6 Vdc rather than 9 Vdc) providing slightly reduced sensitivity. Normally, you will not need D5, so there's no need to get it unless you're certain you need it.

In order to get speed control R2 to increase sensitivity in the expected clockwise direction, its ground terminal is opposite to the usual volume control ground. Facing R1's shaft with its terminals sticking up, the ground terminal is the one on the left.

Meter M1 has five terminals. The one designated "+" and the one adjacent to it are the meter terminals. The three terminals above the meter terminals are the pilot lamps. The extreme end pilot lamp

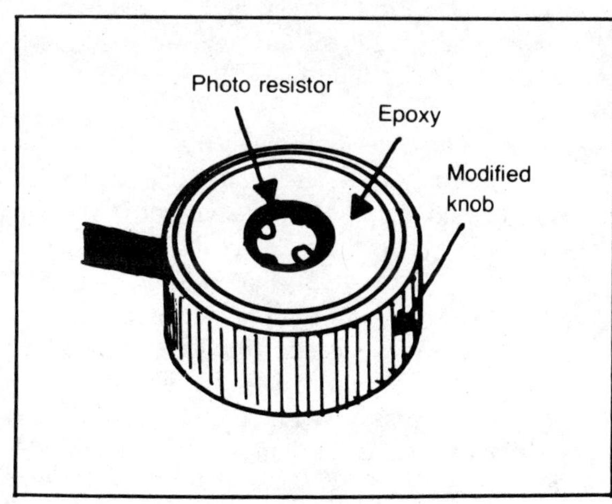

Fig. 6-22. After the sensor is completed, punch a hole in a matching cardboard disc and cement the disc over the sensor. The hole provides a smaller sensitive area required for prints 4 × 5 inches or smaller. Better results with larger prints are also obtained with the mask.

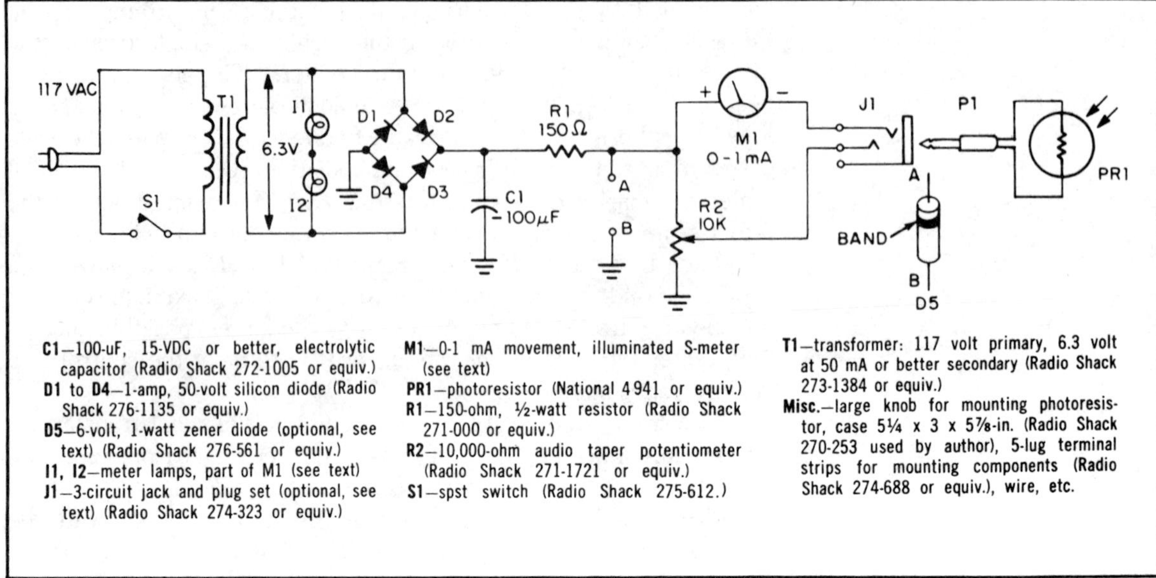

C1—100-uF, 15-VDC or better, electrolytic capacitor (Radio Shack 272-1005 or equiv.)
D1 to **D4**—1-amp, 50-volt silicon diode (Radio Shack 276-1135 or equiv.)
D5—6-volt, 1-watt zener diode (optional, see text) (Radio Shack 276-561 or equiv.)
I1, I2—meter lamps, part of M1 (see text)
J1—3-circuit jack and plug set (optional, see text) (Radio Shack 274-323 or equiv.)
M1—0-1 mA movement, illuminated S-meter (see text)
PR1—photoresistor (National 4941 or equiv.)
R1—150-ohm, ½-watt resistor (Radio Shack 271-000 or equiv.)
R2—10,000-ohm audio taper potentiometer (Radio Shack 271-1721 or equiv.)
S1—spst switch (Radio Shack 275-612.)
T1—transformer: 117 volt primary, 6.3 volt at 50 mA or better secondary (Radio Shack 273-1384 or equiv.)
Misc.—large knob for mounting photoresistor, case 5¼ x 3 x 5⅞-in. (Radio Shack 270-253 used by author), 5-lug terminal strips for mounting components (Radio Shack 274-688 or equiv.), wire, etc.

Fig. 6-23. Darkroom Printing Meter schematic and parts list.

terminals are the 12-volt connections. The center terminal is not used for the 12-volt connection.

The Eye. The only assembly that requires some care is the sensor. The sensor itself is a photoresistor; however, the photoresistor doesn't have enough left to maintain its position on the easel, so it must be mounted in a support that can maintain its position without falling over. The sensor assembly shown consists of PR1 epoxy-cemented into a relatively large knob. The knob must be plastic—not metal, though it can have a metal decorative rim—and it's best if there is a recess on the top even if the recess is produced by a rim. Remove the set screw and drill out the set screw hole with a bit approximately 3/16 inch (not critical). Then, using a 3/8-inch bit, drill through the shaft hole clear through the top of the knob. If the shaft hole has a brass (or other metal) bushing make certain the drill bit removes all the metal.

Pass the PR1 leads through the hole in the knob from the top. Tape it in position. Feed a section of line cord or speaker wire through the setscrew hole and solder the wires to PR1 as close as possible to the knob. Trim away the excess PR1 leads; they should not protrude below the knob. Remove the tape holding PR1, get PR1 as close to the center of the knob as possible, and then pour in a quantity of fast-setting epoxy or liquid plastic from a knob repair kit or plastic modeling kit, and let it set a few minutes until the plastic hardens. Keep the level of the epoxy or plastic below the top of PR1—use less rather than more. Similarly, pack the bottom of the knob with epoxy, plastic or rubber.

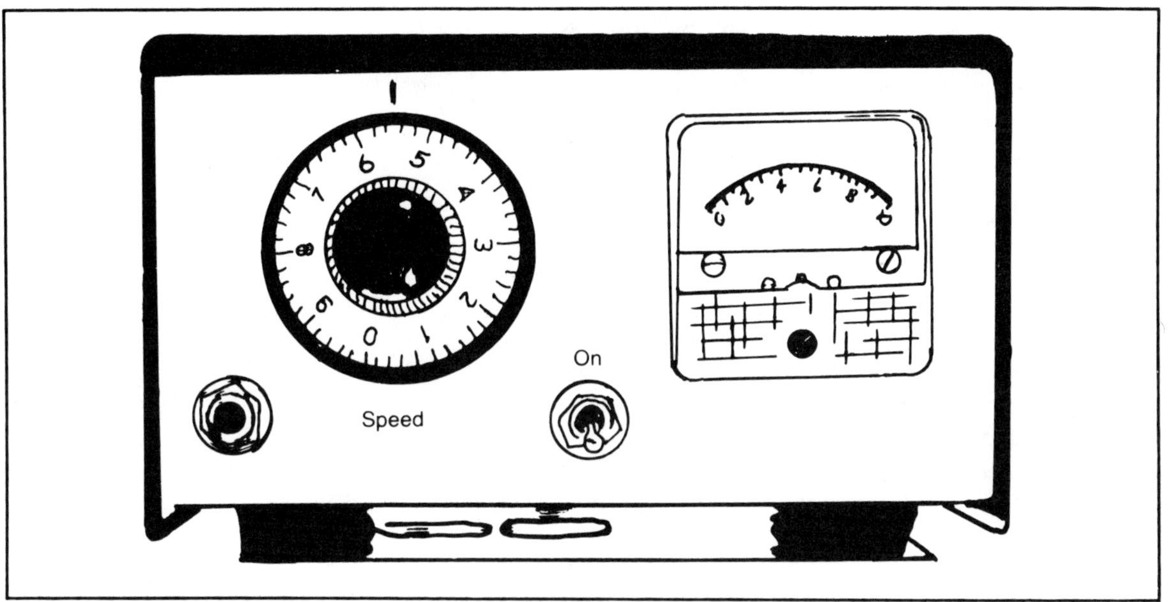

Fig. 6-24. Use the largest calibrated knob you can install without interference by other panel components. The greater the calibration area on the knob is, the easier it is to preset the paper speed with accuracy.

Mask Down. Now, the surface area of the photoresistor is too large for small prints—4 × 5 or smaller—and even some 8 × 10s. So cut a disc the diameter of the knob from skirt cardboard or a manila file folder (but not oak-tag) and using a standard hand punch (such as used in schools) punch a hole in the center of the disc. Apply rubber cement to the rim of the knob and the inside rim of the disc on the knob so the hole exposes a small part of the photoresistor's surface. It's not all that critical; the hole doesn't have to be precisely over the center of the photoresistor. However, the unit is calibrated for a punch-size hole and might not work properly if the disc is not used, or if the hole is a hand-made pinhole. Use the punch.

Using the Meter. The first step is to select a decent reference negative and make a good print using a 10, 15, or 20-second exposure. We suggest 20 seconds as it will become your standard exposure, and will be sufficiently long to allow moderate dodging. When you are certain you have a print exactly the way you want it, and without disturbing the enlarger's controls, place the printing meter's sensor under the *brightest* light falling on the easel—it produces black (maximum shadow) on the final print. Now turn on the printing meter and allow about five seconds for warm up. Adjust speed control R2 so the meter pointer indicates any meter reading you want to use as a reference. It doesn't matter what the reading is as long as you always use the same reference for the standard exposure time. For example, 0.2 on the meter scale is a good choice because it is well illuminated by the meter lamps. But you might just as easily select mid-scale as the reference meter reading. It doesn't make any difference; just be consistent.

Once you have adjusted the speed control for the reference meter reading, note on a piece of paper or in a notebook the dial reading from the speed control's calibrated knob. This is the reference speed value for the particular printing paper. For example, let's say you made the test print on Polycontrast using the No. 2 filter, and the speed knob indicates 5.6. Next time you want to print using Polycontrast with a No. 2 filter you simply set the speed knob to 5.6, put the sensor under the darkest shadow area and adjust the lens diaphragm for a reference meter reading. Everything will be set for your standard exposure time.

Changing Filters. Kodak provides a speed rating for all their papers and you can easily work out the correct (or close) speed control setting without making a "perfect" test print for each type and grade of paper. For example, changing from a No. 2 to No. 4 filter usually means increasing the exposure by a 3.5X factor. If your No. 2 exposure is 10 seconds, the No. 4 exposure will be 35 seconds—somewhat long. You can, however, open up the lens diaphragm for a 3.5X light increase (close enough value) and adjust the speed control for the reference meter reading. The new speed control setting is the speed value for the No. 4 filter. You can do this with variable contrast filters or numbered printing paper.

While the most pleasing print usually has some black, there are times when there can be no black, such as snow scenes and portraits. You can peg the speed control's calibration to a grey corresponding to a skin tone, or any other degree of grey you might desire. The only thing you cannot do is calibrate the meter for highlights, because the meter might not have enough sensitivity for slow papers, and highlights can completely fool the meter.

If desired, you can take a speed control calibration reading for each type of paper (using your standard negative) for both shadow detail and intermediate grey. This way, you can quickly set up for typical snapshots, scenics, or portraits.

Keep In Mind. The sensor has a slight light memory, so turn the sensor face down when not being used and turn the power switch on and off in the dark, though you can keep the darkroom illuminated by a safelight with the power switch on. Meter readings, however, must be taken with all room lights off; only the enlarger should be on and the print meter should be positioned so that its meter lamps do not illuminate the sensor (even slightly).

FLASH TESTER

Even if you spend $18 or $20 for a super-duper professional remote flash tripper, you'll get little more than this two-component circuit. Cost is important if the results are equal.

Transistor Q1 is a light-activated silicon-controlled rectifier (LASCR). The gate is tripped by light entering a small lens built into the top cap.

Refer to Fig. 6-25.

To operate, provide a 6-in. length of stiff wire for the anode and cathode connections and terminate the wires in a polarized power plug that matches the sync terminals on your electronic flashgun (strobelight). Make certain the anode lead connects to the *positive* sync terminal.

When using the device, bend the connecting wires so the LASCR lens faces the main flash. This will fire the remote unit.

No reset switch is needed. Voltage at the flash's sync terminals falls below the LASCR's holding voltage when the flash is fired, thereby turning *off* the LASCR.

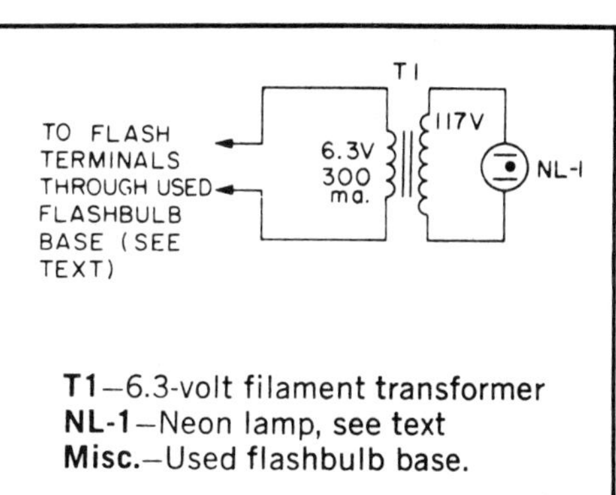

T1—6.3-volt filament transformer
NL-1—Neon lamp, see text
Misc.—Used flashbulb base.

Fig. 6-25. Flash Tester schematic and parts lists.

PHOTOFLOOD DIMMER

Professional quality photographic lighting requires complete control of the studio lights, and that's just what you'll get with the pro-type, full-range 500-watt dimmer. Each one can handle one 500-watt No. 2, or two 100-watt No. 1 photoflood lamps, and the lighting range can be adjusted from full off to full on. Refer to Fig. 6-26.

Triac Q1 must be mounted to a large heatsink, preferably the metal cabinet used to house this dimmer. Make certain you insulate Q1 from the cabinet.

Fuse F1 *must* be used, otherwise, the surge current that occurs

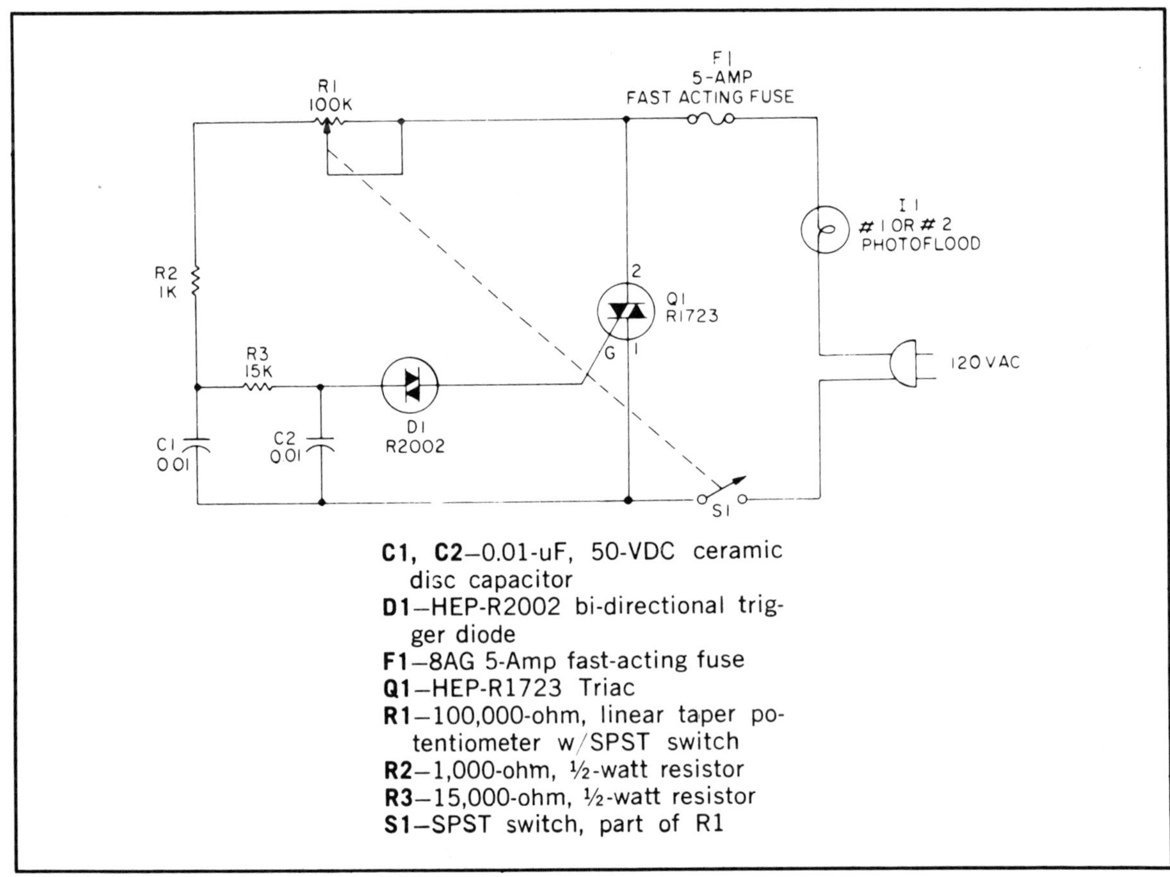

Fig. 6-26. Photoflood Dimmer No. 1 schematic and parts list.

when 500-watt photofloods burn out will instantly destroy Q1. F1 must be a fast-acting fuse such as the type 8AG. The slower fuses such as the 3AG and the slo-blo offer no protection. Switch S1 is part of intensity adjustment R1, and R1 should be wired so it represents maximum resistance just before S1 switches off. (While S1 cannot normally handle a 500 watt load, in this circuit, it switches when the lamp is off and has no trouble handling any size photoflood.)

Chapter 7

Novelty Projects

BUILD THE LI'L WAILER

Slip this small package into your pocket, walk down the street with one of your friends, and push the button; your friend will go crazy trying to figure out where the sound is coming from. You can "raid" your local poker group or let it sound off at a party. We don't recommend using it within earshot of your local gendarme (unless you *want* to attract his attention), however.

Our 'Lil Wailer sounds like a police siren and can be adjusted from a barely discernible cry to a scream that will attract attention for at least 100 feet around. It is rugged, small, and can be built in an evening or two with readily available parts.

Refer to Figs. 7-1 through 7-3.

How it Works. A combination of old and new technology is used in the design of the wailer. The heart of the circuit is the venerable unijunction (UJT) transistor. With power switch S1 on, and with trigger S2 depressed, capacitor C2 charges through Q1 until the level at the emitter of the UJT (Q2) causes it to fire. It discharges through R5 to create the basic siren tone. The voltage applied to the UJT charging circuit is varied to produce the ascending and descending pitch required. As S2 is held, capacitor C1 charges through R1, with the emitter voltage at Q1 "following." Thus, the UJT fires at a faster and faster rate, peaking when C1 is essentially fully charged. When S2 is released, C1 discharges through Q1, causing the voltage to the UJT charging circuit (R2, R3 and C2) to decrease, with the firing rate dropping slowly to zero. R3 provides an adjustment to select the most "authentic" tone.

The sawtooth-like waveform at the emitter of Q2 is then coupled through R6 and R7 to IC1, the LM386 audio amplifier. IC1, a 10-transistor linear amplifier, amplifies the waveform to a level sufficient to drive the tiny 8-ohm speaker through C3. The level at Q2's emitter is fairly high, so R6 and R7 divide it to prevent overdriving the amplifier, with R7 serving additionally as the volume control. The high value of R6 prevents loading down the UJT emitter circuit, ensuring proper operation of the device.

R4 and R5 are values typically associated with the unijunction transistor's characteristics. R4, in particular, was chosen to provide the best temperature stability of the circuit, a dubious requirement in

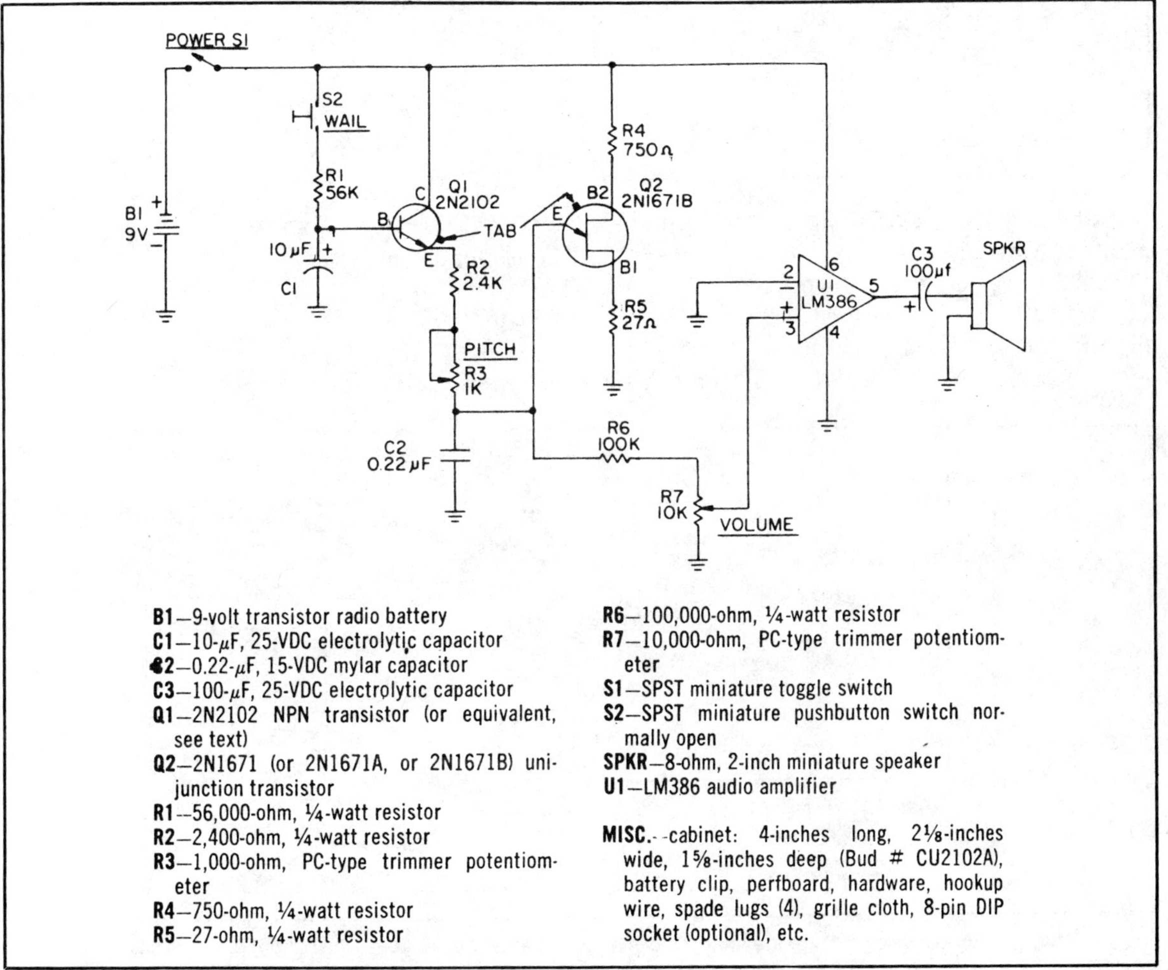

B1—9-volt transistor radio battery
C1—10-µF, 25-VDC electrolytic capacitor
C2—0.22-µF, 15-VDC mylar capacitor
C3—100-µF, 25-VDC electrolytic capacitor
Q1—2N2102 NPN transistor (or equivalent, see text)
Q2—2N1671 (or 2N1671A, or 2N1671B) unijunction transistor
R1—56,000-ohm, ¼-watt resistor
R2—2,400-ohm, ¼-watt resistor
R3—1,000-ohm, PC-type trimmer potentiometer
R4—750-ohm, ¼-watt resistor
R5—27-ohm, ¼-watt resistor
R6—100,000-ohm, ¼-watt resistor
R7—10,000-ohm, PC-type trimmer potentiometer
S1—SPST miniature toggle switch
S2—SPST miniature pushbutton switch normally open
SPKR—8-ohm, 2-inch miniature speaker
U1—LM386 audio amplifier

MISC.—cabinet: 4-inches long, 2⅛-inches wide, 1⅝-inches deep (Bud # CU2102A), battery clip, perfboard, hardware, hookup wire, spade lugs (4), grille cloth, 8-pin DIP socket (optional), etc.

Fig. 7-1. Li'l Wailer schematic and parts list.

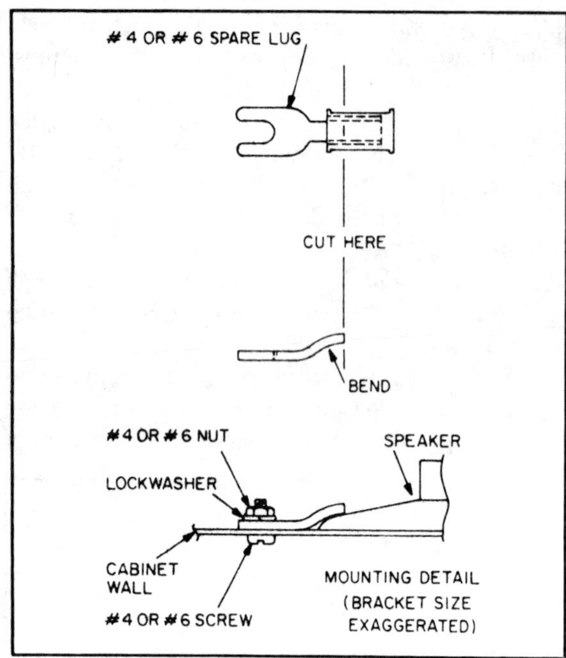

Fig. 7-2. Here's a novel way to mount subminiature speakers without tabs for mounting screws. A few spade lugs bent to the proper angle will solve the problem. You might use this same method (on a larger scale, of course) to flush-mount speakers in your car, where space doesn't allow for running screws through the mounting tabs in the speaker frame, as on door mounts.

this application. If a 2N2646 were to be used, R4 would be 2K. Due to another characteristic of the UJT, its *Intrinsic Stand-off Ratio*, use of the 2N2646 may increase the frequency of the tone, as compared to the 2N1671. It may be necessary to increase the value of R2 to 2.7K, or higher, in order to compensate.

Then 2N2102 and the 2N1671B, were used in the author's wailer

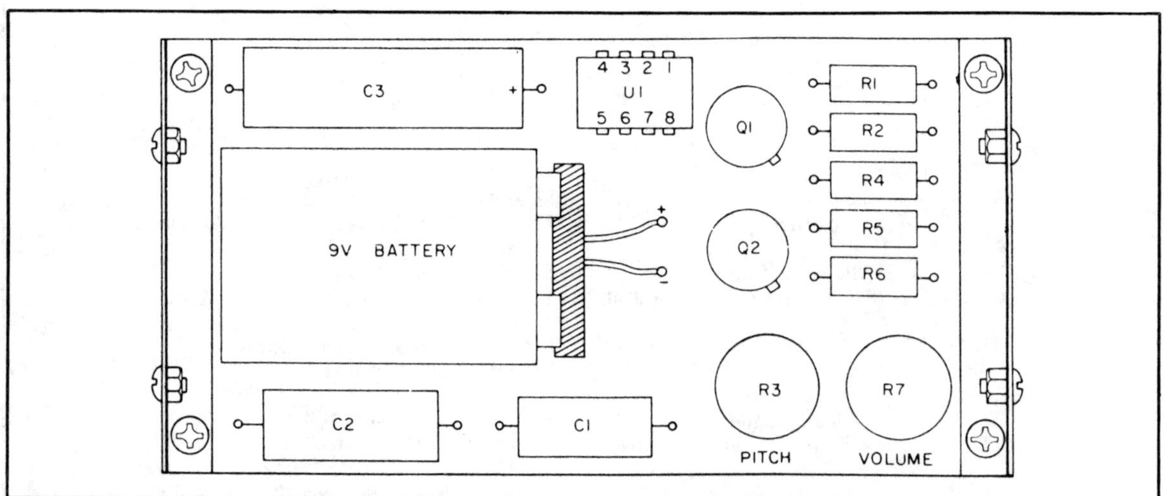

Fig. 7-3. Parts placement, showing how the project looks on perfboard. The diagram is just about drawn to scale, so you can get an idea of how compact the Wailer will be when completed. With a little ingenuity, it can possibly be made even smaller for pocket use.

simply because of their availability from the "junk" box. A 2N2646 may be used, with the value of R4 adjusted as described earlier. Similarly, a 2N3646 or almost any inexpensive NPN silicon transistor will work, although if the device's gain is too high, the descending wail will be prolonged. The builder may wish to try neighboring values around C2, if the peak pitch, within the adjustment range of R3, is not completely suitable to his ears. In like manner, variations in C1 will cause a variation in the tone's rate of ascent and descent; a larger value for C1 slowing both processes.

Construction. Layout is not critical, except for the space allotted by the specified box. Perfboard was used because of the circuit's simplicity. The material used in the unit shown required additional holes drilled to mount the IC. Board material with 0.100-inch spacing would avoid that inconvenience. An IC socket may be helpful in protecting the amplifier during soldering, but since one wasn't handy, the chip was pushed flat against the board, bending over the unused leads to hold it in place.

The perfboard mounting brackets were bent from thin sheet stock and bolted to the enclosure so as to seat the board firmly against the speaker magnet. This keeps the heaviest part, the battery, from bending the board. Foam tape, strategically placed at the end of the battery and also on the cover, secures it nicely, making a rattle-free unit. Care should be taken to ensure that no contact is made between the switch terminals and circuitry on the bottom of the board when the board is in place.

Although glue would likely be adequate to secure the speaker, four mounting brackets (see detail) were made from spade-tongue terminals. A piece of grill cloth was cemented inside the box to cover the five, ¼-inch holes drilled to pass the sound.

Checkout. Before connecting the battery, carefully check all wiring, especially to the LM386. Then, connect the battery and flip S1 to "on." A soft click should be heard from the speaker as IC1 gets power. Set the pitch and volume controls at mid-point and hold S2 down. After a short delay, the siren should wail up, and volume can then be set at the desired level. Adjust the pitch control next for the desired "highpoint" of the wail. Release S2 and the unit should wail down and stop—just like a police siren. Check for suitable wail-up and wail-down times by depressing and releasing S2.

After satisfactory operation is obtained, place the cover on the box, adjusting the foam tape as necessary to ensure a solid, rattle-free unit. Then, have fun with it, and just *try* to keep the kids' hands off of it!

THE ELECTRONIC SLOT MACHINE

The slot machine, or one-armed bandit, is one of the all-time favorite gambling devices. The game can be very captivating and fun. Unfortunately, you can drop a lot of money playing on a real one, not to mention the fact that you have to go to either Las Vegas or Atlantic City to play one. You don't have to lose your shirt to have fun playing our Electronic Slot Machine.

The unit performs virtually like the real thing. Three rows of LEDs simulate the 3 windows of a real slot machine. When you push the button (S2), each of the 3 rows flash the LEDs in a cascading manner that simulates a rolling wheel. After a couple of seconds, the first row window will stop rolling and only one fruit will remain lit. In the same manner, window 2 will stop rolling and only one fruit will remain lit. Another second later, row 3 will stop.

Should you happen to have three LED fruits lit horizontally in a row, you're a winner! If you're lucky enough to land 3 LEDs in the jackpot row, then you've won a 4-to-1 payoff! Hint: It's harder than you think, because naturally, the odds are with the house. However, that's what makes this game so fascinating and worthwhile to build.

Refer to Fig. 7-4.

The total project cost is about $15 or $20. If that seems a bit much, think how fast you'd lose that money playing on a real "one-armed bandit!"

Construction. Before you start building, you should decide on the size of the enclosure that you wish to use. This is largely determined by the battery pack, as four "C" batteries are used, and they take up a bit of room. You don't necessarily have to mount the batteries in the project box itself, but it's nice to have one self-contained unit. The prototype is housed in a cabinet with dimensions of 3¾-inches wide, by 6¼-inches high, by 2-inches deep.

Use the low-power Schottky ICs as recommended in the parts list. Don't worry, these Schottky ICs are as easy to get as standard TTL types.

The project can be built on a piece of perfboard. PC board construction is rather impractical, because of the amount of connection between the ICs.

Notice that IC sockets are used on all ICs. This makes for easy replacement and really makes wiring up the project a lot easier. Using wirewrap IC sockets and a wirewrapping technique, you can wire up the sockets in no time. Point-to-point wiring with solder is also satisfactory, but more time-consuming.

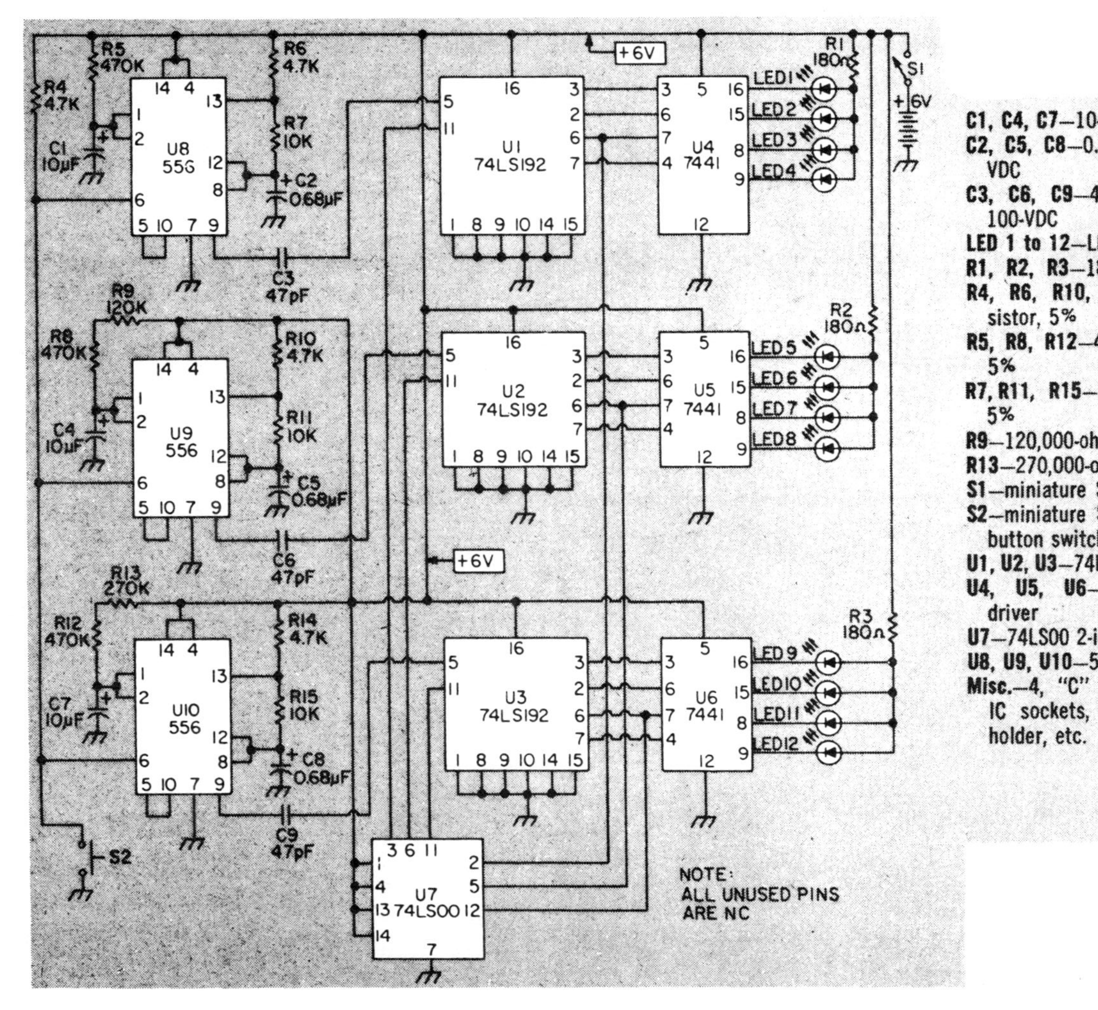

C1, C4, C7—10-uF tantalum capacitor, 16-VDC
C2, C5, C8—0.68-uF tantalum capacitor, 16-VDC
C3, C6, C9—47-pF ceramic disc capacitor, 100-VDC
LED 1 to 12—LED rated 20mA @ 1.75-VDC
R1, R2, R3—180-ohm, ¼-watt resistor, 5%
R4, R6, R10, R14—4,700-ohm, ¼-watt resistor, 5%
R5, R8, R12—470,000-ohm, ¼-watt resistor, 5%
R7, R11, R15—10,000-ohm, ¼-watt resistor, 5%
R9—120,000-ohm, ¼-watt resistor, 5%
R13—270,000-ohm, ¼-watt resistor, 5%
S1—miniature SPST toggle switch
S2—miniature SPST momentary-contact push-button switch
U1, U2, U3—74LS192 decade counter
U4, U5, U6—7441 decimal decoder/LED driver
U7—74LS00 2-input, quad NAND gate
U8, U9, U10—556 timer
Misc.—4, "C" batteries, perfboard, cabinet, IC sockets, solder, hookup wire, battery holder, etc.

Fig. 7-4. The Electronic Slot Machine schematic.

IC sockets, cut in half, were used for mounting the LEDs to the circuit board. While this is not necessary, it *does* make the LEDs easier to mount, especially if you want to have a faceplate that fits flush over the LEDs, as shown in the photo of the prototype. Switches S1 and S2 can be mounted right on the board to eliminate unnecessary stray wires in the cabinet.

Install the ICs in the sockets only after all of your wiring is completed. Make sure that the pin orientation is correct. Observe polarity on the tantalum capacitors. Even though they're less than 1-μF, they are polarized units, and care should be taken with them. When mounting the LEDs, make sure that you get their polarity (anode and cathode) correct as well. Double check all of your wiring carefully.

Testing and Operation. Connect the batteries, and turn the unit on by throwing switch S1. All three rows of LEDs should "roll" sequentially. If they don't, push switch S2. If the LEDs still don't roll, then turn the unit off and check for a wiring error. If the unit is performing correctly, all three rows of LEDs will roll, and then they will stop in order, 1-2-3.

Troubleshooting. If you're sure your wiring is right, but one of the rows still won't flash correctly, check the corresponding chip of the non-working row with one of a working row. (U1 controls row 1, U2 controls row 2, etc.). This switching around can reveal if the trouble is in your wiring or in the chip itself. If the display flashes erratically, fresh batteries are required.

BUG MOTHER NATURE

Englishman George Riley lives in Kent, works in London and goes home to an unusual hobby.

"It all started about a couple of years ago when I borrowed a friend's parabolic directional microphone dish. This type of equipment is hyper-sensitive and can be pinpointed to record a sound without external noise interference. I was using it to record the sound of crickets when I suddenly heard a strange 'slurping crunching' sound. This turned out to be a large snail making the most of some hard grass. From then on I was hooked," says George.

Experts such as zoologist Donald J. Borror have used the parabolic microphone technique to produce 33⅓ rpm records that sonically illustrate ornithology books and booklets.

Refer to Figs. 7-5 and 7-6.

After stumbling over a couple of radar antenna dishes a few years ago, I finally decided to put one of them to work. Since I was no microwave expert, I decided to try an acoustic application. After all, I reasoned, a parabolic dish is a parabolic dish whether it is used for reflecting and focusing microwaves or sound waves. The result is the parabolic microphone described here.

If you want to go all out for added gain, look over the surplus dealers' list for an 18-inch or larger aluminum model. As nearly as I can tell with the test equipment available, the 18-inch reflector adds about 10-dB gain to the microphone.

Construction. It's simple enough as reference to the photos will reveal. The mount for the dish is made of wood and masonite. The dish is held in place by three threaded rods which also serve as the microphone support. Almost any kind of rod material will do, as long as it is or can be threaded. I happened to have some odd pieces of 9-gauge aluminum clothesline which threaded easily with a 10-32 die. Make the rod length about 7½ inches to allow sufficient leeway for adjusting the microphone for optimum focus. A small bracket or block may be added where the dish touches the wooden base to add rigidity, and a hole in the center of the base will make it convenient to mount the whole assembly on a camera tripod.

Any low-priced ceramic or crystal microphone cartridge will work well with this reflector. Mount the microphone cartridge on the rods with rubber bands. The exact method of attaching the rubber bands to the microphone cartridge is left to the ingenuity of the builder, since this will largely depend upon the physical configuration of the microphone.

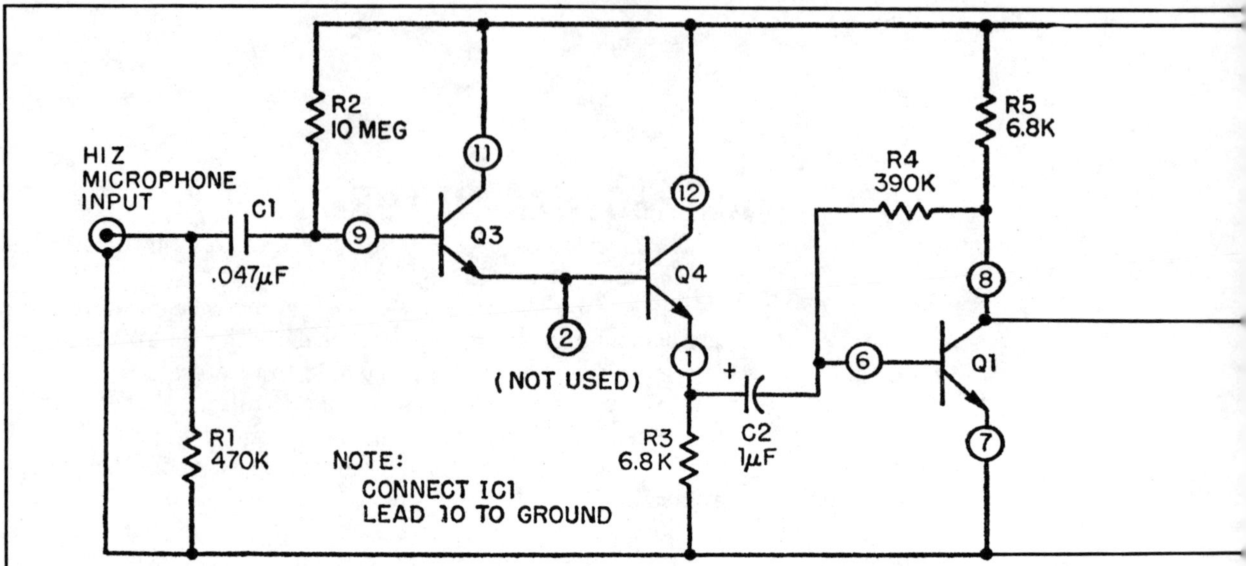

HI Z MICROPHONE INPUT
C1
.047μF
R2
10 MEG
Q3
Q4
R5
6.8K
R4
390K
Q1
(NOT USED)
R1
470K
NOTE:
CONNECT IC1
LEAD 10 TO GROUND
R3
6.8 K
C2
1μF

B1, B2—9-volt battery, 2U6-type (Radio Shack 23-464 or equiv.)

C1—0.047 μF disc or tubular capacitor (Radio Shack 272-1052 or equiv.)

C2 C3, C5, C6—1 μF capacitor, electrolytic (observe polarity) or tubular, 35 volts or better (Radio Shack 272-1055 or equiv.)

C4—0.01 μF capacitor, ceramic disc (Radio Shack 272-131 or equiv.)

IC1—3018 integrated circuit (RCA CA3018) Available from Circuit Specialists Co., Box 3047, Scottsdale, AZ 85257; $2.00 postpaid.

R1—470,000-ohm, ¼-watt resistor (Radio Shack 271-1800 or equiv.)

Fig. 7-5. Bug Mother Nature schematic and parts list.

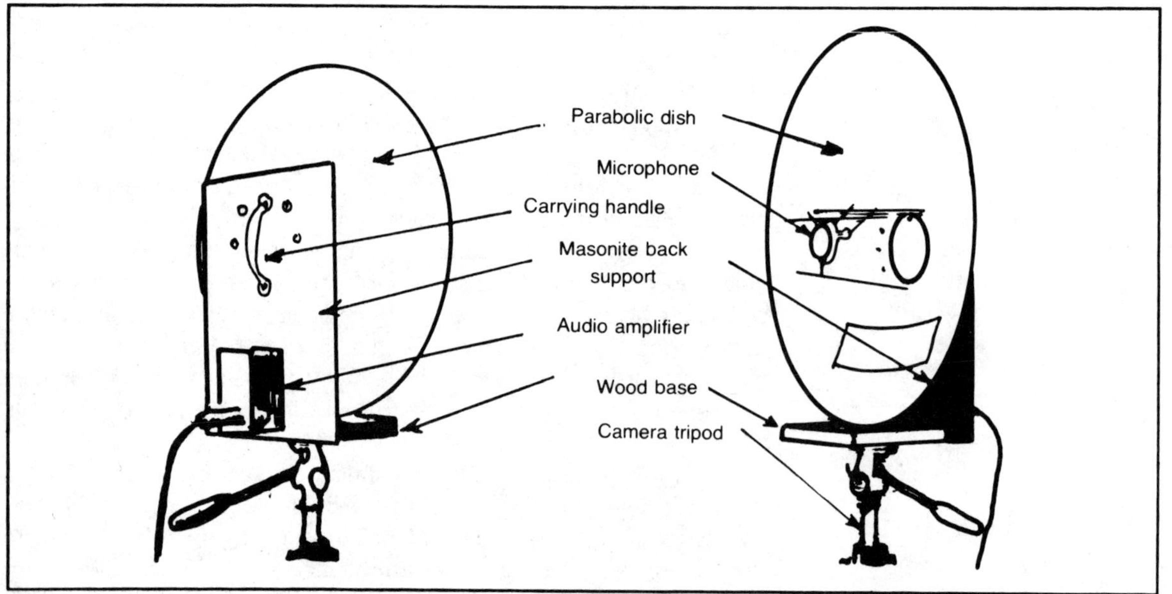

Fig. 7-6. Suspend your microphone from rubber bands that extend to the support rods, a clamp wrapped in foam packing material holds the microphone securely.

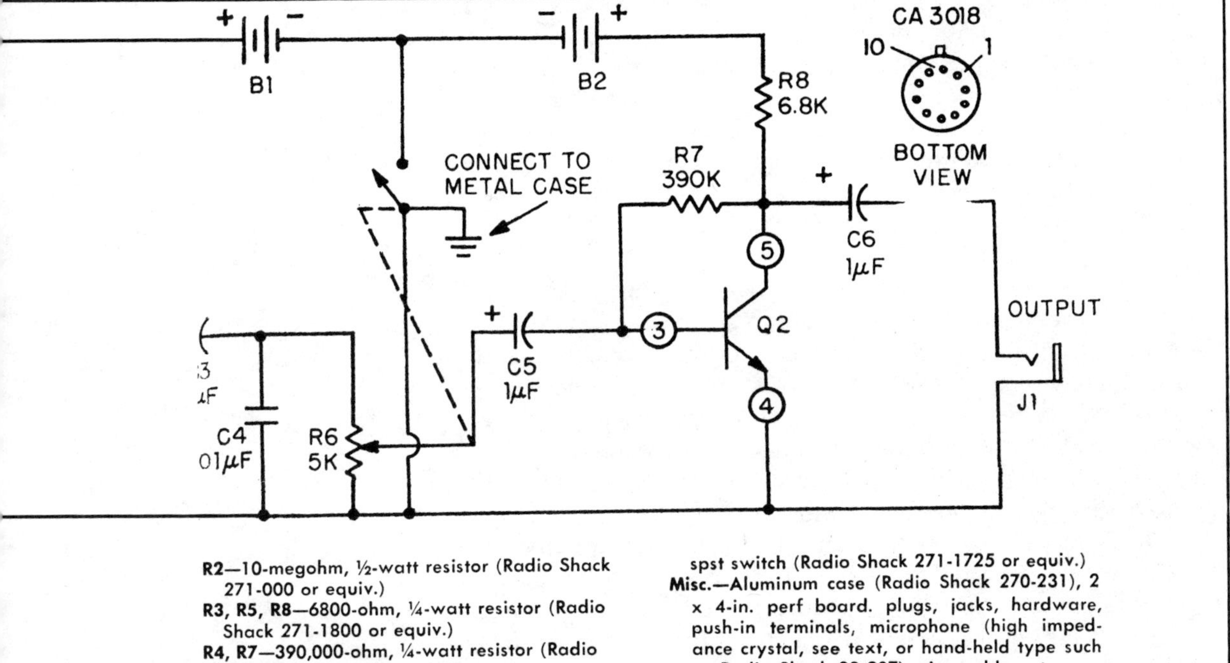

R2—10-megohm, ½-watt resistor (Radio Shack 271-000 or equiv.)
R3, R5, R8—6800-ohm, ¼-watt resistor (Radio Shack 271-1800 or equiv.)
R4, R7—390,000-ohm, ¼-watt resistor (Radio Shack 271-1800 or equiv.)
R6—5,000-ohm potentiometer, audio taper, with spst switch (Radio Shack 271-1725 or equiv.)
Misc.—Aluminum case (Radio Shack 270-231), 2 x 4-in. perf board. plugs, jacks, hardware, push-in terminals, microphone (high impedance crystal, see text, or hand-held type such as Radio Shack 33-907) wire, solder, etc.

Next route a 16-inch piece of shielded microphone cable from the microphone along one of the rods, through the dish (but inside the back plate), and terminate the cable in a phono plug. The cable should have sufficient slack so it may be easily plugged into the amplifier box. Also, be sure to allow sufficient lead slack at the microphone end of the cable so that the shock mount effect of the rubber bands is not nullified. This will complete the microphone reflector assembly, which should be set aside until the amplifier is built.

Electronics. The amplifier is a three-stage affair using an RCA CA 3018 integrated circuit. Transistors Q3 and Q4 are used as a Darlington pair in an emitter-follower circuit in the first stage. This provides the necessary high input impedance required by the crystal microphone. The two following stages utilize Q1 and Q2 respectively as conventional common emitter amplifiers. The average gain per stage is about 38 dB.

Capacitor C4 across audio gain control R6 provides a 3-dB rolloff at 15 kHz, thus limiting amplifier frequency response to the desired audio range. In addition to limiting the frequency response, this capacitor also reduces the tendency of the amplifier to oscillate at higher frequencies, which could result in instability and low output. The 3-dB point at the low frequency end is about 70 Hz, sufficient for this application.

Two 9-volt transistor batteries are used to power the amplifier; not because of high current drain, but, rather, to avoid common coupling between the output stage and earlier stages of the amplifier. An RC decoupling network could, of course, be used instead of two batteries, but it was found that oscillation would occur in spite of the decoupling network after the batteries had been in service for awhile. Two batteries absolutely against amplifier instability during the useful life of the batteries. The total current drain of the amplifier, by the way, is only 1.5 mA.

No trouble should be experienced with the amplifier if the original layout is followed. All amplifier components are mounted and wired on the perfboard as shown. The volume control, capacitor C4, and the earphone jack are mounted on the part of the minibox that serves as a cover and battery holder. All connecting wires are soldered to push-in terminals on the perfboard, and the perfboard is mounted above the batteries with small bolts and spacers. After assembly, connect the microphone to the amplifier input with a short piece of cable.

Checkout. When testing the amplifier on the bench, either have the microphone connected to the input terminals or substitute a half-megohm resistor for the microphone input. If you have a hum problem it is probably caused by nearby ac wiring. (I had to turn off power to the workbench whenever I tested the amplifier out of its case.) Alternatively, you may find a place in the house that is hum free; make your tests there. With the amplifier completely enclosed in its case, there is absolutely no hum pickup problem.

When you are satisfied that the amplifier is stable and working properly, solder the short microphone cable to the input terminals and mount the amplifier in its case. You are now ready to set up the microphone for maximum gain. To do this, you will need a code practice oscillator or other source of audio signal and an ac voltmeter with 10-volt range connected to the amplifier output.

Set the equipment up in a clear area. Enable the CPA and adjust the audio gain so that the voltmeter reads two volts or less. Next move the microphone cartridge towards and away from the center of the dish to find the microphone position giving the greatest output. Do not let the voltmeter reading go above three volts because overloading the amplifier will make it difficult to find the point of maximum gain. After finding the best position for the microphone, secure the rubber bands on the support rods with dabs of cement.

The parabolic snooper may be used in several ways. As a portable field instrument, just plug in a set of 2000-ohm earphones and be on your way through the woods. The unit will also work as a combination microphone-preamplifier with any amplifier or tape recorder. However, if you are using a speaker for monitoring outside noises, be sure to have sufficient acoustic isolation between the microphone and speaker, such as closed doors and windows. If you don't, all the world will know by your feedback howl that you are listening. When using the unit with an audio power amplifier it is best to run the gain quite high on the amplifier and adjust the system gain as needed with the preamp gain control. Now you're ready for a new world of close up sound.

THE THIRD EAR

For many creatures on this earth, two ears are the norm. The two-eared arrangement does more than allow listening to your mother-in-law and wife at the same time. Thanks to some special neural circuitry (which, among other things, performs phase and magnitude comparisons between left and right ear signals) a two-eared individual can quite accurately tell where a sound is coming from. You know how marvelously well the present system works, but think of the possibilities afforded by a third ear.

Wait a second now! No one is advocating surgery as a hobby (á la Frankenstein). The Third Ear in this instance is a versatile, electronic, sound-actuated system. It can spy on your friends, mind the phone, babysit, thwart would-be burglars and much more. Later on, the Third Ear's applications will be explored in detail, but first let's examine its circuit.

Refer to Figs. 7-7 through 7-13.

The Circuit. The heart (better yet, the eardrum) of the Third Ear is a tiny module, the ETCO S-210U sound trigger. This little device originally formed the nervous system of an electronic turtle. The species is now extinct, unfortunately, but its innards are available as a great surplus bargain. As you can see from the schematic, the S-210U contains a crystal microphone, a transistor amplifier, and an SCR. The module's black lead goes to the minus side of a battery, while the red and green leads will be shorted together in this application. The shorted leads connect to one side of a low-resistance load (like a relay), and the opposite end of the load goes to battery positive. Sound picked up by the microphone is amplified by the transistor and fed to the SCR's gate. If the sound is sufficiently loud, the SCR latches in a conducting state, thus drawing a relatively large current through the connected load. Power must be removed in order to turn off the device again.

A more versatile system should operate in either of two modes, latch or pulse. After the first triggering impulse of sound, the latch-mode system remains active. A pulse-mode system, on the other hand, remains active only for some pre-determined time interval after triggering. It then returns to its inactive state, where it rests until re-triggering occurs. Then, the process repeats itself.

Construction. Adding pulse-mode capability to the S-210U is a simple matter. All it takes is some auxiliary circuitry to sense the latching of the SCR and to unlatch it again after a user-selected time delay. Unlatching an SCR can be accomplished by opening a switch in series with the anode or closing a switch to short the anode and cathode

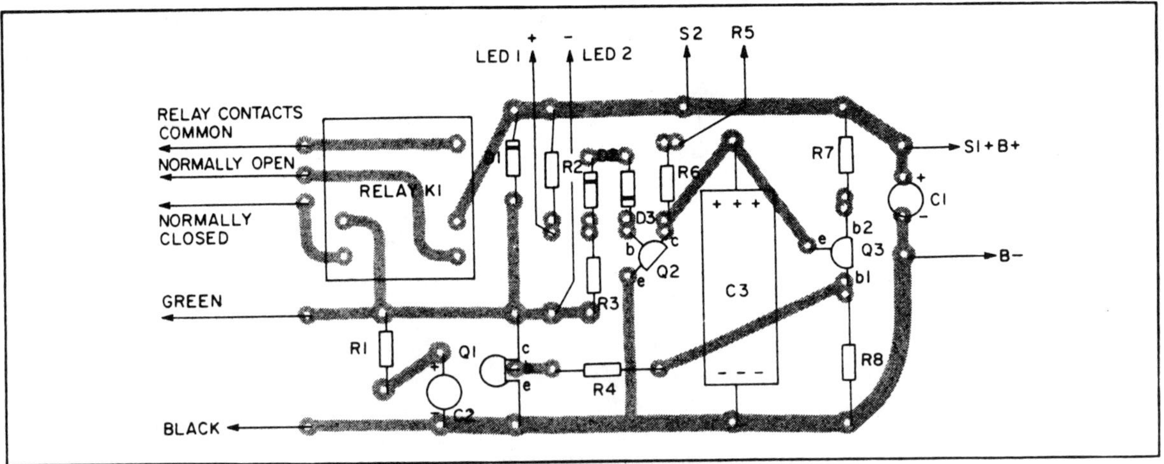

Fig. 7-7. This is a view of the component side of the main printed circuit board. The foil pattern is on the reverse side. If your K1 relay has mounting tabs, you may have to make some holes.

together. The latter method is the one used in the Third Ear, but before getting any further into that, there are a few easy modifications that must first be made to the S-210U.

The figure shows the four necessary modifications in detail. First, remove the 5000-ohm trimmer by unsoldering it. This device is unusual in that it has two mounting pins, not three. Wire a 2500-ohm potentiometer in series with a 680-ohm resistor so that the net resistance is a minimum (680 ohms) when the pot is fully clockwise. The two wires from the pot/resistor combination should be soldered into the holes vacated by the 5000-ohm trimmer. This new pot will function as the Third Ear's sensitivity control (with maximum sensitivity in the clockwise position).

The second modification requires that the 0.1-μF disc capacitor in the upper left-hand corner of the S-210U be unsoldered. In the holes vacated by the capacitor, install and solder a jumper of bare, solid hookup wire.

The third step is to cut the red wire in the lower lefthand corner completely off at the point where it joins the PC board. Finally, unsolder the 32-μF electrolytic capacitor from the board, and replace it with a similar unit having a higher working voltage; 16 Vdc or higher. In general, your replacement may have a value anywhere between 22 and 47-μF, with 33-μF being about optimum. Remember that since you are dealing with electrolytic devices, the orientation must be correct. In the pictorial you can see that capacitor positive (+) must be pointing upward.

Now, let's see how the modified S-210U mates with the rest of the circuitry in our Third Ear. As the schematic diagram shows, the green and black leads of the module are its only connection to the external circuitry. Capacitor C2 bypasses the module's supply leads in order to keep the sensitivity high, while R1 isolates C2 to reduce its effects on the performance of the rest of the circuit. Whenever the module's SCR latches into conduction (due to sonic triggering), current will be drawn through relay K1 and the LED1/R2 combination. As a result, the normally open relay contacts will close, and the LED will light simultaneously. These two conditions will persist as long as the SCR remains latched.

Note how switch S3 selects either the normally open (N.O.) or normally closed (N.C.) contacts of K1. This allows a load to be turned on or off, respectively, when the circuit is activated. Diode D1, connected across K1's coil, is normally reverse-biased (not conducting). When the SCR is forced to unlatch, however, K1's coil generates an inductive kickback voltage which could cause trouble if D1 were not there to clip it.

In order to see how unlatching is accomplished, let's assume that the SCR in the module is initially unlatched, and that mode switch S2 is closed in its "pulse" position. Since the SCR is not conducting, the voltage at the green lead of the module must be high (about 7 volts above ground). This potential drives sufficient current through R3, D2 and D3 into the base of transistor Q2 to ensure that Q2's collector is conducting current heavily. This prevents the voltage on C3 from rising, and nothing of interest happens.

Suppose, however, that a sound triggers the module into conduction. The potential of the green lead drops to less than one volt, which

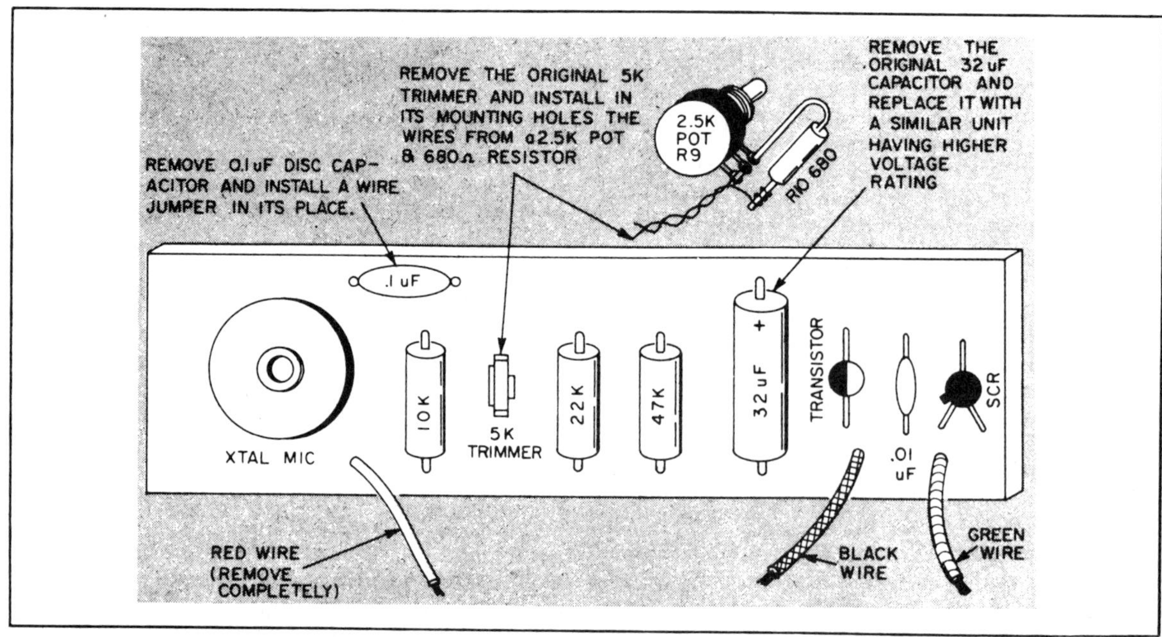

Fig. 7-8. This diagram shows all the modifications needed on the S-21OU module. The 32-μF capacitor should be replaced by one with a 16-Vdc or higher voltage capacity.

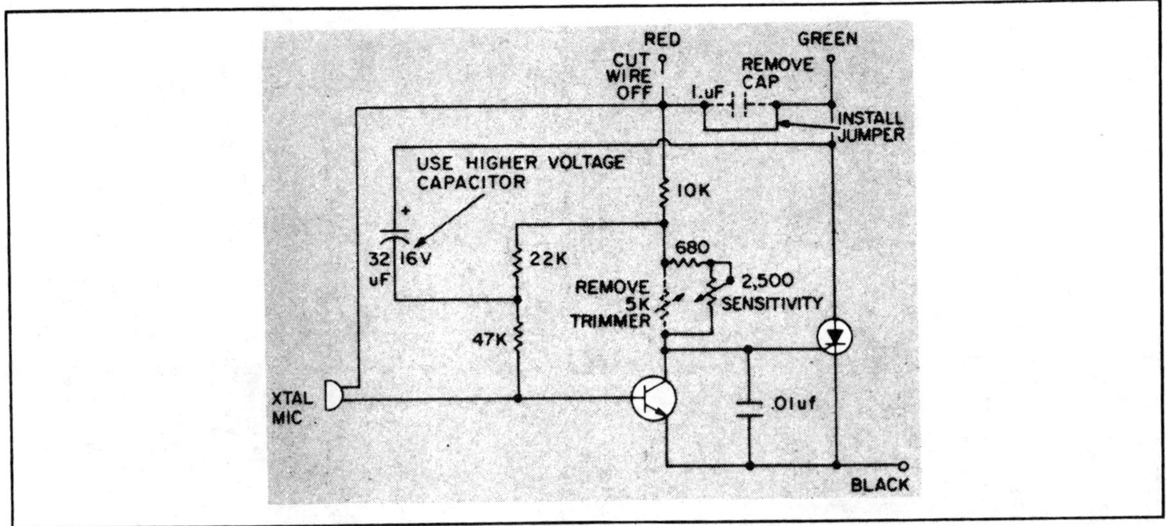

Fig. 7-9. Schematic of the S-210U. The module can be obtained from the company listed in the parts list.

is less than the 2-volt minimum needed to turn on the D2/D3/Q2 combination. Consequently, Q2's collector no longer conducts current, and the potential across capacitor C3 rises as charging occurs through R5 and R6. The rate of ascent is controlled by potentiometer R5; higher resistance causes the potential on C3 to climb more. Eventually, the voltage on C3 will reach a critical level, at which point unijunction transistor Q3's emitter-to-base 1 impedance will break down to a very low level. This rapidly discharges C3 and causes the appearance of a voltage spike across resistor R8.

This voltage spike drives current through R4 into the base of transistor Q1. As a result, Q1's collector conducts current heavily, thus shorting the module's green and black leads together. This deprives the SCR of anode current, causing it to unlatch. Because the voltage spike lasts only a brief instant, less than 0.1 second, Q1 soon loses base drive and ceases to conduct. When this happens, current can no longer activate K1 or LED1, and both will remain off until another sound triggers the module. As you can see, the circuit has returned to the state it was in at the beginning of this discussion.

If mode switch S2 had been opened to its *latch* position, no current would have been able to flow through R5 and R6 to charge C3. Since the charging of C3 is an essential part of the unlatching process, it is clear that the module would have remained latched indefinitely. In fact, in the latch mode, the only way to reset the circuit to its inactive state is by opening power switch S1 for at least five seconds. This gives the various capacitors time to discharge completely, thus ensuring that the circuit will be inactive when S1 is again closed. Similarly, should you wish to manually unlatch the module in the pulse mode before the time delay elapses, the same procedure applies.

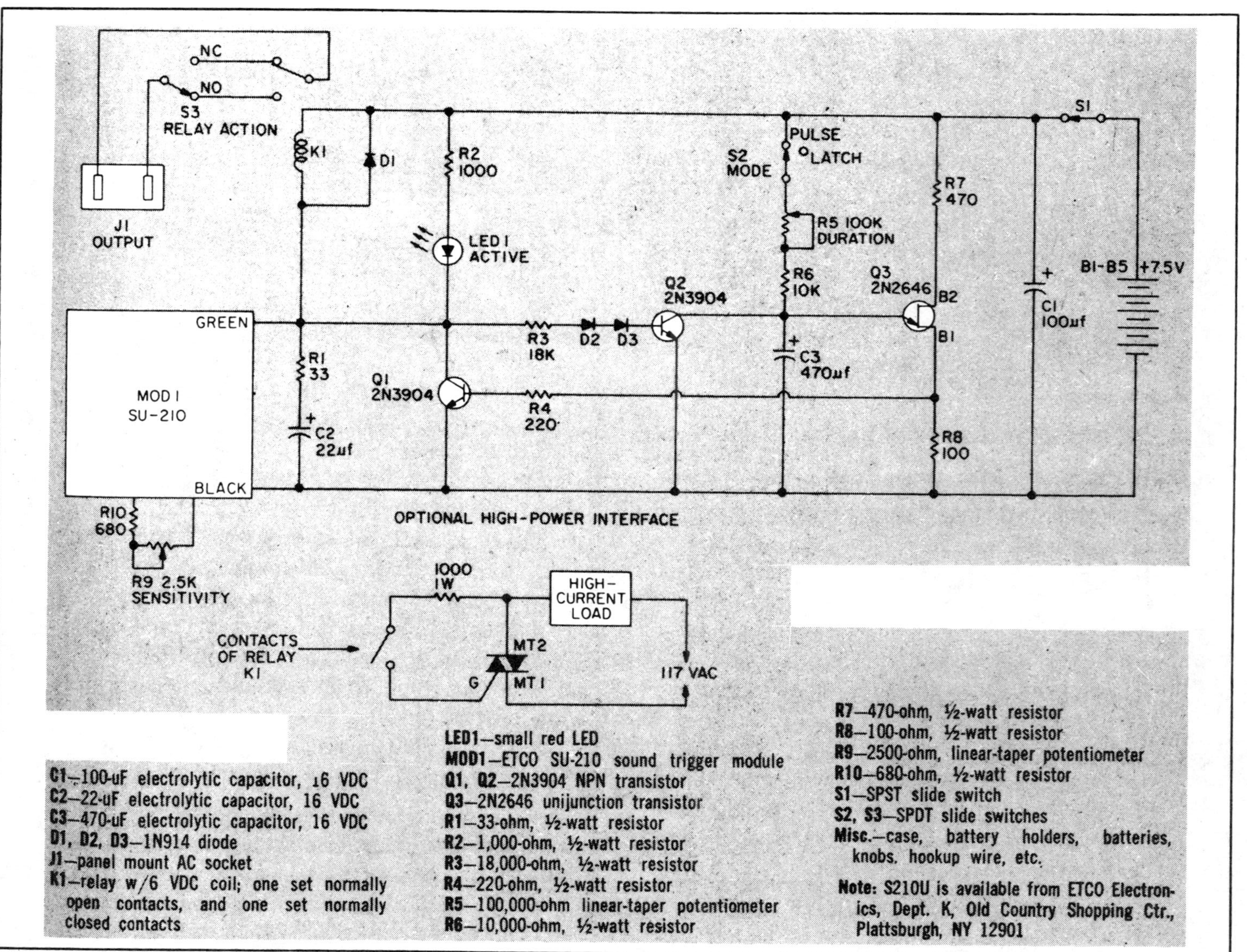

C1—100-uF electrolytic capacitor, 16 VDC
C2—22-uF electrolytic capacitor, 16 VDC
C3—470-uF electrolytic capacitor, 16 VDC
D1, D2, D3—1N914 diode
J1—panel mount AC socket
K1—relay w/6 VDC coil; one set normally open contacts, and one set normally closed contacts
LED1—small red LED
MOD1—ETCO SU-210 sound trigger module
Q1, Q2—2N3904 NPN transistor
Q3—2N2646 unijunction transistor
R1—33-ohm, ½-watt resistor
R2—1,000-ohm, ½-watt resistor
R3—18,000-ohm, ½-watt resistor
R4—220-ohm, ½-watt resistor
R5—100,000-ohm linear-taper potentiometer
R6—10,000-ohm, ½-watt resistor
R7—470-ohm, ½-watt resistor
R8—100-ohm, ½-watt resistor
R9—2500-ohm, linear-taper potentiometer
R10—680-ohm, ½-watt resistor
S1—SPST slide switch
S2, S3—SPDT slide switches
Misc.—case, battery holders, batteries, knobs, hookup wire, etc.

Note: S210U is available from ETCO Electronics, Dept. K, Old Country Shopping Ctr., Plattsburgh, NY 12901

Fig. 7-10. The Third Ear schematic.

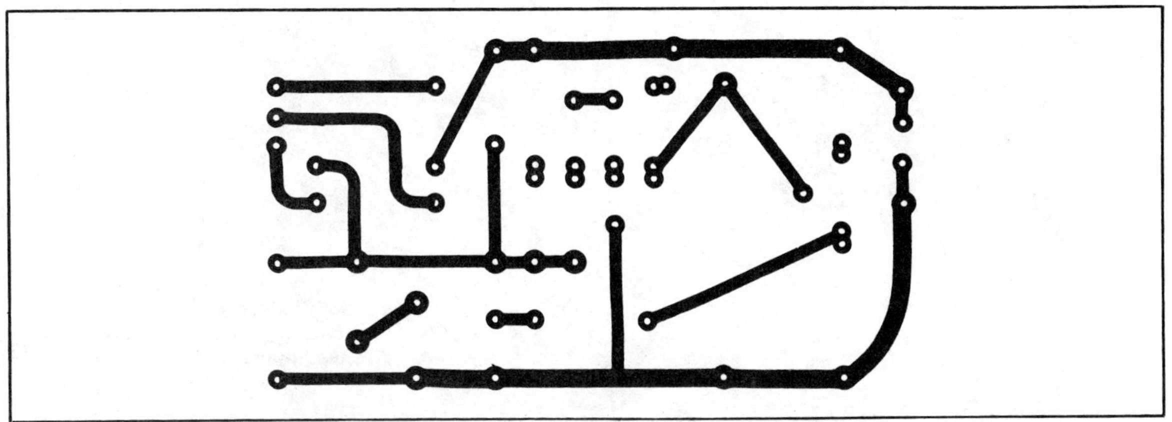

Fig. 7-11. The printed circuit board for The Third Ear is easy to make. You can use this template for a photographic copy or just duplicate the pattern with a resist pen.

In the prototype's pulse mode, duration control R5 was able to produce time delays between 9 and 130 seconds. The actual control range obtained in your model is likely to be somewhat different because of variations in the characteristics of Q3 and C3. Furthermore, any leakage within C3 will exert yet another influence on the time delay; the leakier the capacitor, the longer the charging time. With this in mind, it is wise to use a new, high-quality electrolytic capacitor for C3.

Power Supply. Power for the Third Ear comes from five "D" cells in series, yielding 7.5 volts. Electrolytic capacitor C1 keeps the power supply's impedance low. Inactive, the Third Ear draws only 2 mA, but current consumption jumps to 22 mA when the circuit is active. At these small rates of discharge, "D" cells will last a long time. Some readers might prefer to see the Third Ear powered by an ac supply; however, transformers hum at 60 Hz, and the Third Ear is

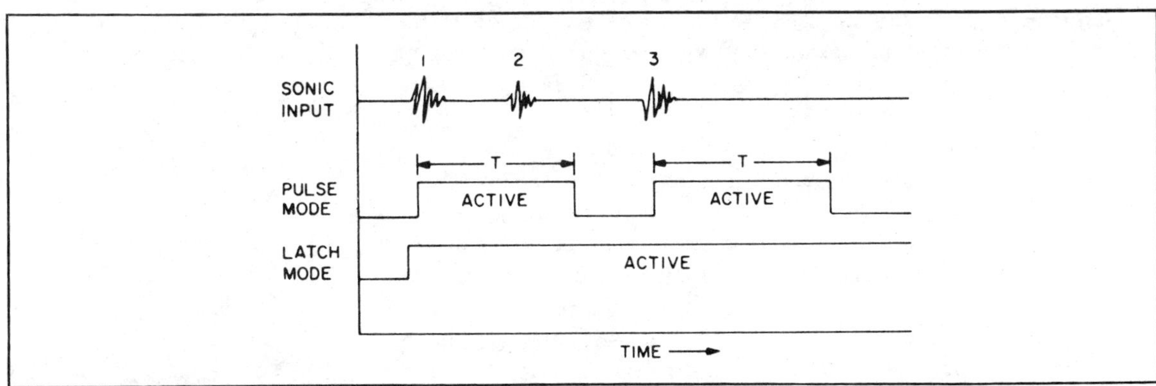

Fig. 7-12. The Third Ear can be adjusted so that it triggers and holds, or so that it triggers and resets. This chart shows how the output level varies with incoming sound pulses.

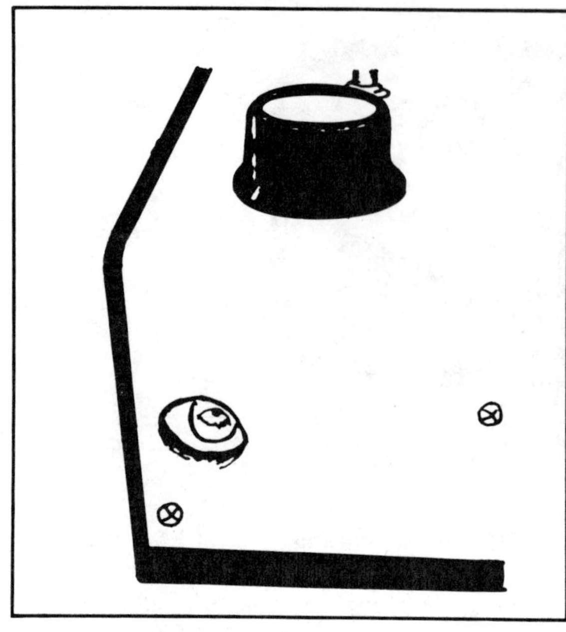

Fig. 7-13. Make a ¾-inch diameter hole in the front panel of the cabinet or wherever you want to put the microphone. Adjust the sensitivity so that it triggers correctly.

sensitive enough to be triggered if a transformer is mounted inside its case. If you want to use an ac supply, a 6- to 9-volt dc unit will work well, but it *must not* be mounted inside the Third Ear's case.

Construction. Construction is easy because you don't need to worry about the layout. Anything will do. A PC board is not absolutely necessary, but if you like to give your projects that professional look, use the PC patterns provided.

You should test LED1's sensitivity before wiring it into the circuit. Bargain LEDs especially may not be sensitive enough to be used here. Hook your LED in series with a 1000-ohm resistor, and connect the combination to a 7.5 Vdc source. (Get the polarities right). If you do not obtain an easily visible red glow, try another LED. Red LEDs are more sensitive than green or yellow ones, so stick with red.

When wiring duration control R5, make sure you obtain maximum resistance in the fully clockwise position. This will then give you a maximum time delay.

When building your Third Ear, you would be better off with slide switches. When a toggle is snapped quickly, the click of the switch can activate your system, regardless of the sensitivity setting. Slide switches require very little operating force and are practically silent.

The contacts of relay K1 are rated for a load of up to one ampere, which is more than adequate for most applications. Sometimes, however, you may wish to control a high-power load, such as a flood lamp. One method of doing this would be to substitute a relay with a higher contact rating for K1, but high-current, good-quality relays are expensive. Besides that, all relays arc, especially with high-power loads, so a relay's lifetime under such conditions is limited. A cheaper, better solution is the high-power interface. Note that the triac controls the ac load, but the relay contacts control the triac. In this way, the relay contacts carry only the small gate current of the triac, and your Third Ear remains isolated from the ac line (and shock hazards) by the relay. Choose a triac with a current rating high enough for your load, and heatsink it. Mount the triac and heatsink in a well-ventilated *plastic* case to prevent accidental shocks.

Checking It Out. After construction is complete, you should check out the operation of your project. Set your Third Ear into the pulse mode, with R5 set for a minimum duration, and sensitivity control R9 placed at the midpoint of its range of rotation. Now, turn on power switch S1. LED1 should flash momentarily as power is applied. Snap your fingers directly in front of the microphone, and note the length of time that LED1 remains lit.

Next, rotate the duration control to maximum. Snap your fingers, and again make a note of how long LED1 stays illuminated.

Finally, turn the power switch off, and flip S2 to the latch mode. After five seconds, re-apply power. Snapping your fingers should now cause the LED to light and stay lit for as long as power is applied. You can do some experimenting with the sensitivity control, too. In the prototype, operation at maximum sensitivity was impossible because even the faintest ambient noise would trigger the circuit.

The applications for the Third Ear are only limited to the uses your imagination can find, and with its switching flexibility, it can control almost anything you may wish to operate around the home or office.

LIGHT-BEAM COMMUNICATOR

Have you ever tried to communicate when skip was coming in over the CB band? Interference from local stations was only exceeded by interference from long-distance stations. The channels were so crowded that stations were as tight as packed sardines. Communication was impossible. There is an intriguing solution to this common problem. Leave the roaring CB crowd behind, and escape up to the light waves.

The Light-Beam Communicator described here demonstrates how a light beam can be used for voice transmission. This communicator is also useful to trap intruders at a remote location, and as a top secret communications link between two stations.

The clarity and quality of audio reproduction is crystal clear, with more than enough pick-up sensitivity and modulation power than normally would be needed. Range of the units should be line of sight up to 1000 meters or better. Alignment is easily accomplished by sighting along the barrels of the units. Short-range communication (several hundred meters) is easily accomplished by simple sighting to one another's respective units. Long range setups are more conveniently obtained using a camera tripod. Units are built in a pistol-type configuration with all power and optics self-contained. A rear panel contains the necessary controls for operating along with jacks for headsets and a built-in microphone. The device is designed so that it also can be used for actual "listening" to other light sources such as TV pictures, scopes, fluorescents and many other infrared and invisible radiation sources.

Refer to Figs. 7-14 through 7-23.

Normally, the units are built using a visible red transmitter for ease and convenience in nighttime alignment. For serious longer range, low/noise performance, they can be equipped with optional filters for invisible infrared transmitting capabilities.

Looking at the Circuit. The light beam transmitter-receiver consists of a phototransistor receiver which picks up modulated light that is fed through a high gain amplifier and then to headsets or a loud speaker. When in the transmit mode, the amplifier becomes a sensitive mike pre-amp that drives a current amplifier modulating a light emitting diode as the transmitter.

The receiver section consists of a phototransistor (Q4) positioned at the focal point of lens LE2 inside enclosure EN2 (a separate enclosure, lens and phototransistor for transmitting and receiving enhances the flexibility and performance of the device). This, however, adds to the cost. Duplicating these components for both functions could be done; however, overall performance is sacrificed.

Q4 is mechanically secured to a sliding dowel DO1 that is adjusted to its proper distance from LE2 and secured with a screw. The signal from Q4 is fed into J1 via a shielded cable to keep hum and other electrical pickup to a minimum. Switch S2A now selects J1 in the *receive* mode and feeds the signal to the amplifier via C1. The signal is now matched and amplified via the integrated circuit U1. A gain control R7 controls the sensitivity of the amplifier and also serves as an ON/OFF switch for the receiver section. The output of U1 is now further amplified by Q1 and impedance matched via transformer T1. S2B now connects T1 and J2 for feeding 8-ohm headsets or an external speaker.

The transmitter section consists of a narrow beam visible red or optional infra-red light emitting diode LED1 located at the focal point of lens LE2 inside enclosure EN1. EN1 also contains the electronics and controls for hand grip EN3. A mike M1 is located on the rear panel RP1 and is fed to the amplifier U1 through C1 via mode select switch S2A. The amplifier now becomes a pre-amp for the mike. The output of the preamp is further amplified by Q1 and impedance matched by T1. The output of T1 is fed to Q2 via S2B. Q2 is dc coupled to Q3 whose quiescent state is selected via R15 in determining the dc current through LED1. A modulation signal is ac coupled to Q2 via C8. The hole is covered to minimize random light or noise in the circuit.

Power for the transmitter section Q2, Q3 and LED1 is from battery B3, is controlled by S2C and is used only during transmit mode. This enables the device to be used as a receiver.

Construction. Begin by making the following parts. The parts list supplies details on how to purchase the items already fabricated. The main enclosure EN1 is an 8-inch long piece of PVC tubing, with a 3½-inch outside diameter. This is sometimes called schedule 40 PVC tubing. PVC tubing is obtained in plumbing supply stores, hardware stores or building supply outlets.

Cut a 2-inch hole for the handle, using a hole saw. The assembly diagram shows where this hole is placed. File a ¼-inch slot on each side of the hole that is 2½-inches from the back of EN2. Remember to curve the slot to take into account the pivoting action of EN2's other screw when optical alignment is later attempted. Next, drill a ¾-inch hole approximately 3 inches from the rear end of the receiver enclosure. This hole is for optical alignment. EN2 is a 6½ × 2⅜ inch (outer diameter) piece of schedule 40 tubing. There are two mating holes to secure the receiver enclosure to EN1. These holes are on top of the piece to allow access with a screwdriver. The bore axis of these two tubes must be parallel. The large ¾-inch hole can be covered with a plug or a piece of tape.

Transformer T1 has a 3:1 turn ratio of 1500 to 500 turns. The

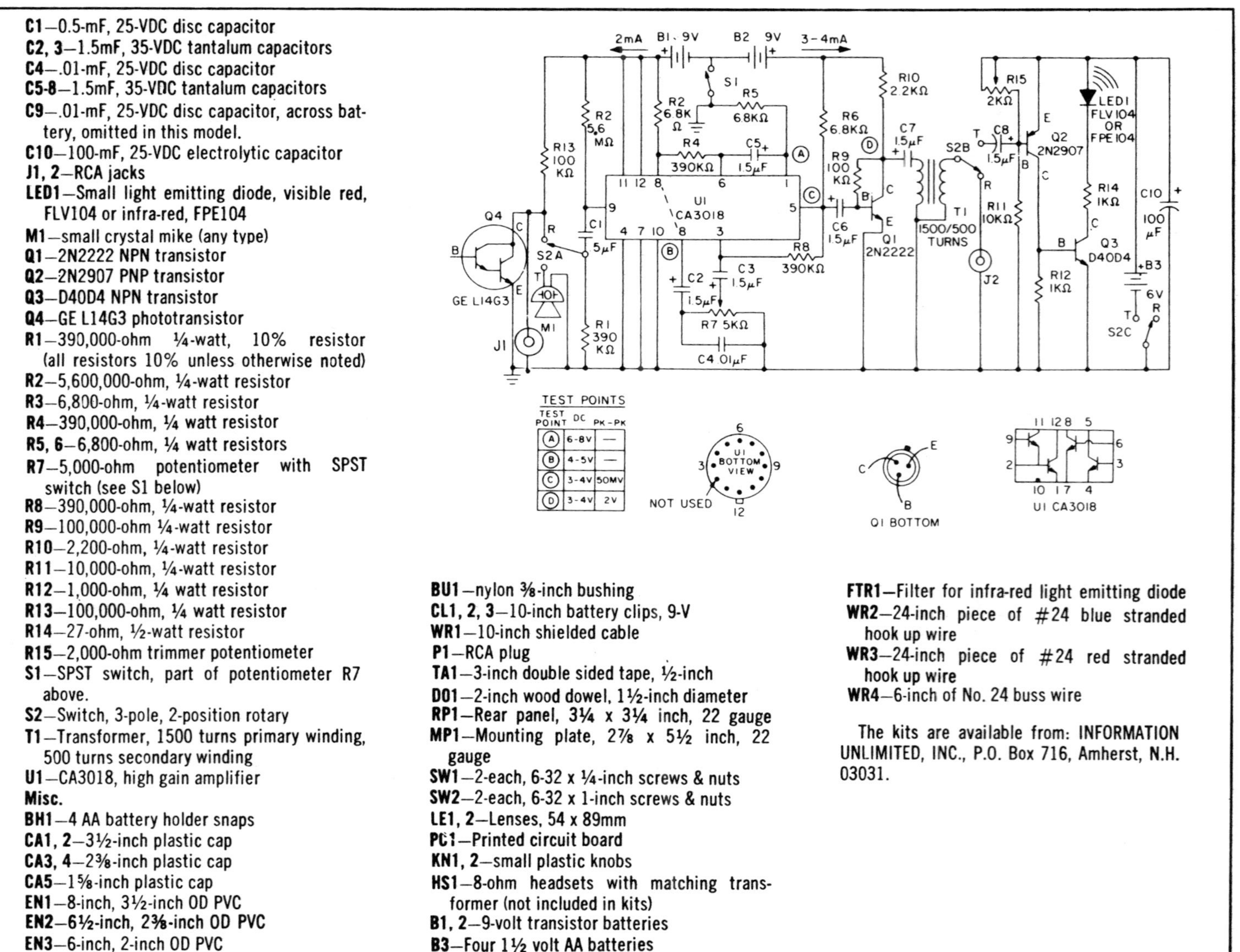

TEST POINT	DC	PK-PK
Ⓐ	6-8V	—
Ⓑ	4-5V	—
Ⓒ	3-4V	50MV
Ⓓ	3-4V	2V

C1—0.5-mF, 25-VDC disc capacitor
C2, 3—1.5mF, 35-VDC tantalum capacitors
C4—.01-mF, 25-VDC disc capacitor
C5-8—1.5mF, 35-VDC tantalum capacitors
C9—.01-mF, 25-VDC disc capacitor, across battery, omitted in this model.
C10—100-mF, 25-VDC electrolytic capacitor
J1, 2—RCA jacks
LED1—Small light emitting diode, visible red, FLV104 or infra-red, FPE104
M1—small crystal mike (any type)
Q1—2N2222 NPN transistor
Q2—2N2907 PNP transistor
Q3—D40D4 NPN transistor
Q4—GE L14G3 phototransistor
R1—390,000-ohm ¼-watt, 10% resistor (all resistors 10% unless otherwise noted)
R2—5,600,000-ohm, ¼-watt resistor
R3—6,800-ohm, ¼-watt resistor
R4—390,000-ohm, ¼ watt resistor
R5, 6—6,800-ohm, ¼ watt resistors
R7—5,000-ohm potentiometer with SPST switch (see S1 below)
R8—390,000-ohm, ¼-watt resistor
R9—100,000-ohm ¼-watt resistor
R10—2,200-ohm, ¼-watt resistor
R11—10,000-ohm, ¼-watt resistor
R12—1,000-ohm, ¼ watt resistor
R13—100,000-ohm, ¼ watt resistor
R14—27-ohm, ½-watt resistor
R15—2,000-ohm trimmer potentiometer
S1—SPST switch, part of potentiometer R7 above.
S2—Switch, 3-pole, 2-position rotary
T1—Transformer, 1500 turns primary winding, 500 turns secondary winding
U1—CA3018, high gain amplifier
Misc.
BH1—4 AA battery holder snaps
CA1, 2—3½-inch plastic cap
CA3, 4—2⅜-inch plastic cap
CA5—1⅝-inch plastic cap
EN1—8-inch, 3½-inch OD PVC
EN2—6½-inch, 2⅜-inch OD PVC
EN3—6-inch, 2-inch OD PVC
BU1—nylon ⅜-inch bushing
CL1, 2, 3—10-inch battery clips, 9-V
WR1—10-inch shielded cable
P1—RCA plug
TA1—3-inch double sided tape, ½-inch
DO1—2-inch wood dowel, 1½-inch diameter
RP1—Rear panel, 3¼ x 3¼ inch, 22 gauge
MP1—Mounting plate, 2⅞ x 5½ inch, 22 gauge
SW1—2-each, 6-32 x ¼-inch screws & nuts
SW2—2-each, 6-32 x 1-inch screws & nuts
LE1, 2—Lenses, 54 x 89mm
PC1—Printed circuit board
KN1, 2—small plastic knobs
HS1—8-ohm headsets with matching transformer (not included in kits)
B1, 2—9-volt transistor batteries
B3—Four 1½ volt AA batteries
FTR1—Filter for infra-red light emitting diode
WR2—24-inch piece of #24 blue stranded hook up wire
WR3—24-inch piece of #24 red stranded hook up wire
WR4—6-inch of No. 24 buss wire

The kits are available from: INFORMATION UNLIMITED, INC., P.O. Box 716, Amherst, N.H. 03031.

Fig. 7-14. Light-Beam Communicator schematic and parts list.

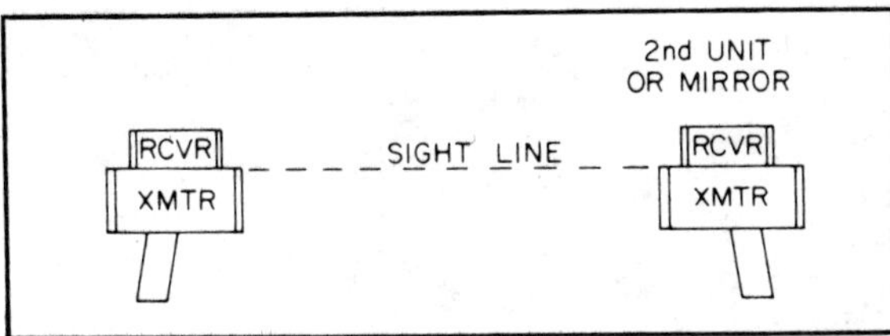

Fig. 7-15. Positioning Light-Beam Communicator for two-way communication is line-of-sight. With ideal conditions, it is possible to send and receive up to one mile.

handle and battery enclosure EN3 is a 6-inch long piece of PVC tubing (schedule 40) with a 2-inch outside diameter. Insert the handle after everything has been assembled, and glue it with PVC cement.

The rear plate RP1 can be fabricated from a 3¼ × 3¼-inch square piece of No. 22 galvanized sheet metal or .035 aluminum. Use the RP1 template shown to locate the holes. The mounting plate MP1 is made from a 2⅞ × 5½-inch square piece of galvanized sheet metal (No. 22) or .035 aluminum. This time, use the MP1 template to locate the holes in this piece.

Centering dowel DO1 has a 2-inch length and an outside diameter of 1½-inches. It should fit smoothly in to EN2. The cable, WR1, is fed to the phototransistor, Q4, through a slightly off center feed hole in the dowel. The connection is made by soldering to the exposed leads of Q4. The leads should be as short as possible, and glued with RTV cement. The leads should be only long enough to allow touch up, repositioning to the true optical axis.

The plastic cap CA1 is 3½ inches, with a 1⅝-inch hole in the center. Use a sharp knife or small snips. If you are not neat in this procedure, the appearance of the device can be ruined. Four small pieces of double sided tape (TA1) are used for securing the lens LE1 to the cap. Be sure that the tape does not contact the ridge of EN1, otherwise it will be difficult to remove for checking. The other plastic cap CA2 is 3½-inch. It has a ¼-inch lip to hold the subassembly into EN1.

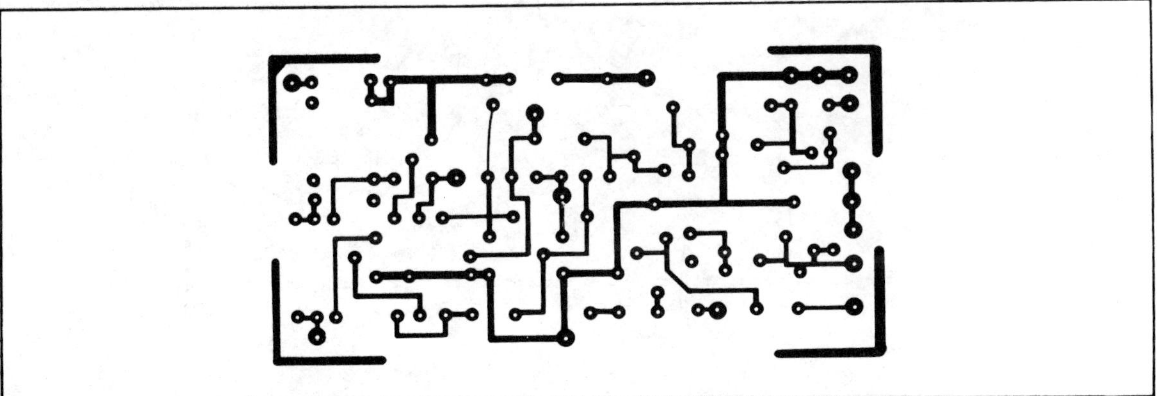

Fig. 7-16. If you are not an expert at etching your own circuit boards, we suggest you buy the kit from Information Unlimited, which includes PC board and parts.

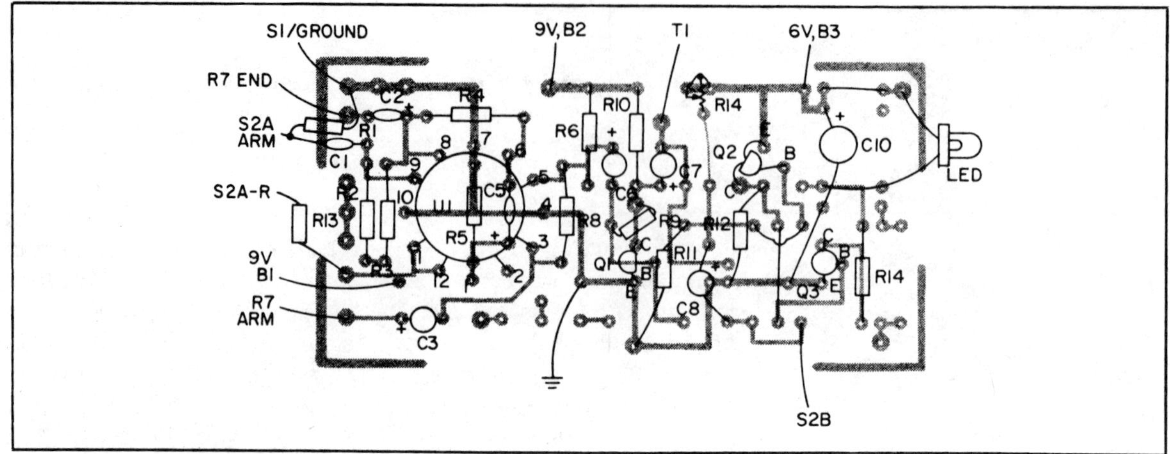

Fig. 7-17. The parts overlay shows components on top of PC board. Note that all of the electronic components fit handily on board, making for easy assembly.

CA3 is the 2⅜-inch plastic cap. Remove the ⅜-inch lip, to retain lens, LE2, and optional filter FTR1. These are fitted against the end of EN2. CA4 is a 2⅜-inch plastic cap. Place a small hole for cable WR1, to create a friction hold and prevent DO1 from sliding once set. After alignment, secure with RTV cement. These plastic caps are available from Information Unlimited (refer to parts list).

It would be preferable to construct all circuits on a board, unless

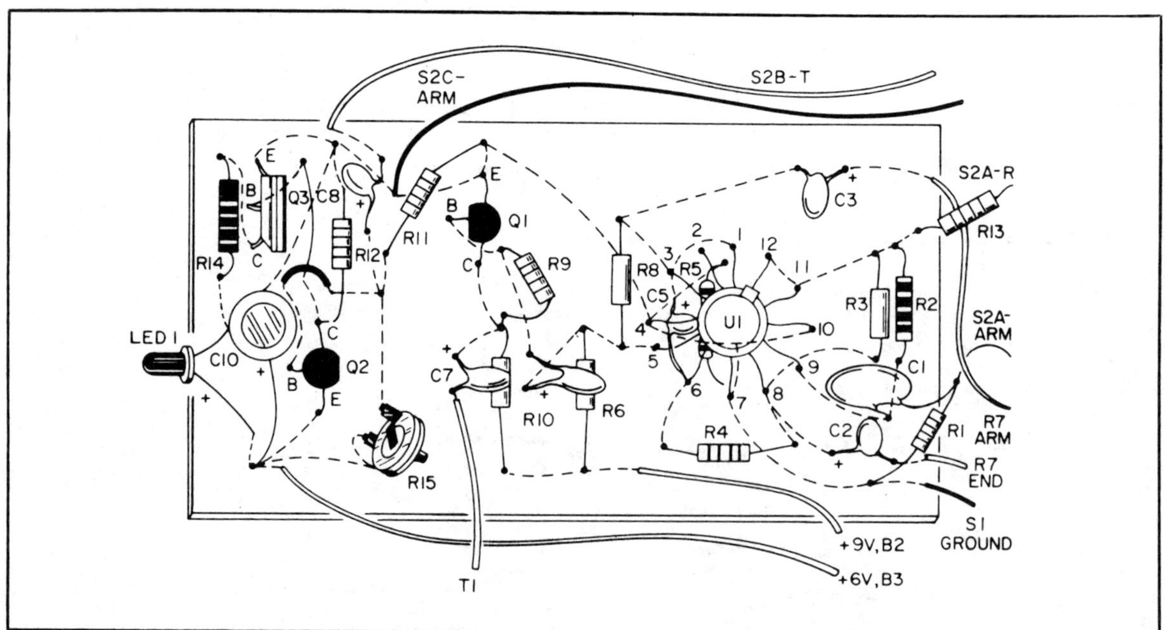

Fig. 7-18. Actual parts placement on the Light-Beam Communicator's board is not critical; however, you must identify all of the leads going to the components to avoid confusion.

you are very familiar with perfboard assembly. If so, a 3 × 1½-inch perf-board, with a .01-inch grid, may be used. Follow the layout as shown in the printed circuit board layout. As always, use resin-core solder when you build this project. Remember to remove any excess solder from the PC board to prevent shorting.

Assemble RP1 as shown in Fig. 7-22. Atach R7 to RP1 as illustrated in Fig. 7-20; the leads from R7 must be twisted and as short as possible. Connect C4 to S2 and then attach the assembly to RP1. Twist all wires together leading to S2. Route these leads close to metal. Switch leads are all identified in the schematic. Attach microphone M1 using RTC cement. Position and wire as shown. Assemble jacks J1 and J2. Attach RP1 to MP1 via screws SW1 (6-32 × ¼-inch).

Construct PC1 according to the circuit foil pattern shown. Assemble the board by using the schematic diagram given and Figs. 7-18 and 7-20. Note polarity of tantalum and electrolytic capacitors and the position of U1, Q1, Q2, Q3 and LED1.

Connect the wires from RP1 assembly to board noting identification of leads shown in sketches. Position as shown in Fig. 7-20, but do not adhere to tape at this time. Carefully position LED1 into B3 as shown. Attach battery clips CL1, 2 and 3 to respective points, as shown in Fig. 7-23. It is advantageous at first to allow the board freedom to be moved for total access during preliminary troubleshooting and testing. Leads may be further shortened after several minutes of proper operation have been verified.

Next, attach T1 to MP1, bending tabs in the small holes. Solder one lead from primary to secondary and sandwich between core of T1 and plate. This makes the ground contact of the transformer. You may want to solder these wires directly to the plate for a positive contact. Use a heavy iron for this. Note ungrounded 500-ohm lead going to S2B and ungrounded 1.5K ohm (winding marked P) going to C7 on board. Connect PL1 phone plug to WR1 cable from receiver section.

Testing the Units. It is assumed that the assembled unit, to this point, has been wired correctly, with no shorts, and good solder connections. You will note that the complete working unit is conveniently built on a single removable assembly. This assembly should have the battery clips CL1, 2 and 3 connected to their respective batteries.

Testing the Receiver. Turn S2 and R7 fully counterclockwise. Connect one terminal of a fresh 9-volt battery to CL1 and connect a 100-mA ammeter between the contact of the battery and the clip. Turn on R7. Current reading should be approximately 2-mA. Fully connect battery and designate B1.

Repeat above using a second battery connected to CL2. Turn on R7/S1. Current reading should be 3 to 4-mA. Fully connect battery designated B2.

Fig. 7-19. Side view of the Light-Beam Communicator.

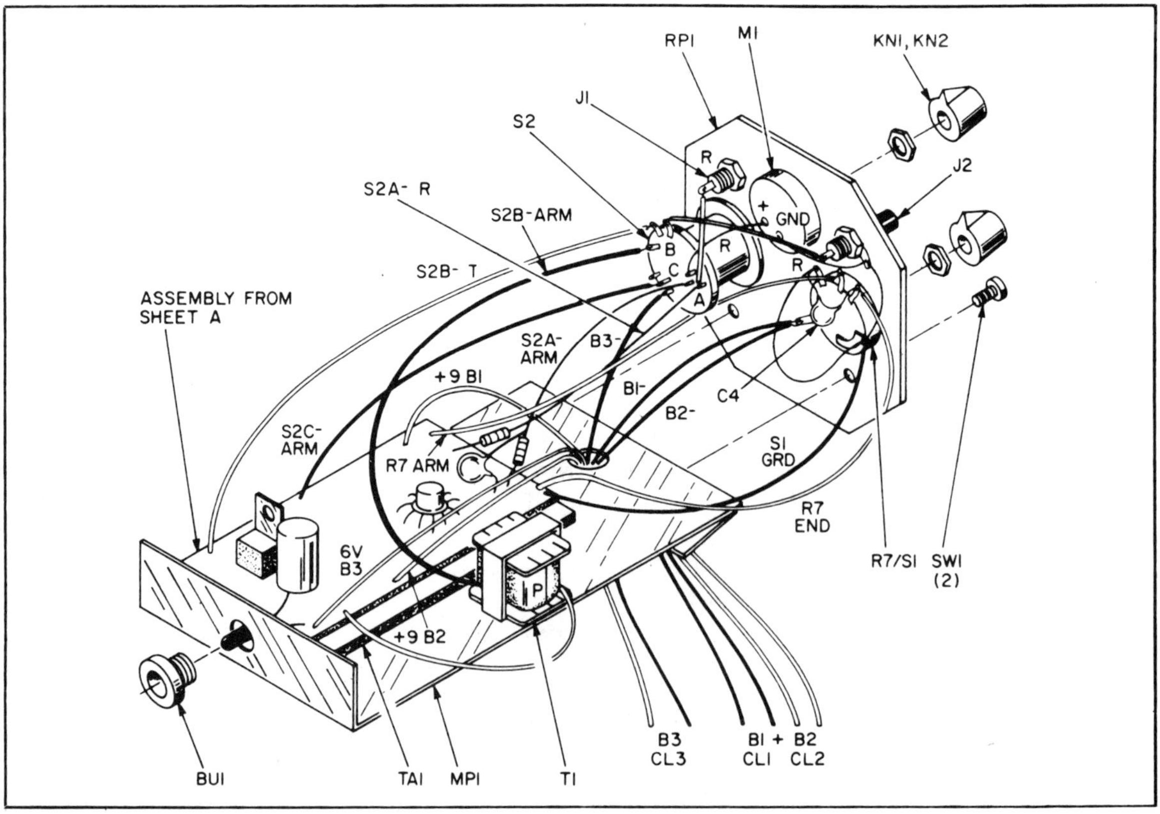

Fig. 7-20. Leads from the board to the rear plate assembly (RP1) are positioned as shown. Leave plenty of free play in the leads to allow for adjustments to the device.

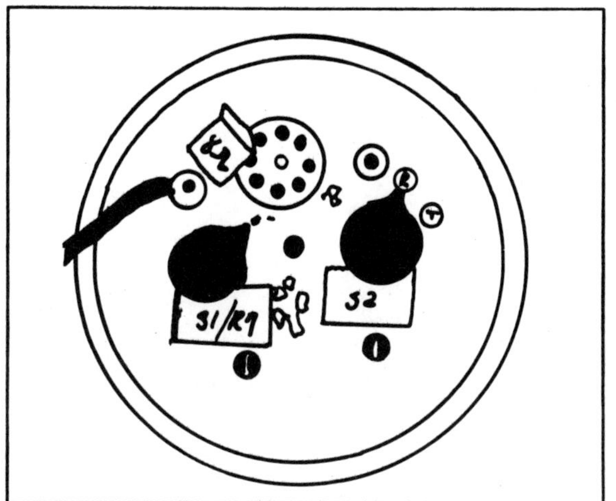

Fig. 7-21. The back end of the transceiver shows mike, mode, and volume controls.

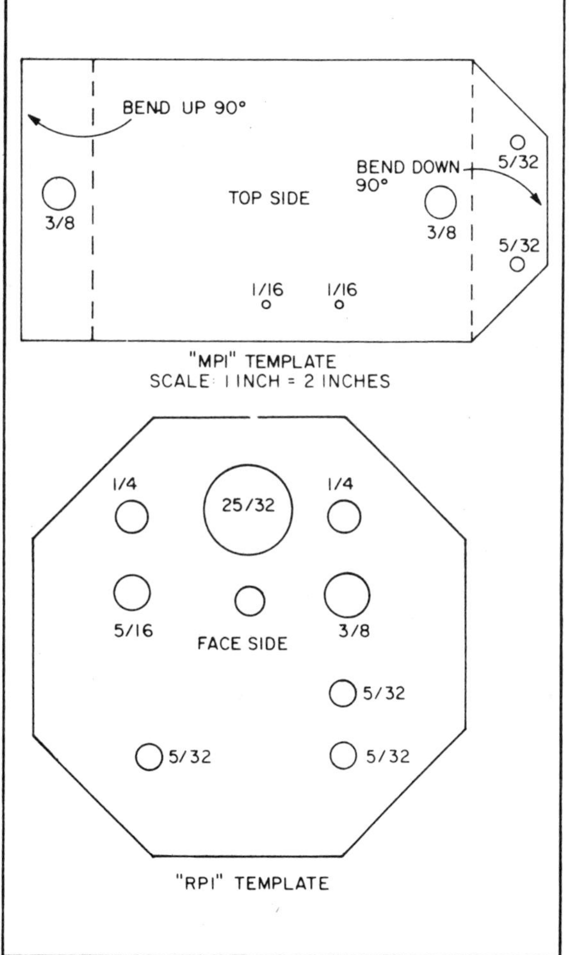

Fig. 7-22. The dimensions of the board and back plate mountings for Light-Beam Communicator unit assembly.

Plug a high-impedance set of headphones into J2. HS1 is a standard 8-ohm headset with a spliced in matching transformer that steps up to 1000 ohms. This is suggested, as high-impedance headsets are scarce and uncomfortable to wear. Plug PL1 from Q4 (receiving phototransistor) into J1.

Turn on R7/S1 and slowly turn up gain until a loud 60-Hz hum is heard. This is the normal ac room lighting frequency being picked up by Q4. At normal ambient lighting conditions it will completely block the amplifier. Reduce the gain and attempt to point Q4 at various objects indicating different levels of signal, depending on reflection characteristics of surfaces. You will note that the circuit is relatively prone to power line hum pick up. It is assumed that testing will be done in normal electrical lighting for this step. If not, you may not obtain the 60-Hz hum.

If everything above checks out OK, you can proceed to the transmitter section. If not, troubleshoot the faulty circuit. It may be convenient to use the test points shown on the schematics and thoroughly familiarize yourself with the circuit description given in the beginning of the plans.

Testing the Transmitter. With all switches full counterclockwise, connect CL3 to 6-volt B3 as done with B1 and B2. Connect a 100-mA meter in series and turn S2 clockwise to the transmit position. Adjust R15 to read 25-50-mA. LED1 should light to about one-half maximum brilliance. Turn R7/S1 and note LED1 changing brilliance with sounds. Whistle. The current meter should jump to nearly 100-mA. *Note:* LED1 increases in brilliance with sound indicating upward modulation. The device seems to work all right with the LED1 downward modulating, but we recommend upward modulating. Certain diodes may require less current than 50-mA for good upward modulation.

You are now ready for final assembly of the unit, and optical alignment. Note that LED1 should automatically center itself inside of EN1.

Optical Alignment. In order to obtain maximum performance from your light-beam communicator, it is necessary to properly optically align both transmitter and receiver according to the sight line diagram. You will note that the receiver tube EN2 is secured to EN1 via screws. The rear hole is slotted to allow a side to side movement of the receiver enclosure in respect to the transmitter enclosure. The up and down position usually self-adjusts simply by the abutting of the two enclosures. Remember that both receiver and transmitter sections must be optically aligned to view the same area for maximum two-way communications.

The method we demonstrate here is not necessarily the only way to align these devices and is only suggested as a possible means. The builder may have his own ideas and methods for accomplishing the above.

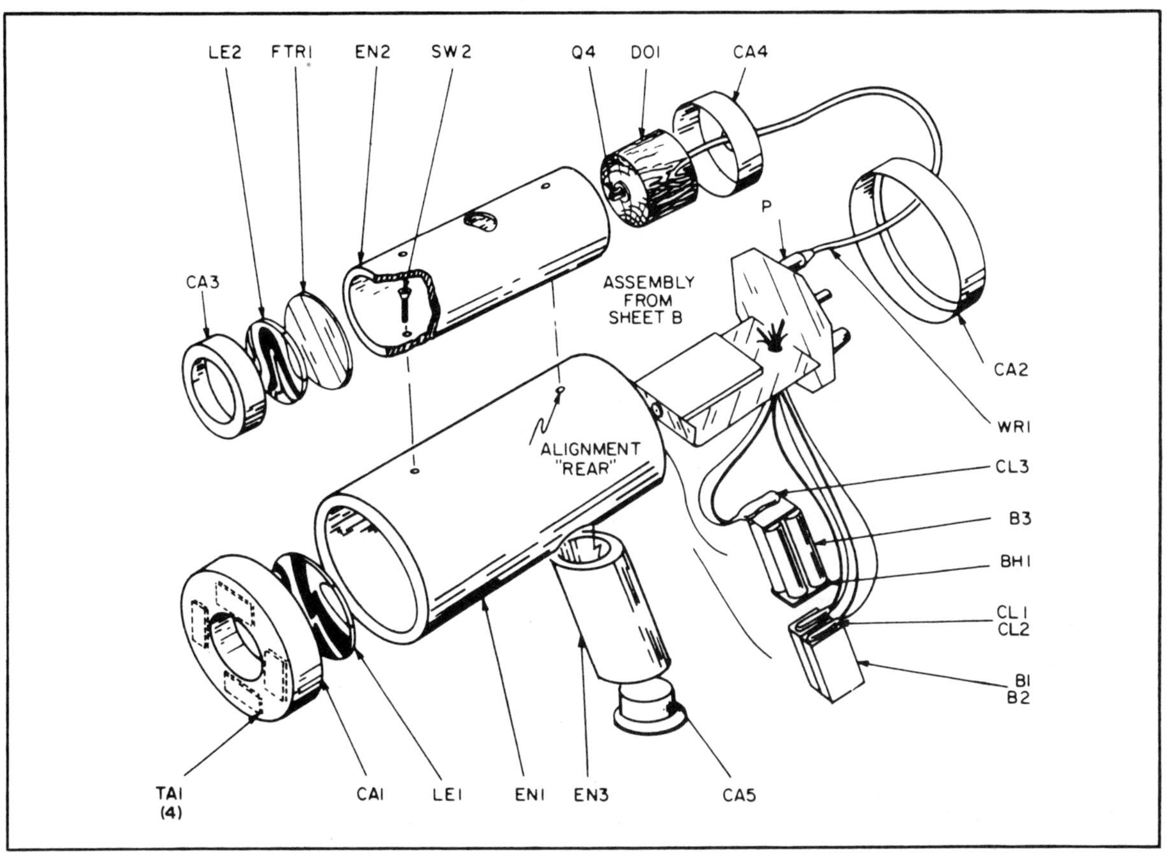

Fig. 7-23. This exploded view shows the placement of the boards, housing, lenses, and other parts necessary for construction and operation of the Light-Beam Communicator.

The following steps were used at our lab and found to be relatively easy in accomplishing acceptable alignment.

☐ Remove transmitter lens and cover and place some thin paper over open end. Adjust LED1's output to the center of paper (this is the bore sight of enclosure). Secure and replace the lens and cover.

☐ Secure communicator in vise or other similar holding attachment.

☐ Locate a mirror about 20 feet from the device.

☐ With transmitter properly aligned and secured, adjust mirror for reflection of output light occurring in receiver lens. This is adjusted by sighting mirror reflection along sight line at surface intersection of the two enclosures (note drawing).

☐ With the device and mirror aligned as above, carefully adjust the position of the receiver enclosure so that the focused received beam is centralized (bore sighted) inside the tube. This is best accomplished by placing a thin strip of paper through the adjust hole over the phototransistor and adjusting the dowel to the focal length of the lens. This should place the focused received light directly on the lens of the phototransistor. Further touch up can be done by careful positioning of the phototransistor with needle nose pliers. Secure dowel, enclosures, and other parts to eliminate movement and improper alignment.

☐ Repeat with the other unit. You should now be able to hand-sight units along sight lines for medium range use. Good, reliable long range use should be done with a camera tripod. Night use, with the visible red transmitting diodes is easily accomplished at ranges up to 1000 meters or so by noting the reflection of the transmitting light in the receiving lens as noted at the transmitter station. Daytime operation is best with filter and IR transmitter. Securing of optical components via permanent means should only be done when optimum optical alignment can be assured.

Operation. For both transmitter and receiver to be in the OFF mode, S2 must be at R position and R7 to the OFF mode.

To use receiver only, plug in headsets to J2 jack and turn on R7 and adjust to desired level (usually no more than ⅛ turn). Point unit at a normal 60-Hz lamp, TV or other light source and note hum.

To use in transmit mode, all that is necessary is to place S2 from the R to the T mode. The modulation level is preset by R7 when used in the receiver mode.

One way to test is to look into the transmitter section and note the LED flickering with audio signals. R7 can then be readjusted if necessary by this indication. Note that the LED only has to change ever so slightly for sufficient modulation.

You will see there is a trimpot R15 on PC board. This adjusts the quiescent current through this LED and should be set just where the LED is emitting with no audio signal. This saves batteries and prevents downward modulation. This probably should be reset as batteries weaken. Also, note that the units pick up 60-Hz hum from power lines and normal lighting. The visible red LED (supplied in these units) obviously operates best in darkness. For normal daylight operating, the infrared LED and filter must be used.

In general, reception is possible as long as the transmitter output light can be seen by the naked eye.

Applications. Aside from the line of sight communications possibilities, this communicator is extremely useful for surveillance applications. Install one of the communicators in a location that is vulnerable to trespass by intruders. You can buy an inexpensive sound-activated alarm, with its microphone taped to an earphone. The alarm will be activated whenever there is noise at a normally quiet location which is under surveillance.

TAPE TESTER

Virtually no two brands of recording tape, particularly the cassette tapes, deliver similar results at slow speeds. Even within the same price range, tape A might sound best on machine X, while tape B is the right one for machine Y. Fact is, because there is so much variation between tapes, many of the better cassette decks provide some means to optimize the bias for optimum frequency response from the tape playback.

The trouble is, not all machines have user-adjustable bias, and some models with the adjustment have no test system. The user must try for an adjustment that sounds right to the ear, a somewhat dubious method. Basically, most of us simply try a wide assortment of brands and types hoping to find the one that works best on our recorder.

Refer to Figs. 7-24 through 7-26.

With our easy-to-build Tape Tester, however, you can check out tapes in seconds to find the one that gives the best response in your sound system. And the same thing goes if your deck has a bias adjustment; again, it takes just a few seconds to find the optimum adjustment for peak performance.

Easy to Use. The tape tester provides two switch-selected test tones of approximately 1000 and 12,000 Hz at 0 dB reference output level—for adjustment of the recording level—and at approximately −20 dB, the standard cassette test level. (Of course, it can be used for reel-to-reel machines, though their test level is generally −10 dB.)

To determine whether you are using the best tape for your machine, or to make a bias adjustment, you simply record the −20 dB 1000 Hz signal for a few seconds, followed by 12,000 Hz. On playback, the 12,000 Hz tone should be within 3 dB of the 1000 Hz reference for "average" performance, and within 1 dB for optimum sound reproduction. If you're using metal tape and want to see how well the tape, or the machine itself, handles high-frequency high-level situation, simply use the 0 dB output before running the standard −20 dB level test.

As an example, assume you've checked a tape and the 12,000 Hz playback is 7 dB below the 1000 Hz playback. Obviously, this isn't hi-fi, so try a tape with "hotter" highs. Alternately, if the 12,000 Hz output was 6 dB above 0 dB, you need a tape with less high frequency output—generally a less expensive tape. If your machine has a bias adjustment but no test system, simply adjust the bias until the 12,000 Hz playback level reads within 1 dB of the 1000 Hz playback level on your deck's VU record/playback meters.

Construction. The Tape Tester shown here is assembled in a

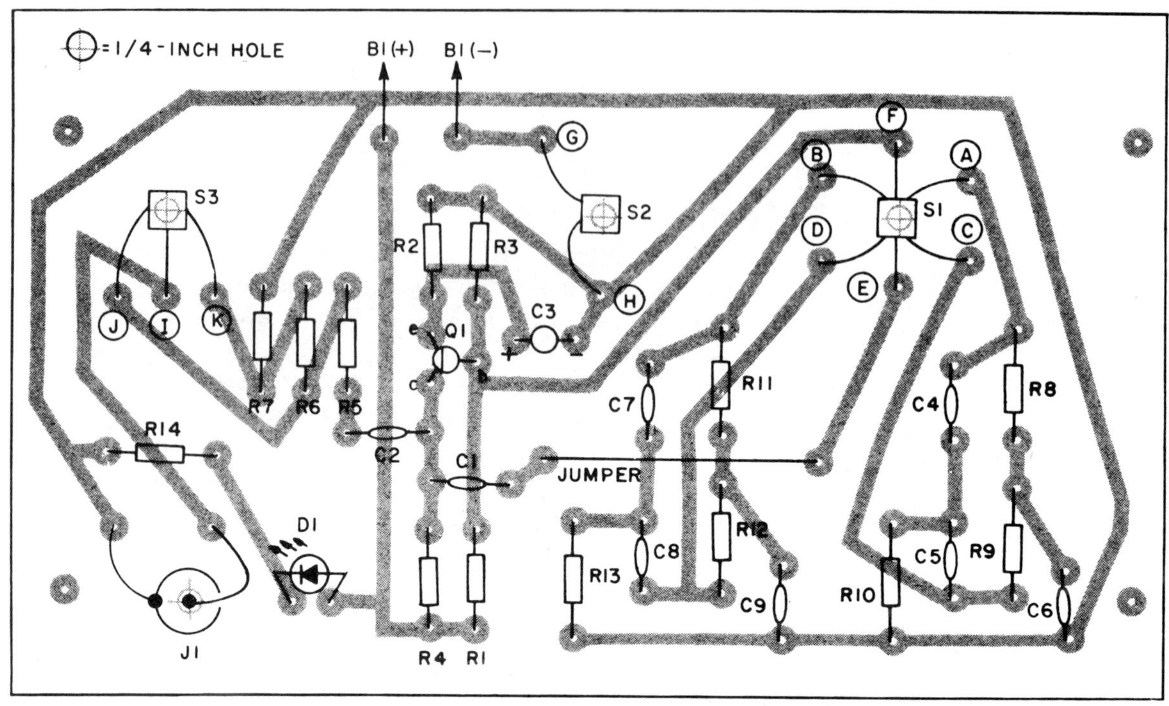

Fig. 7-24. In this expanded view of the parts arrangement on the printed circuit board, note that ¼-inch holes are required for mounting S1, S2, S3, and J1. Be sure to remember to wire in the lone jumper between C1 and point E at switch S1. Also, observe polarity when installing D1 and C3 on the PC board.

plastic cabinet sized approximately 2⅝ × 1⅝ × 5 inches, usually available from Radio Shack. Some cabinets have square corners; some are round. It doesn't make any difference unless you contemplate problems in rounding the corners of a printed circuit board.

The cabinet cover (which replaces the supplied metal cover) is the printed circuit board itself. This arrangement eliminates the hassle of trying to mount the PC board inside the cabinet.

Before etching the printed circuit, make certain the board is cut to the exact size of the metal cover supplied with your particular cabinet. Not only is there the problem of round or square corners, but the overall dimensions and the location of the mounting holes vary from cabinet to cabinet. The holes for the switches and jack are ¼-inch diameter, and all other component holes are made with a No. 58, No. 59, or No. 60 drill.

All oscillator values have been selected so the project will work with low-cost, normal tolerance components. There is no need to go through the extra expense of purchasing precision resistors and capacitors. However, take extra care that you do not change any specified value. For example, do not change R1 to 270K, or R13 to 1200 ohms.

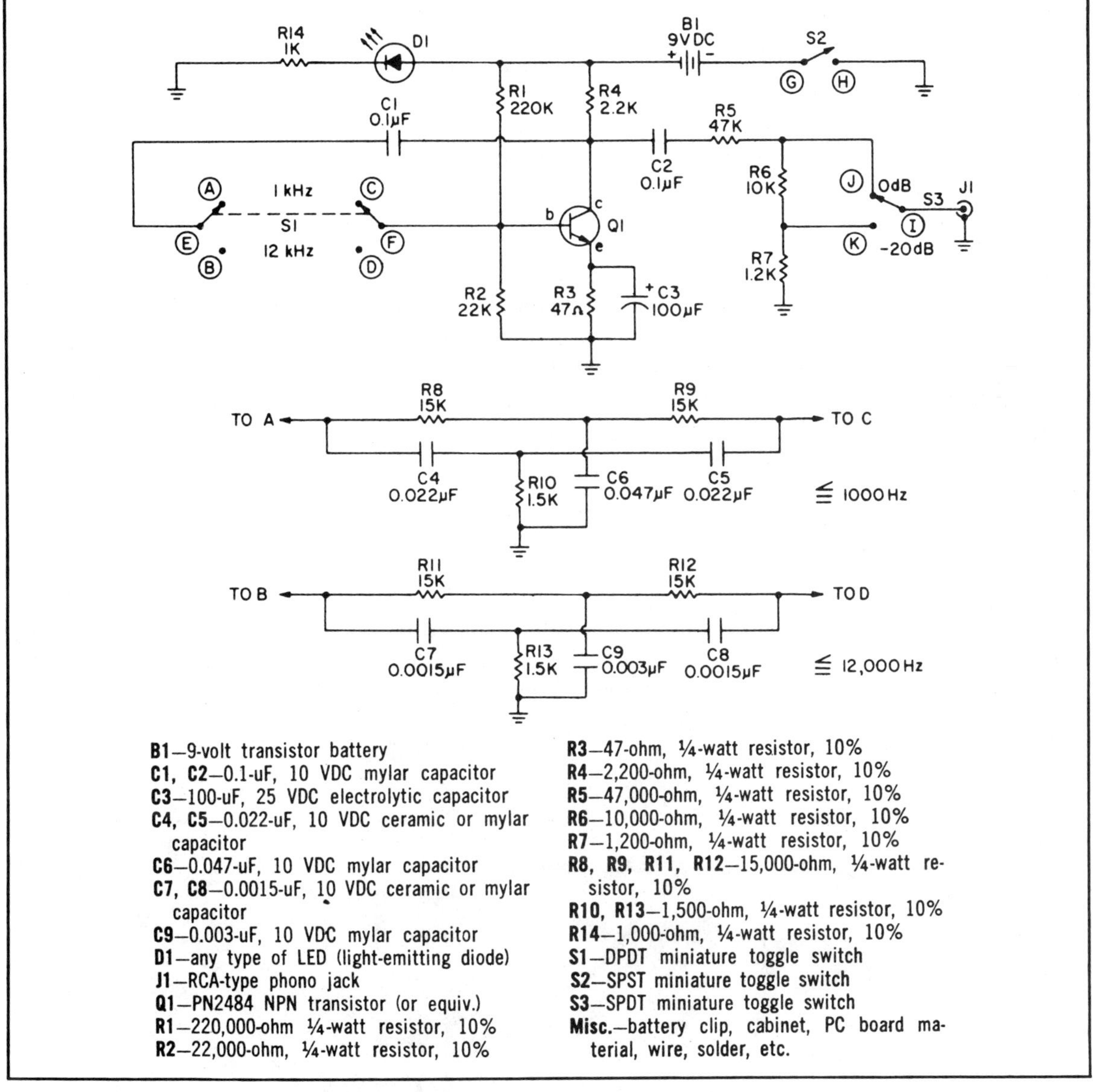

Fig. 7-25. Tape Tester schematic and parts list.

If you cannot obtain the specified values, make them. For example, the 0.003-μF capacitor used for C9 can be a 0.002-μF and a 0.001-μF connected in parallel by simply twisting their leads together. Similarly, C7 and C8 can be a 0.001-μF in parallel with a 0.0005-μF (500 pF). The 0.022-μF capacitor is a standard value available just about anywhere. Similarly, all resistor values are standard in the sense that they are generally available types.

Since the switches are installed from the foil side of the board, take care that the connecting wires don't extend up *through* the board where they can act as a hazard to fingers working the switches. First, solder the wire to the switch contact (or jack). Before soldering to the board, trim the wire flush with the top (non-foil side) of the PC board. Next, using long nose pliers or tweezers, back the wire very slightly towards the switch so the end is just below the top of the PC board—but still within the hole—then solder the wire to the foil. If done correctly, there will no sharp edges sticking up through the PC board.

Check the LED polarity before you install it on the PC board, because some of the instructions supplied with hobby-grade LEDS are incorrect. Temporarily tack-solder a 1000-ohm resistor to either LED wire. Clip one LED wire to either battery terminal. Touch the free end of the resistor to the other battery terminal. If the LED doesn't light, reverse the wires to the battery. When the LED lights, the LED wire attached to the resistor is the same polarity as the battery terminal to which the resistor connects. (If the resistor connects to the positive battery terminal, the LED wire on the opposite end of the resistor is the anode). If the resistor is connected to the negative battery terminal, the LED wire on the opposite end is the cathode. If the LED doesn't light with either polarity, it is defective (not unusual in LED hobby assortments.)

Using the Tape Tester. The output of the calibrator at 0 dB is nominally 0.5-volts, and should be connected to a recorder's LINE/AUX input, not to the microphone input. Because of normal component tolerance, the oscillator might not "start" when power is first applied if the frequency switch is set for 12 kHz. If this occurs with your model, rather than trying to "trim" resistor and capacitor values in the twin-T

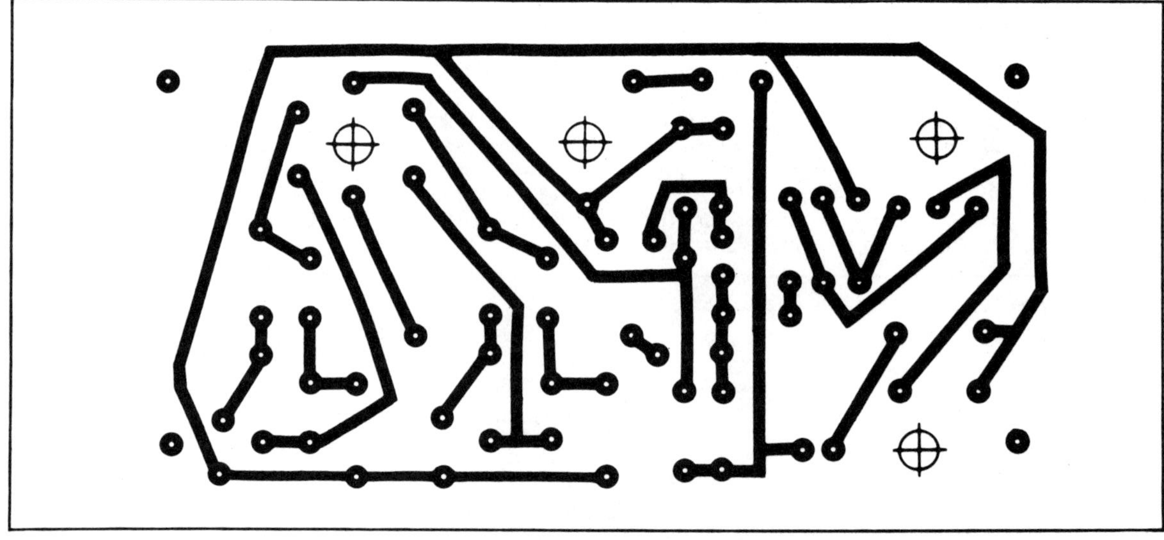

Fig. 7-26. Full-scale etching guide for the Tape Tester's PC board. Note that the holes marked for the switches and J1 are not full-size. You will need a ¼-inch drill bit.

feedback networks—R8 through R13 and C4 through C9—simply set frequency selector S1 to 1000 Hz before applying power. The oscillator will always start at the lower frequency. If the oscillator does not start at all, either you have made a wiring error, or the capacitors in each twin-T network are not approximately equal to the specified values. Generally, the C4/C5 and C7/C8 combinations don't have to be the precise specified value, but each capacitor in a pair must be close in value to the other.

Set S3 for 0 dB output, S1 for 1000 Hz, and turn power switch S2 on. Adjust the record level for a 0 dB record level, or whatever is the maximum meter-indicated level for your recorder. Set switch S3 for a −20 dB output; you will see the recorder's meter drop 18 to 20 dB. (The exact amount depends on the components used in your project. It doesn't matter what the drop is, as long as it's near 20 dB.)

Start the recorder in the record mode and record a few seconds of 1000 Hz, followed by a few seconds of 12,000 Hz. On playback, the output of the two frequencies should be within 3 dB. Adjust the recorder's bias, or select a tape type that delivers this performance. Because of the 20 dB input attenuation, you will probably need an external level indicator because the usual −20 dB indication on a recorder's meter is not all that easy to read for precise value on playback. Use any external indicator, such as the meters on another recorder (feed the output of the recorder being tested to the input of a second recorder), the meters of a power amplifier or an ac/audio meter. You can even use an oscilloscope if you have one. To check for metal tape saturation at maximum recording level, set S3 for 0 dB and record at 0 dB record level, or whatever value is maximum for your recorder.

THE OBNOXILLATOR

This little audio oscillator emits a sound that's obnoxious to both man and beast, which is why we call it the Obnoxillator. The tone starts out at a relatively high pitch which, over a period of about one second, swoops downward in frequency. Then, the signal jumps abruptly to its initial pitch and commences its downward plunge once more. The effect is approximately as pleasant as running your nails over a blackboard, and as such it will get people's attention—if not their admiration.

Q1 together with R1, R2, R3 and C1 comprise a conventional UJT relaxation oscillator with a period of approximately one second. The roughly sawtoothed voltage developed across C1 drives current source Q2, the output of which charges capacitor C2. Adjustment of R4 affects the magnitude of the current and, hence, the rate at which capacitor C2 charges. Unijunction transistor Q3 discharges C2 when the voltage on the capacitor reaches 4.2 volts, or so. The rate at which the voltage on C2 oscillates is in the audio range and is much faster than that of the waveform developed across C1.

Refer to Fig. 7-27.

After C2's discharge, the capacitor once again gets charged up by current from Q2. Since Q2's charging current is a function of the voltage across C1, the rate at which C2 charges will vary (in fact, diminish) over the 1-second interval it takes for C1 to charge. Once Q1 discharges C1, Q2's charging current returns to a high value, and the frequency of the sawtooth waveform across C2 jumps back to its initial high value.

Emitter follower Q4 reads the signal developed on C2 and provides a buffered audio output with a maximum peak-to-peak amplitude of about 1-volt. Volume control R9 can be used to vary the magnitude of the output, which should drive an audio amplifier through its high-level input.

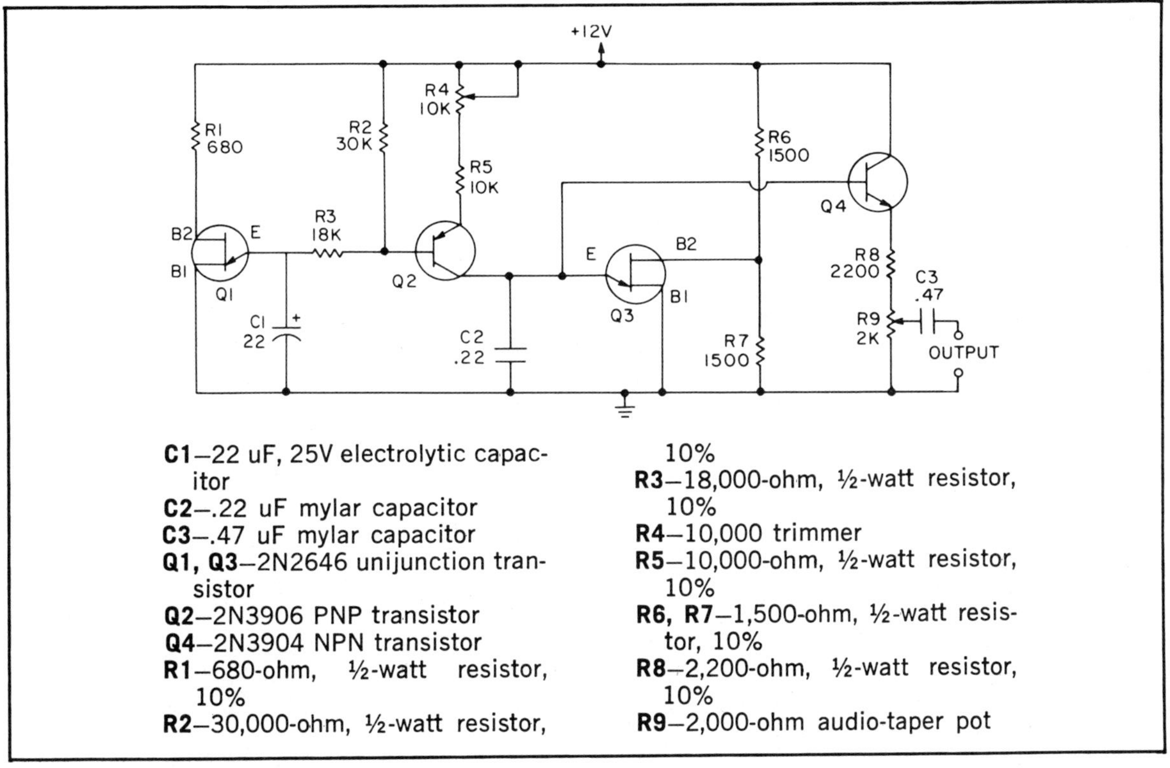

C1—22 uF, 25V electrolytic capacitor
C2—.22 uF mylar capacitor
C3—.47 uF mylar capacitor
Q1, Q3—2N2646 unijunction transistor
Q2—2N3906 PNP transistor
Q4—2N3904 NPN transistor
R1—680-ohm, ½-watt resistor, 10%
R2—30,000-ohm, ½-watt resistor, 10%
R3—18,000-ohm, ½-watt resistor, 10%
R4—10,000 trimmer
R5—10,000-ohm, ½-watt resistor, 10%
R6, R7—1,500-ohm, ½-watt resistor, 10%
R8—2,200-ohm, ½-watt resistor, 10%
R9—2,000-ohm audio-taper pot

Fig. 7-27. Obnoxillator schematic and parts list.

SIGNAL-OPERATED SWITCH

If a VOX is a voice-operated switch, is this signal-operated switch a SOX?

You can take a signal, like the earphone jack output from a radio or tape player, and use it to trigger the relay operation. If used with an FM wireless mike, an FM radio and a cassette recorder, for example, this circuit could start the recorder whenever the FM radio receives the wireless mike signal. D1-R1-C1 form an R-C delay network that delays the turn-off of the relay until some time (the number of seconds of delay is roughly the number of ohms of R1 times the number of microfarads of C1 divided by a million) after the signal stops.

Refer to Fig. 7-28.

The signal charges C1 through D1, which keeps it from discharging back through the signal source. C1 then holds the base of Q1 high until it discharges enough through R1 and the base-emitter circuit of Q1 to reach a turn-off point. Q1 completes the circuit for K1's coil, and you can do whatever you want with the contacts (turn on a light, start a motor, honk a horn, fire up a computer, light up your TV).

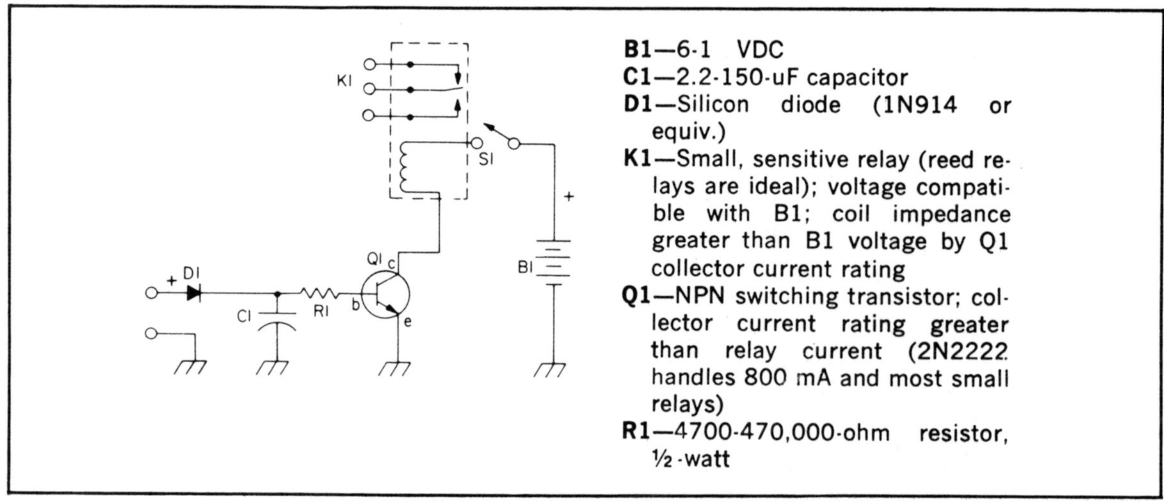

B1—6-1 VDC
C1—2.2-150-uF capacitor
D1—Silicon diode (1N914 or equiv.)
K1—Small, sensitive relay (reed relays are ideal); voltage compatible with B1; coil impedance greater than B1 voltage by Q1 collector current rating
Q1—NPN switching transistor; collector current rating greater than relay current (2N2222 handles 800 mA and most small relays)
R1—4700-470,000-ohm resistor, ½-watt

Fig. 7-28. Signal-Operated Switch schematic and parts list.

MAGNETIC LATCH

Refer to Fig. 7-29.

Here is a circuit that enables you to turn a device ON and OFF using a small, permanent magnet. Bring your magnet close to S2, and the reed switch will close. This shunts all current away from Q2's base and sends it collector potential high. As a result, base drive is available for Q1 and Q3. Transistor Q3 obliges by conducting and thereby causing relay K1's contacts to close. Q1 conducts too, and this sends Q1's collector to ground potential. Therefore, when you remove your magnet from S2, Q2 remains latched in a non-conducting state, since Q1's collector is low.

Now, suppose you approach S1 with your magnet. Once the reed switch closes, it shunts base current from Q1, thus causing Q1's

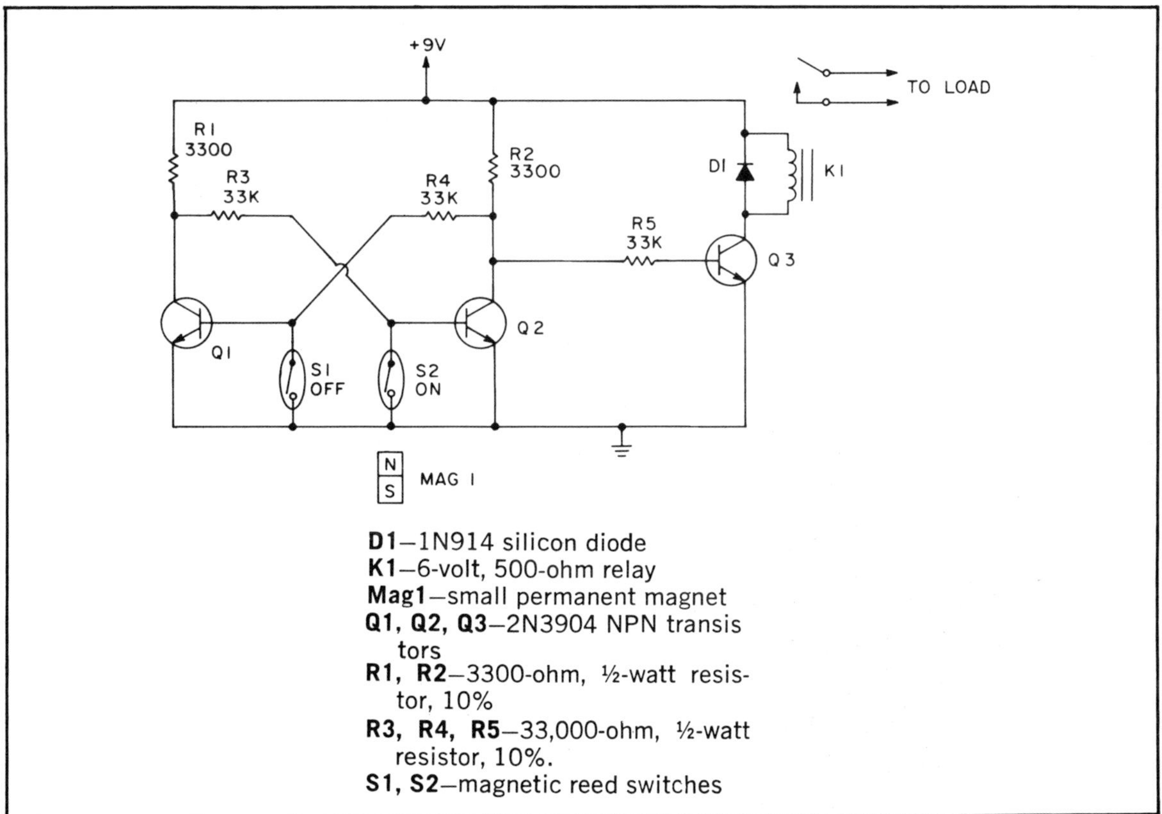

D1—1N914 silicon diode
K1—6-volt, 500-ohm relay
Mag1—small permanent magnet
Q1, Q2, Q3—2N3904 NPN transis tors
R1, R2—3300-ohm, ½-watt resis- tor, 10%
R3, R4, R5—33,000-ohm, ½-watt resistor, 10%.
S1, S2—magnetic reed switches

Fig. 7-29. Magnetic Latch schematic and parts list.

collector potential to go high. As a result, Q2 receives base current that causes it to conduct, which sends its collector to ground potential. This removes base drive from Q3, so K1 is no longer energized. When you remove your magnet, the circuit remains latched in this OFF condition, since Q2's grounded collector cannot supply base current to transistor Q1.

PHOTOELECTRONIC ANNUNCIATOR

Refer to Fig. 7-30.

Momentarily interrupt the beam of light shining on Q1, and you get a one-second beep from this circuit. Most likely you've encountered circuits of a similar nature in retail stores, where the buzzing sound signals your entrance and alerts salesmen to their prey. Obviously, a great many other applications are possible as well.

With light shining on Q1's sensitive face, the phototransistor conducts heavily and shunts current away from the base of Q2. But

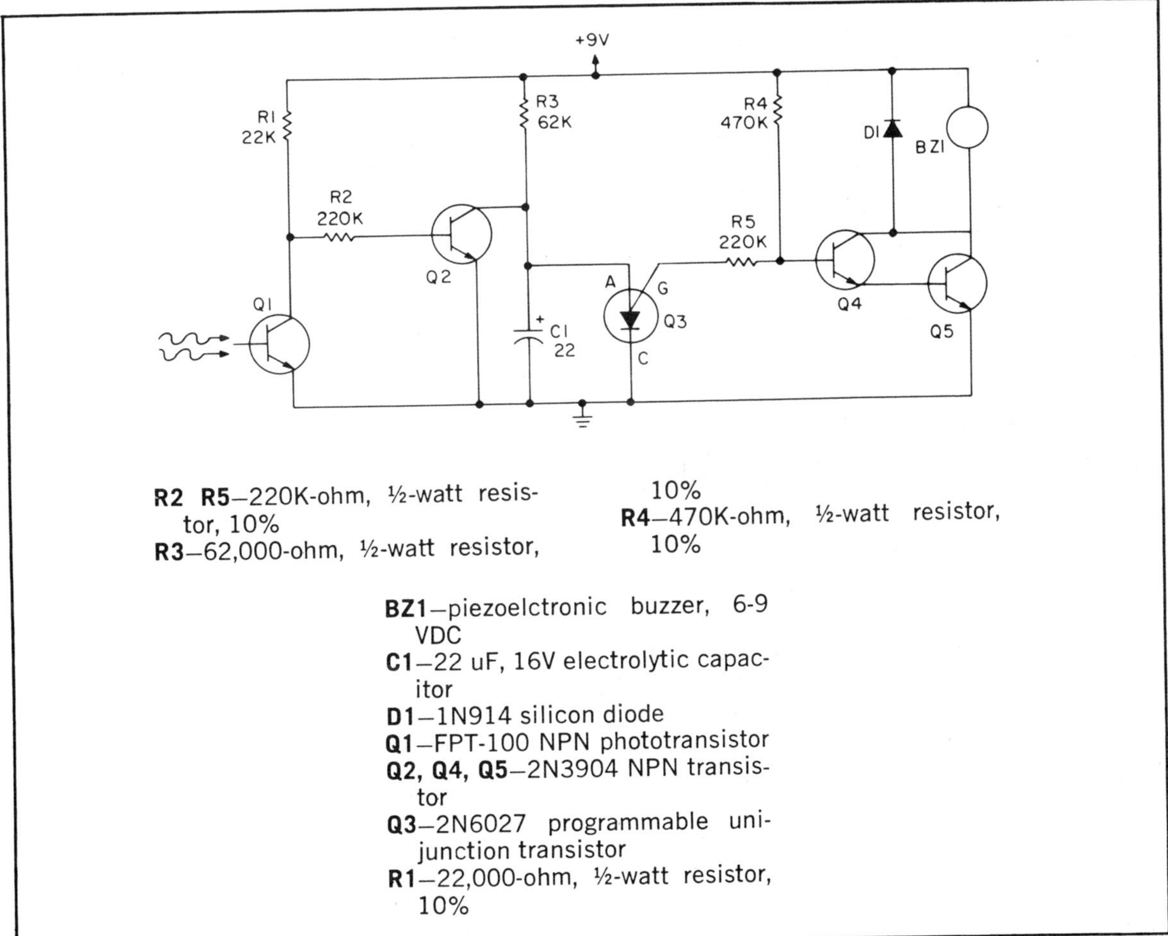

Fig. 7-30. Photoelectronic Annunciator schematic and parts list.

when the beam of light is interrupted, Q1 ceases to conduct—thus allowing current to flow through R1 to R2 into Q2's base. The collector of Q2 then conducts current and rapidly discharges capacitor C1. This allows Q3's gate lead (G) to swing high, thereby turning on Q4, Q5 and the buzzer.

Assuming that the interruption of the beam was only temporary, Q2's collector will now have ceased to conduct current. This allows C1 to charge until it reaches a level sufficient to trigger Q3, a programmable unijunction transistor (PUT). When that happens (in about 1 second), Q3's gate potential drops, which turns off Q4, Q5 and the buzzer. Another interruption will repeat the whole process and yield one more beep.

IDIOT'S DELIGHT

Refer to Fig. 7-31.

Sometimes the dashboard idiot lights aren't warning enough that something's gone awry. Bright sunlight, a burned-out lamp or simply a lack of attention can obviate Detroit's brilliant efforts. But this simple gizmo adds a buzz to their blink, plus a luxurious extra. R1, C1 and Q1 give you about 7 seconds when you first get into the car to get yourself going and let the idiot lights douse before the buzzer can sound.

D1-D5 can be added to or subtracted from to fit the number of dashboard dimwits on your car. You can use something other than a buzzer, if you wish, to help you keep from getting confused about your door being ajar, your key being in, or your lights being left on.

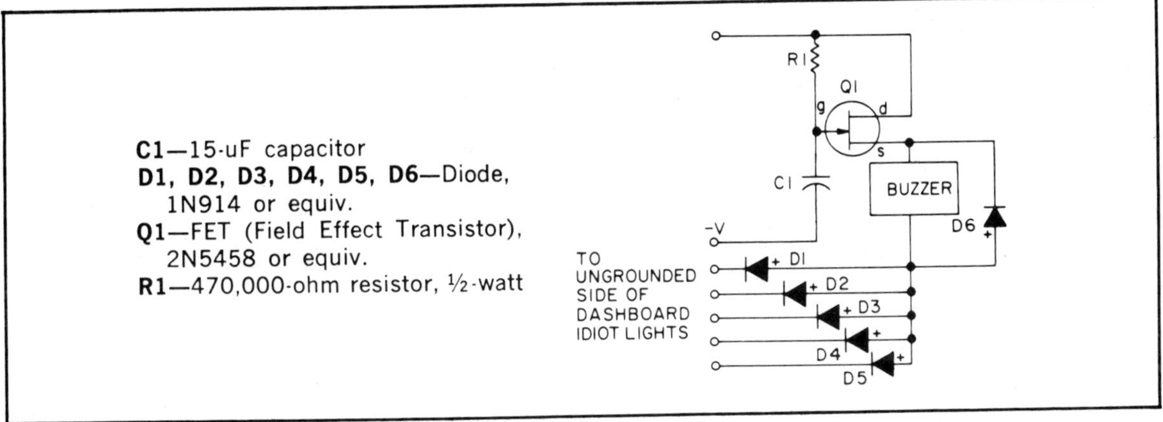

Fig. 7-31. Idiot's Delight schematic and parts list.

DIODE PUZZLE

This innocuous-looking little circuit will provide a good indication of how well you really understand the rectifier diode and the light-emitting diode. Your task is to determine which of the five LEDs will light up when 6.3 volts ac is applied to the circuit. We won't give you the answer; to find that out, just breadboard the circuit. However, we will supply you with a couple of hints. First, the forward voltage drop of a rectifier diode is approximately .8 volts, while that of an LED is about 2 volts. Naturally, rectifiers conduct current in one direction only. LEDs will light up only when their anodes (arrows) are 2 volts more positive than their cathodes (bars). Finally, you can expect to find 3 LEDs lit and 2 LEDs dark. Pencils sharpened? OK, begin.

Refer to Fig. 7-32.

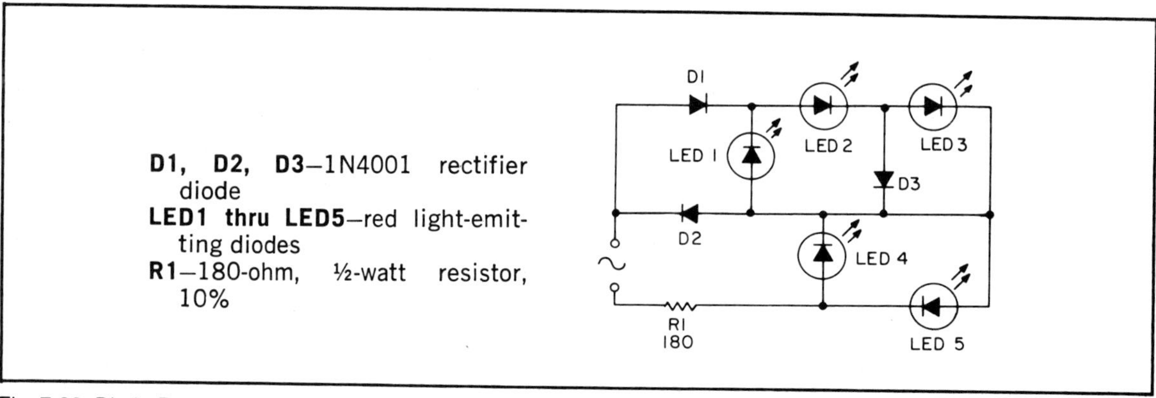

Fig. 7-32. Diode Puzzle schematic and parts list.

Chapter 8
Science Projects

THE MOVING COIL

Learning electronic theory is dull, dull, dull. Right? Not necessarily. In some cases you can learn a lot and still have a lot of fun. How? Leave your theory textbook behind and build a current-measuring gadget that will allow you to see electronic theory in action.

Moving coil meters are used universally in all types of test instruments to measure dc current because they are highly sensitive, rugged and reliable, and relatively inexpensive. They may also be used to measure ac current by first rectifying the ac with a small meter rectifier.

Refer to Figs. 8-1 through 8-4.

Our project uses the moving coil principle in a simplified form of indicating meter. You can learn more about this form of meter by building our easy-to-construct project.

How It Works. The operation of the moving coil type of meter is based on the attraction between a permanent magnet and the magnetic field of a movable coil of fine wire. The coil is pivoted in the center of the space between the poles of the magnet, and has a return spring (to return the meter pointer to zero). This spring is usually of a spiral form in commercial meters, but our meter unit uses a rubberband coil suspension in place of both the pivot and spiral zero return spring.

When dc current flows through the coil, it produces a magnetic field that is attracted to the permanent magnet's magnetic field. This

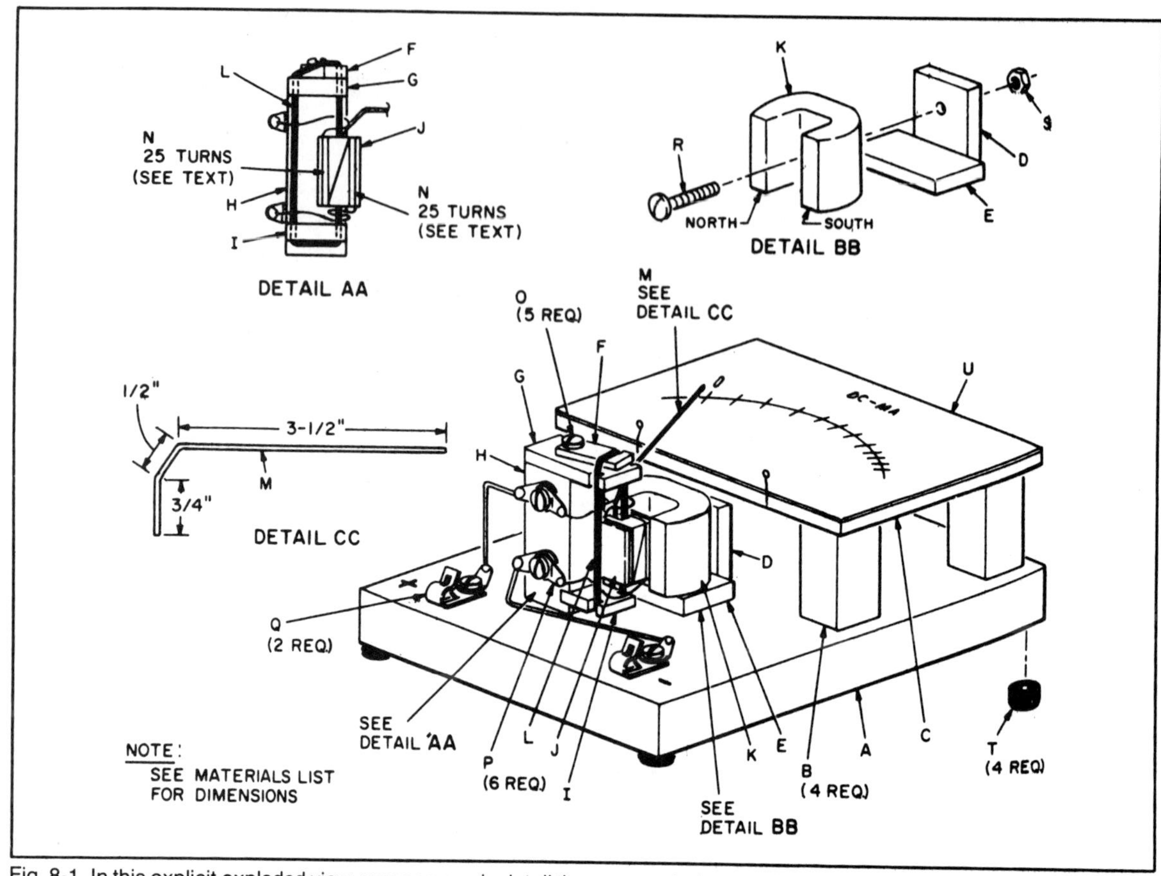

Fig. 8-1. In this explicit exploded view, you can see in detail the more varied steps of the Moving Coil Meter assembly. To find the name of each part, refer to the Parts List.

attraction imposes a turning force on the coil, that rotates the coil and moves the attached pointer over the meter scale. The amount of turning force is dependent on the amount of dc current flowing through the coil. The more dc current, the more the coil rotates. The meter is calibrated in dc current indications.

Construction. As can be observed from our drawings, our model of a moving coil meter has been made in a very simplified form to facilitate construction. It's a taut-band moving coil instrument by virtue of the rubber band suspension of the moving coil assembly. We used wood for the various supporting structures because it's easier to work with and most everyone has the few hand tools required.

Size is relatively unimportant. However, we suggest you follow the dimensions and construction details given in the drawing and Parts List. In this way you should have no difficulty in making the meter and you won't have to fiddle with changing the number of turns of wire for the moving coil to compensate for a change in physical size.

You should cut out all of the various pieces of wood and sand them smooth before actually starting to assemble the meter. Mount rubber feet on the bottom four corners of the base (A) and then glue the 4 supports (B) to base (A), as shown in the pictorial drawing. Next cement the scale platform (C) to these supports. We used our electric glue gun, but almost any similar adhesive can be used with equal success.

Now you're ready to cement the magnet and moving coil assembly supports to base (A). Pieces D, E, G, and I are made from ¼-inch plywood. First step is to cement D and E in their respective locations and fasten the magnet in place. The magnet used in our unit has a mounting hole. If the magnet you use isn't drilled at the bottom center of the *U* to allow a bolt to go through it to hold the magnet in place, it too can be cemented to D and E.

At this point the main support block (H) should be readied for cementing. But first you must notch it out so that piece I can be properly fastened to it. Hold H on the base (A) near piece E and mark H so that the top of the notch will be even with the top of E. The notch should be about ¼-inch deep. The best way to determine its depth is to hold piece G in position at the top of H and place piece I so that its notched end is even with the notched end of G. Mark the depth of the notch in block H based on the position of the end of piece I that will be inserted in the notch where its end is matched with piece G as mentioned above. Be sure that the notch in block H is cut square so that the surface of piece I will be square with the surface of block H where I is cemented in place. The notches in the free ends of I and G are required only to hold the rubber band in position. Cement block H in position, and also piece G to block H as shown in drawing.

The form block J for the moving coil is made from balsa wood, which is lighter in weight than any other wood and therefore contributes to the sensitivity of the instrument. Cut a notch in the center of J as shown in our drawing. The rubber band (L) is cemented in this notch. We used a rubber band approximately 6¾ × ⅜ × 1/16 inch. The coil is made in two sections by winding 25 turns of No. 38 enameled magnet wire on one half of J and by repeating this winding process in the same direction on the other half of J. Put a touch of cement to the ends of each coil to hold the wire to the ends of each coil to hold the wire in place and have 6-inch lengths of the start and finish of the two-section coil for future connection to it. Mount the coil assembly by stretching the rubber band over pieces G and I, centering it vertically within the height of the pole pieces of the magnet.

Now for a Pointer. Straighten out a 4¾-inch length of No. 18 gauge bare copper wire and then form it as shown in the drawing. The pointer is cemented into the notch in block J so that it rests near the O end (left side) of the scale platform. Piece F is used to make final O rest position adjustments after a scale has been cemented in position.

Fasten two double-solder lugs to block H; these are intermediary

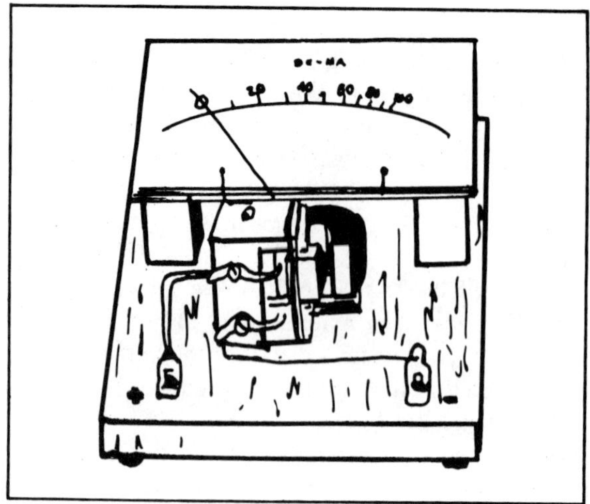

Fig. 8-2. Here's how your finishing MCM will look. Its innards are very similar to a bought meter.

connecting points for the two wires from the coil. Form a helix like a hairspring with each of these leads so that they will wind up as the coil assembly moves clockwise. Solder the end of the wire from the top helix to the top lugs and the bottom helix to the bottom lugs. Mount two Fahnestock clips or binding posts along the front edge of the meter baseboard and connect them to the solder lugs on H, using No. 18 solid base wire. Since meter polarity is determined by magnet polarity and the direction of current flow, established by how the coil is wound, the correct polarity markings of the meter should be determined when you calibrate the instrument.

Calibration. In order to calibrate this instrument, you'll need a potentiometer having roughly 20 ohms resistance, a 1½-volt battery, and a dc milliammeter, preferably a multi-range one available as part of a VOM.

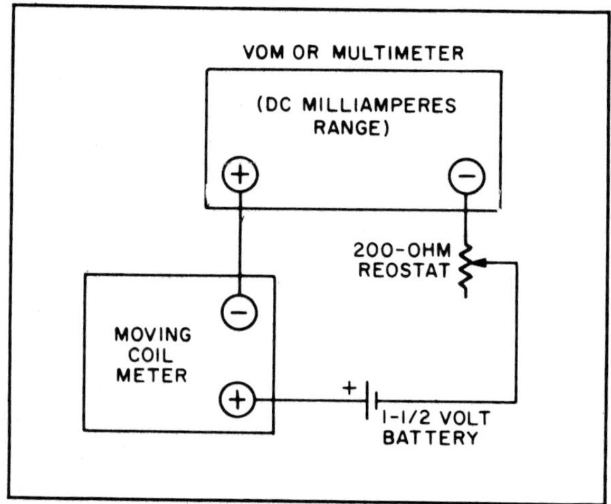

Fig. 8-3. Any 0-100 mA or higher range milliameter will test your MCM just as well as a VOM or multimeter, provided its accuracy is fairly good.

Fig. 8-4. Moving Coil Meter parts list.

A—6 x 7 x ¾-inch white pine
B—¾-inch square x 1½-inch white pine (4 required)
C—4 x 5 x ¼-inch plywood
D—1¼ x 1 x ¼-inch plywood (see text)
E—1¾ x 1 x ⅜-inch plywood
F—⅜ x 1 x 3/16-inch white pine with end notch
G—¾ x 1½ x ¼-inch plywood with end notches
H—¾-inch square x 2-inch white pine (see text)
I—¾ x ⅞ x ¼-inch plywood with end notches
J—½-inch square x ¾-inch balsa wood with slot
K—Alnico magnet
L—Rubberband (see text)
M—Meter pointer (#18 copper wire—see text)
N—50 turns #38 enameled magnet wire (see text) (Radio Shack #278-011, #32 gauge wire, may be substituted)
O—Five #4 x ½-inch wood or sheet metal screws
P—Solder lugs (6 required) (Radio Shack #64-3029 or equiv.)
Q—Fahnestock clips (2 required) (Radio Shack #270-393 or equiv.)
R, S—8-32 x 1 or 1¼-inch machine screw and nut
T—4 rubber bumpers
U—4 x 5½-inch heavy white paper for meter scale
Misc.—Hookup wire, 200-ohm rheostat, 1½-V. battery, VOM or DC milliameter

Now you are ready for the calibration scale that's mounted on the platform C made during the framework construction. The scale is drawn on a piece of heavy white paper (U) which will be cemented to the platform after the calibration marks have been drawn. (Rub-on numerals, such as Datak, make a neat scale.) Temporarily fasten this white paper (U) to platform C, draw an arc, and place a mark on the left-hand side for a zero reference point.

Connect a 1½-volt battery, a 200-ohm potentiometer (used as a rheostat) a multimeter set on dc milliamp ranges (or a milliammeter), and the moving coil meter you have just built, as shown in the calibration diagram.

Set the potentiometer for maximum resistance and at the start use the highest milliamp range of the multimeter. If the pointer on your moving coil meter deflects to the left, below the established O point, reverse connections to it and then mark the binding posts + and −. Use the connection diagram to determine their polarity markings after connecting the meter so that the pointer moves to the right.

Slowly turn potentiometer to reduce resistance in the circuit and note the readings of the multimeter milliamp range selected. Mark your moving coil meter with the same readings shown on the milliammeter. We divided the 0-100 scale into 10 mA divisions. In the manufacture of dc moving coil meters spring tensions, spacing and coil weight are carefully controlled so that these meters are linear. For this reason commercial milliammeters have uniform spacing between divisions. Our moving coil meter doesn't have such uniformity because of the variations in the rubber band used for suspension and tension, and because it's difficult to maintain accuracy of positioning the various pieces and to be assured of the strength of magnetic field developed by the magnet. Once you have established the calibration points they can be considered accurate.

Now that you have marked the scale in pencil you can remove it from the platform and apply the permanent markings. Permanently fasten the scale in position and stand back to admire your work. If you used reasonable care in following the instructions, you'll have good reason to be proud of your handiwork.

DIGITAL WINDSPEED METER

Increasing energy costs have driven many people to thinking of alternate sources of power, such as solar energy and water power. But the technology for these natural energy sources is still quite expensive and complicated to install. It'll be at least several years before the cost of most natural energy systems comes down enough and the parts are easy enough for most people to install. Wind power for generating electricity, on the other hand, has been available for many years. For several decades, farmers and others in rural areas have used windmill generators as standby electricity and in some cases, as their main power supply. Windmills and wind-driven electrical generators can be bought off the shelf by anyone, and require no expertise other than the usual home mechanic skills to set up.

Refer to Figs. 8-5 through 8-13.

Have you wondered if there's enough wind where you live to drive a windmill electrical generator? Do you know if there's enough wind to fly that big kite you've often thought of constructing? Is there enough wind coming over the hills near your area so you can get into hang-gliding? Or do you live in an area where tornadoes or hurricanes sometimes strike? If so, it could literally be a matter of life-and-death for you to read the wind-speed easily, with an accurate, easy-to-install anemometer (windspeed meter). That's what the Digital Windspeed Meter is—an accurate anemometer using the latest digital TTL (transistor-transistor logic) integrated circuits.

Though this project isn't recommended for someone who's never built any solid-state projects before, it should be easy enough for anyone who has built one or two simpler projects. In addition, it's the sort of project which will get you started easily in digital logic circuitry, the circuits and components which are the basic building blocks of computers and most other advanced electronics today.

How Anemometers Work. There are two types of electronic anemometers in general use. One type uses air cups or a wind turbine to turn a tiny electric generator whose output is directly connected to a milliammeter. The faster the wind blows, the faster the generator turns and the higher the meter reads. This type of anemometer is simple and reliable but it usually is not accurate.

A more sophisticated type uses air cups to turn a shaft to produce electric pulses. The pulses are integrated by a capacitor and related circuitry to produce a dc voltage whose magnitude is directly proportional to the wind speed. This voltage is also displayed on a meter. This method is easier to calibrate, and thus is more accurate than the

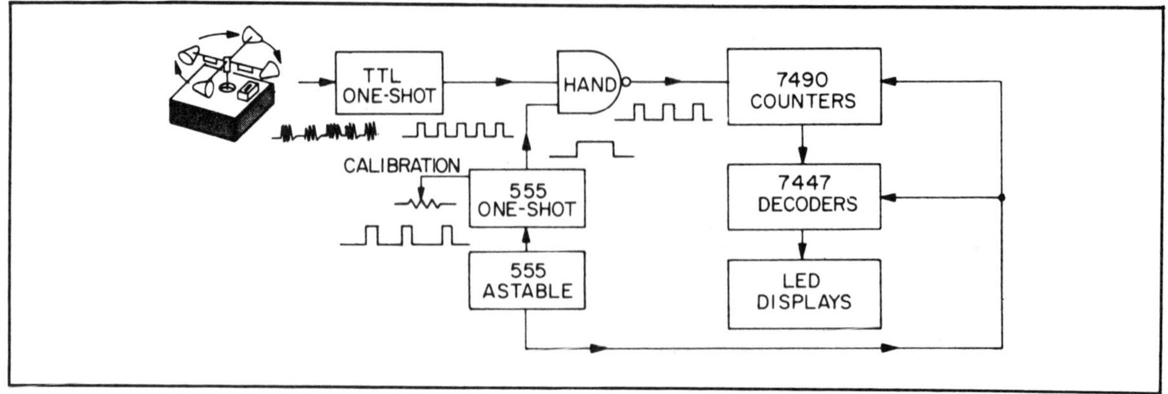

Fig. 8-5. Block diagram for digital anemometer. As the calibration control is varied, it changes the duration of the pulse put out by the 555 one-shot. This acts as a variable window for the pulses coming from the windspeed sensor permitting accurate readout of the LEDs.

simple generator method. By using state-of-the-art digital electronics, improvement can be made upon this method of measuring the wind's speed. Instead of the round-about method of adding up the electric pulses by charging up a capacitor, why not just count them directly? The digital anemometer described here does just that. The result is a more accurate sophisticated instrument that is easier to read and cheaper to build.

How It Works. The theory behind the digital anemometer is simple. See Fig. 8-5. The wind turns a shaft which has streamlined plastic cups attached to it. On one rod that holds two oppositely directed cups are placed two small magnets. A reed switch is mounted on the stationary base beneath the rotating cups so that it will be operated by the rotating magnets above. Each time the cups make a full revolution, the reed switch opens and closes twice. The pulses generated by this reed switch trigger a one-shot multivibrator (TTL-7412) which cleans up the pulses, eliminating contact-bounce and other error pulses. The cleaned-up pulses are then fed to a TTL NAND gate which is controlled by the 555 one-shot multivibrator. The 555's one-shot output pulse is manually adjustable to let us calibrate the anemometer.

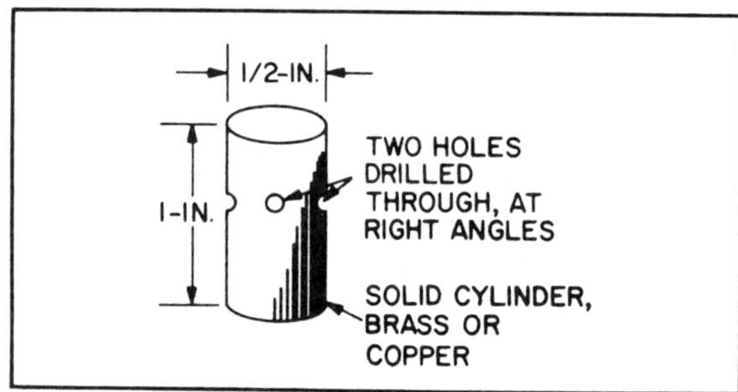

Fig. 8-6. Centerpiece of windspeed sensor.

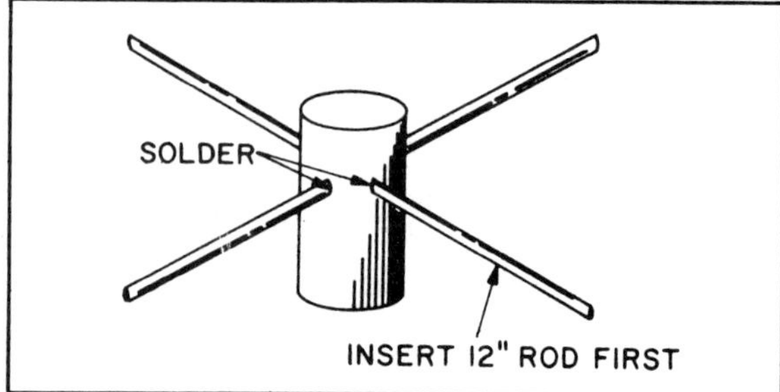

Fig. 8-7. Assembly of rods and centerpiece to form rotor.

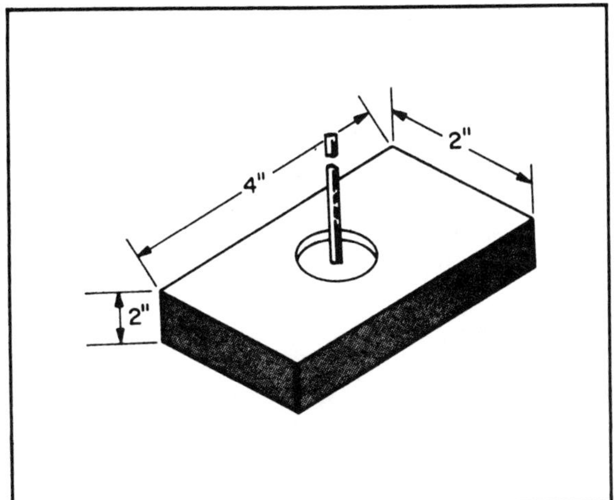

Fig. 8-8. Wood block mount with bearing.

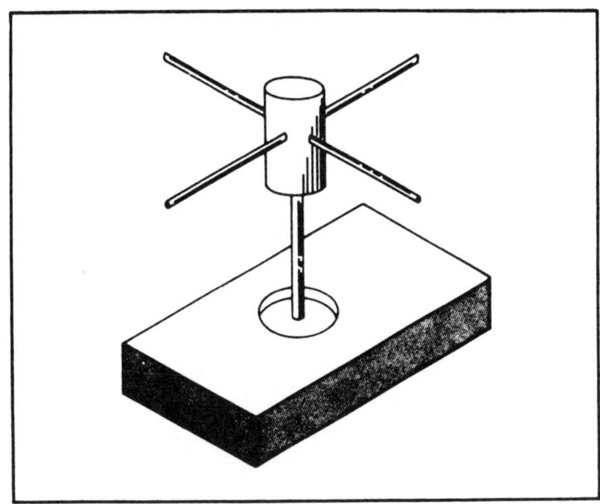

Fig. 8-9. Rotor in place on bearing.

Another 555 astable multivibrator provides automatic triggering pulses for the 555 one-shot as well as supplying reset and blanking pulses for the counters and decoders. The resulting controlled and cleaned up pulses (which originated in the reed switch) are counted on two TTL decade counters and displayed on two LED displays.

Construction. The rotating wind sensor is made up of four plastic cups, mounted with 3/32-inch or ⅛-inch rods to a slot-car motor or similar cheap and readily available bearing. (The brushes of the motor can be removed if desired.) The egg-shaped containers in which *Leggs* nylons are sold are ideal for the plastic cups which catch the wind.

The rods which support the cups can be steel welding rods or (better) copper or brass. One rod should be one foot long and the other two should be six inches long.

Next, obtain a small cylindrical piece of a solid metal that is easily solderable—brass or copper is best. Drill two holes, using bits the

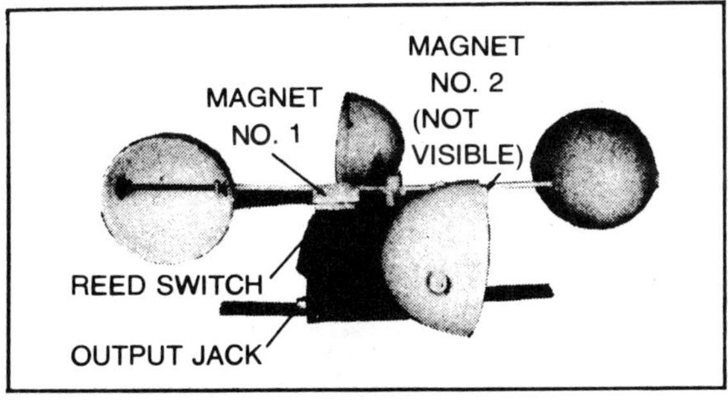

Fig. 8-10. Completed unit. Adjust height of reed switch so magnets pass about ¼-inch over it (or less).

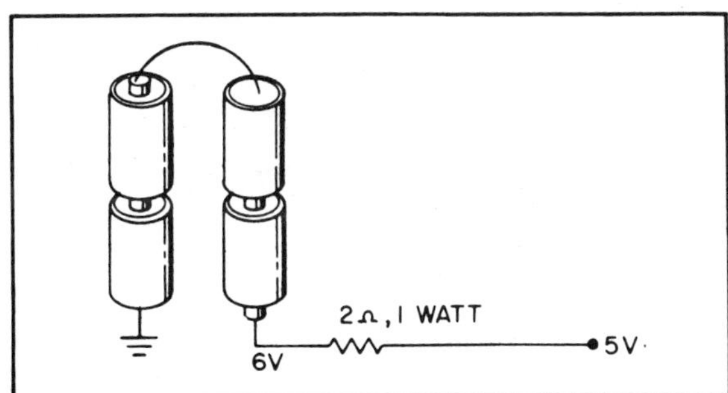

Fig. 8-11. Temporary battery power supply for use when calibrating the instrument in an automobile.

same size as the rods, at right angles to each other through this cylinder of metal as shown in Fig. 8-6.

Now center the 12-inch rod in the cylinder. Insert the two shorter rods in the remaining two open holes in the cylinder, as shown in Fig. 8-7. Using acid-flux, solder the rods to the cylinder.

Mount the motor, which is used as the bearing, in a 2-inch long piece of two-by-four. To mount the motor, drill and file a hole in the wood large enough to take the motor. Cover the motor's case with epoxy glue and insert it in the hole as shown in Fig. 8-8.

Using a bit as close to the diameter of the motor's shaft as possible drill a hole about ½-inch deep in the bottom of the cylinder (see Fig. 8-7). which now has rods soldered to it. Insert the motor shaft into this hole and solder it, using acid-core flux. (If steel is used, secure with epoxy glue.)

Now mount the four plastic cups to the rod, taking care to correctly orient the cups. Drill holes in the cups and insert the rods in the holes. Keep the cups in place with epoxy or other good glue.

Next we mount the magnets on the rods. If copper or brass rods are used, great, just solder or glue the magnets to the undersides of two opposite rods, centering them one inch from the pivot. The reed switch is then mounted on the wood base so the magnets pass a quarter of an inch above it.

If the rod is iron or steel, we have a problem because it will distort the magnet's magnetic field. This problem is overcome by using a non-magnetic spacer between the magnet and the rod—¼ inch is enough space. A ¼ × 1-inch piece of wood is glued to the rod and then the magnet glued to the wood. Since there is very little weight involved, a good glue will hold the magnet fine. This completes the construction of the wind sensor.

Circuit Assembly. To build the circuit, use any convenient layout on a perfboard. The position of the components is not critical. If you haven't worked with ICs before, you'll be better off soldering IC sockets in place on the perfboard and connecting the other components to the pins of the IC sockets. If you've had a fair amount of experience and can solder ICs directly into a circuit without overheating the pins (using a pair of long-nose pliers as a heatsink while soldering to each pin), do it that way.

The main job in wiring the anemometer lies in making the printed circuit board. The pattern shown can be made by using the simple resist method. Simply draw the pattern with a felt-tipped resist pen on the foil side of the printed circuit board, place in etching solution for an hour or so and drill the holes marked. The somewhat more sophisticated, yet still easy, non-camera photo method can also be used.

If a small 25-watt soldering iron is used, the *ICs* can be soldered directly to the board, although IC sockets are less risky. Be sure to orient the notch on the *ICs* as shown in the component layout diagram.

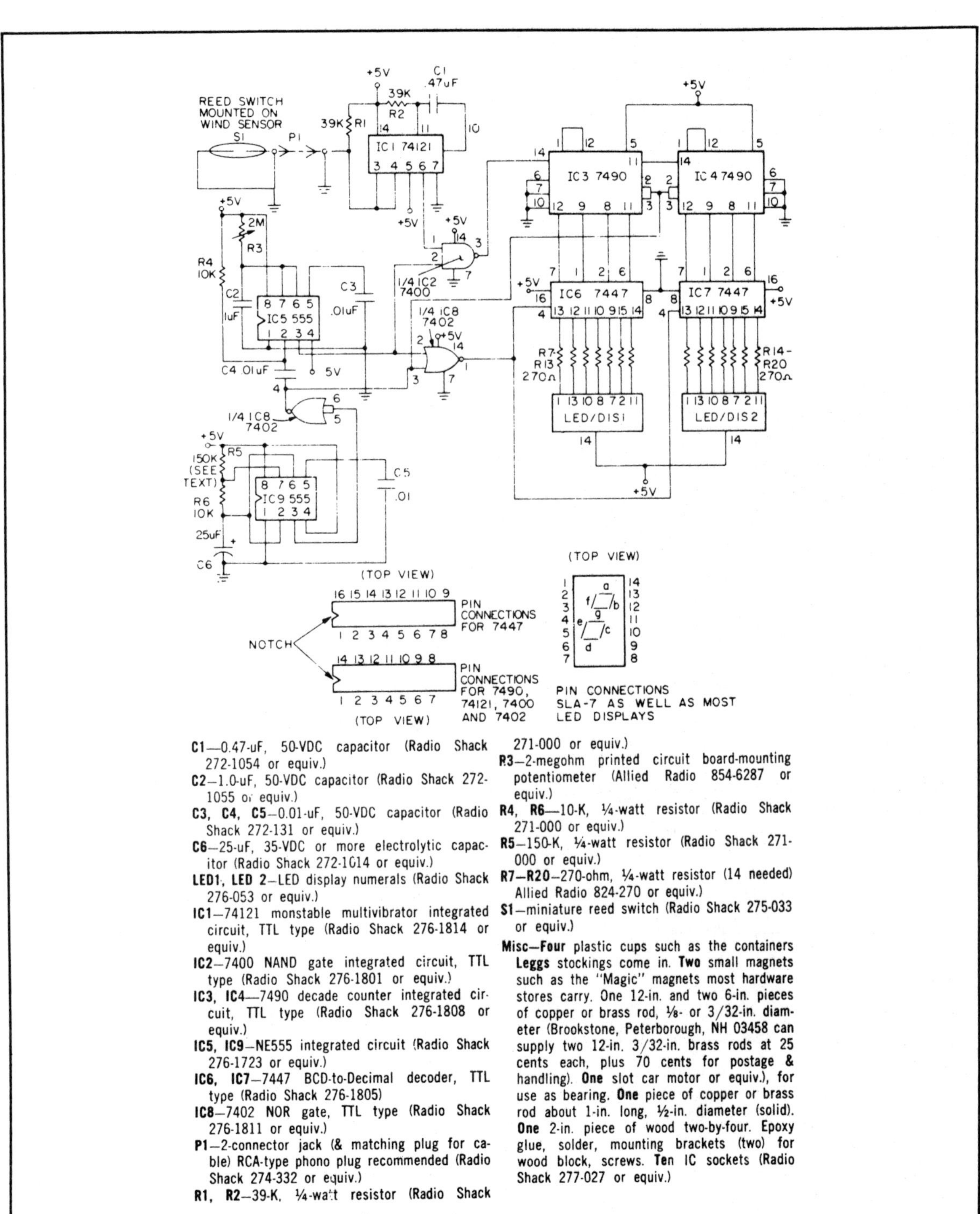

C1—0.47-uF, 50-VDC capacitor (Radio Shack 272-1054 or equiv.)

C2—1.0-uF, 50-VDC capacitor (Radio Shack 272-1055 or equiv.)

C3, C4, C5—0.01-uF, 50-VDC capacitor (Radio Shack 272-131 or equiv.)

C6—25-uF, 35-VDC or more electrolytic capacitor (Radio Shack 272-1014 or equiv.)

LED1, LED 2—LED display numerals (Radio Shack 276-053 or equiv.)

IC1—74121 monstable multivibrator integrated circuit, TTL type (Radio Shack 276-1814 or equiv.)

IC2—7400 NAND gate integrated circuit, TTL type (Radio Shack 276-1801 or equiv.)

IC3, IC4—7490 decade counter integrated circuit, TTL type (Radio Shack 276-1808 or equiv.)

IC5, IC9—NE555 integrated circuit (Radio Shack 276-1723 or equiv.)

IC6, IC7—7447 BCD-to-Decimal decoder, TTL type (Radio Shack 276-1805)

IC8—7402 NOR gate, TTL type (Radio Shack 276-1811 or equiv.)

P1—2-connector jack (& matching plug for cable) RCA-type phono plug recommended (Radio Shack 274-332 or equiv.)

R1, R2—39-K, ¼-watt resistor (Radio Shack 271-000 or equiv.)

R3—2-megohm printed circuit board-mounting potentiometer (Allied Radio 854-6287 or equiv.)

R4, R6—10-K, ¼-watt resistor (Radio Shack 271-000 or equiv.)

R5—150-K, ¼-watt resistor (Radio Shack 271-000 or equiv.)

R7–R20—270-ohm, ¼-watt resistor (14 needed) Allied Radio 824-270 or equiv.)

S1—miniature reed switch (Radio Shack 275-033 or equiv.)

Misc—Four plastic cups such as the containers **Leggs** stockings come in. **Two** small magnets such as the "Magic" magnets most hardware stores carry. One 12-in. and two 6-in. pieces of copper or brass rod, ⅛- or 3/32-in. diameter (Brookstone, Peterborough, NH 03458 can supply two 12-in. 3/32-in. brass rods at 25 cents each, plus 70 cents for postage & handling). **One** slot car motor or equiv.), for use as bearing. **One** piece of copper or brass rod about 1-in. long, ½-in. diameter (solid). **One** 2-in. piece of wood two-by-four. Epoxy glue, solder, mounting brackets (two) for wood block, screws. **Ten** IC sockets (Radio Shack 277-027 or equiv.)

Fig. 8-12. Digital Windspeed Meter schematic and parts list.

It is always wise to use IC sockets when mounting display LEDs. Be sure to either bend back or cut off pin 12 on the socket which is used to mount Display No. 1.

Unless double sided PC boards are used, jumpers made up of hookup or bare wire are needed. Place jumpers between the two J1s, J2s, J3s, J4s, J5s and J6s. In addition, interconnect the +5 Vdc points on the PC board with jumpers (6 needed).

Connect the two leads from the remote mounted reed switch to points P1 and to one of the two GNDs. Connect the plus power supply lead to the +5 point at the top of the board. Connect the other supply lead to the other GND point which is also located at the top of the board.

The entire circuit can be mounted in any convenient size bakelite or aluminum case with aluminum panel. For a smart appearance, spray paint the panel with some auto-touch-up white lacquer. Use dry transfer decals for the lettering.

Cut a slot in the panel so the two digit LED display can be readily seen. If desired, the switch to turn on the power can be an inexpensive

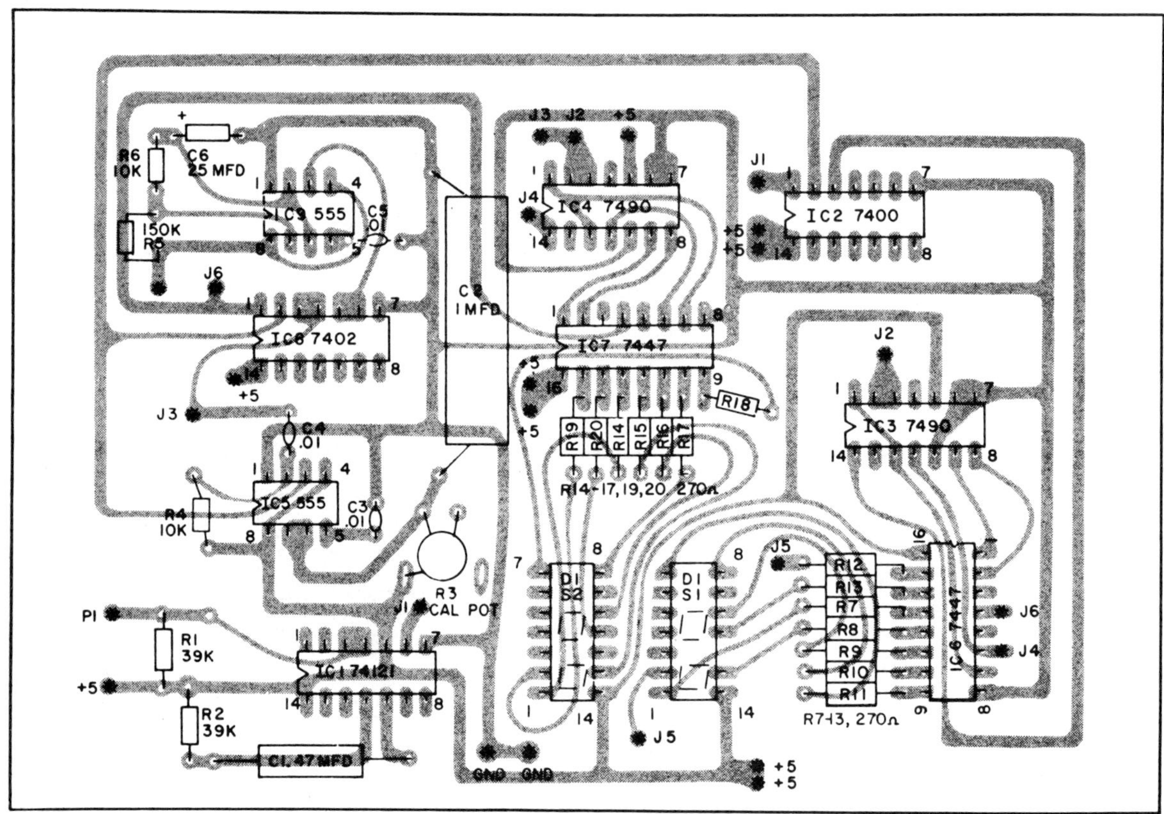

Fig. 8-13. Location of components as seen from the bottom (through the foil pattern) as they would be viewed in soldering to the foil.

slide switch but a toggle switch is more reliable and easier to mount. The circuit board and all other components should be mounted to the back of the front panel for ease of accessibility.

If one desires a longer display time, increase R5 from the recommended 150k to 220k or even 270k ohms.

Any type of two-conductor connecting jack can be mounted on the front panel (I used an RCA-type jack) as long as the appropriate plug is used. The two wire cable which connects the rotary wind sensor the electrical unit must be long enough to reach from the roof to the place in your home where you want to keep the display unit. Any kind of shielded cable, including audio cable or microphone cable is all right. Coax such as RG-59/U is fine, but don't buy it special for this job because it costs much more than other (audio) cable.

Calibration. This anemometer is easily calibrated since there is just one pot to adjust. As an initial test, plug the unit in and connect the wind sensor to the display unit. After a few seconds warmup, the unit should show 00 and then go momentarily blank. Turn the cups by hand and a number should appear on the display for a second or two and then disappear for a second. Now turn the cups as fast as you can by hand and adjust the calibration pot to read as close as possible to 20. If everything so far works all right, it is time to take the anemometer for a ride. If not, go back to Square One and check your wiring and the seating of the LED display modules.

The anemometer should be calibrated against an accurate automobile's speedometer. Since the anemometer will be away from the regular house supply, you will have to take along a 5-volt battery supply. In order to drop the voltage to the required 5 Vdc, you must connect a 2-ohm resistor in series with a 6-volt battery.

With someone else driving, take the unit in an auto on a nearly calm day and drive as steadily as possible at a certain definite speed, say 30 mph. Drive up and down a quiet road, with the wind sensor held out the window and adjust the calibration pot so the display will read an average value of 30.

Use. The wind sensor should be mounted on a roof or other location where there are few obstructions. Because of the one-shot ahead of the NAND gate, the anemometer may suddenly go blank when winds are of hurricane speed. So if the display one minute shows 75 mph and the next minute 00, don't stick your head out the window to see if something happened to the wind sensor on your roof, a tree might just be sailing by.

A simple way of checking your speedometer is to drive down an expressway at 55 and have someone time you between two mileposts. Then get your hand calculator out and divide 3600 by the number of seconds it took you to travel the mile. The result is your true speed.

PRECISION ELECTRONIC THERMOMETER

If you've always wanted a really precise electronic thermometer to measure the temperature outdoors at a distance, continuously check the operation of your heating system at home, monitor the cold in your refrigerator or ice compartment, or check the temperature down deep in the water for fishing up the big ones, the *Precision Electronic Thermometer*—PET, is for you.

If the cost, or the supposed complexity has scared you off—or if you've heard it's difficult to calibrate an electronic thermometer—take heart, PET is for you. Most electronic thermometers have calibration direction which instruct one to fiddle with 2 or even 3 adjustments while alternately sticking the thermometer's probe in ice water and lukewarm water. Even after minutes or hours of fussin' and cussin' and water everywhere, the accuracy of the thermometer often leaves much to be desired. But not so with PET.

Refer to Figs. 8-14 through 8-22 and Tables 8-1 through 8-3.

National Semiconductor has brought out a relatively inexpensive temperature transducer, the LX5700, which makes electronic thermometers fun to build and calibrate. The LX5700 has a temperature-versus-output curve as flat as a pancake. Its non-linearity is less than ½ of one percent, compared to over one percent for *good*-quality mercury thermometers. Its only real deviation from perfection is an easily-corrected offset error of about ±4 degrees Celsius. This means you can build PET, a simple highly-accurate electronic thermometer with this transducer, yet have only one simple adjustment.

The heart of the LX5700 transducer is the sensor, which is made of two identical transistors fabricated on the same silicon chip but operating at different current densities. The 10-millivolt-per-degree Kelvin output of the sensor is proportional to the difference in emitter-to-base voltages, which is linearly related to temperature. This sensor was impossible to construct before integrated circuit techniques were perfected since it depends upon making two identical transistors right next to each other on the same chip.

In addition to its temperature linearity this transducer has two other features which make it a really neat device for people who love simplicity. First, it has its own built-in voltage regulator (see Fig. 8-14, showing the zener diode in the block diagram) which makes "it great for accurate, battery-operated thermometers. Second, the transducer also includes in its tiny case that marvelous device, the op-amp. By adding two resistors you can take that 10 mV-per-°Kelvin output of the sensor and amplify it to any practical output.

(Note that 1 mV is one-thousandth of a volt.) K is degrees Kelvin, which is the absolute temperature in the metric system. Kelvin degrees are the measure of temperature universally used by physicists. All you have to do to get the absolute—Kelvin—temperature in any system is to put a plus sign on its absolute zero and add it to the regular (Celsius) temperature. That is, degrees Kelvin = degrees Celsius + 273°. As an example, the room temperature of 25°C (77°F) is actually 289°K. Other equivalents:

$$Tc = (40 + Tf)\,\frac{5}{9} - 40 \text{ and}$$

$$Tf = (40 + Tc)\,\frac{9}{5} - 40$$

where, of course, *T*f stands for degrees Fahrenheit and *T*c for degrees Celsius.).

Figure 8-15 shows an electronic thermometer that uses only four components (including a voltmeter). Assuming the transducer in Fig. 8-15 is at the room temperature of 298°K (25°C-77°F), if you connect an accurate voltmeter to the output of the transducer, you will get a voltage reading of 2.98 volts (±.04 volts). The problem with such a thermometer, however, is the same problem some people find with life in general—a lack of *meaning*. For instance, wouldn't it sound funny to hear a disk jocket say on the radio, ". . . so folks another bitter cold night tonight with a low near 260° Kelvin . . ." Another problem arises if a standard 0-5 Vdc meter is used as a thermometer's readout, for even if the thermometer is taken from Niagara Falls to Miami in January, the needle's movement will hardly be visible. If cost were no

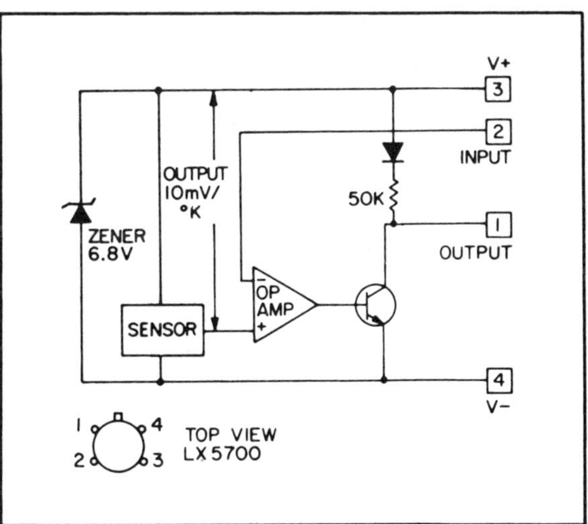

Fig. 8-14. Block diagram shows what's inside the LX5700 temperature-sensing transducer by National Semiconductor. The heart of the device is the sensor itself, which produces an output of 10 millivolts per degree Kelvin (see text). Op amp and transistor raise output, and voltage divider (external to transducer) determine meter deflection due to output at terminals 1 and 4.

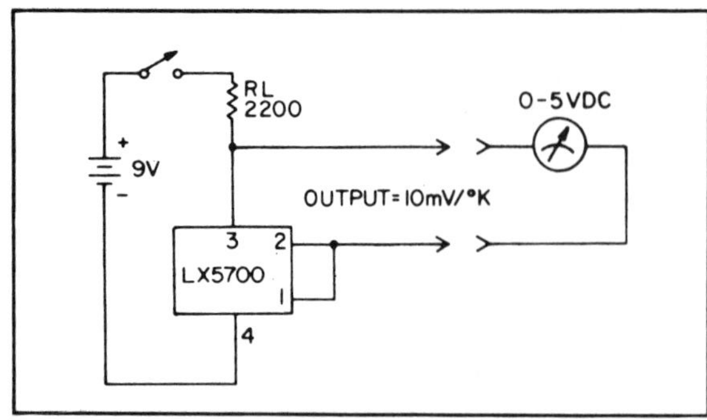

Fig. 8-15. Basic electronic thermometer circuit shows voltage divider is in the PET case, sensor and op amp contained in the LX5700 transducer package and readout meter.

problem, a digital voltmeter could solve this problem but the issue of "meaning" would remain.

In order to make a simple, useful thermometer with "meaning" we must add a few more resistors and a capacitor. Although our prototype PET measures temperatures between −20° and +105°F, I also explain how to make thermometers with ranges from 0-100°F and −50°C+50°C. The only variation between the three thermometers are three resistors. If you plan to use your thermometer to measure the outside air temperature, the range you choose depends on where you live. If you live on the East, West, or Gulf Coasts the 0-100°F range would do nicely. If you live in the northern great plains, desert southwest or similar areas or you are into metric, the −50°C to +50°C range is for you. For most of the rest of the country the −20°F to +105°F range is ideal. See the tables for the resistors required for each temperature range.

Circuit Operation. Although the full schematic diagram is fairly simple it isn't much help in understanding exactly how the circuit works. To help clarify the matter, a simplified schematic is shown in Fig. 8-16. R1, R2 and R3, are the same resistors shown in the full schematic.

From basic operational amplifier theory we know the following: The differential input voltage is zero, so V1 = V2. It can also be shown that

$$V_o = \frac{R2 + R3}{R3} \times V2.$$

From basic electricity,

$$Im = \frac{V1\text{-}V_o}{R2} + \frac{6.8 - V_o}{R1}$$

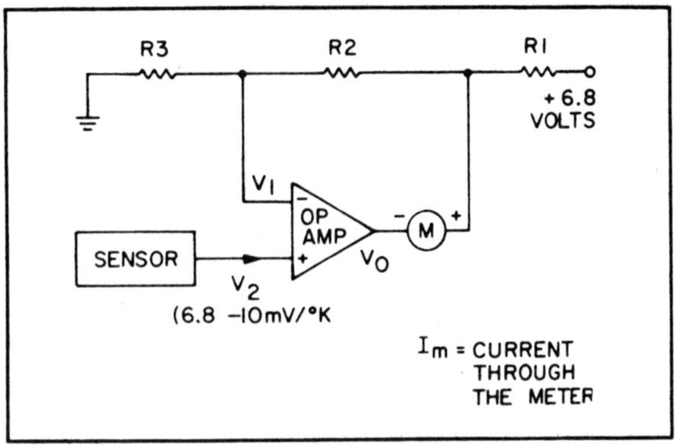

Fig. 8-16. Simplified schematic of the PET precision electronic thermometer shows basic thermometer circuit adapted to LX5700 sensor.

To simplify the equation further, to understand the circuit, we can say that Im is approximately equal to

$$\frac{6.8 - V_o}{R1}.$$

We can do this since we know that R2 is much greater than R1. Since $V_o = KV2$, where K is a constant and $V2 = 6.8 - 10$ mV/°K, we see that the higher the temperature the smaller will be V_o. Thus, the current through the meter, Im, increases as the temperature increases.

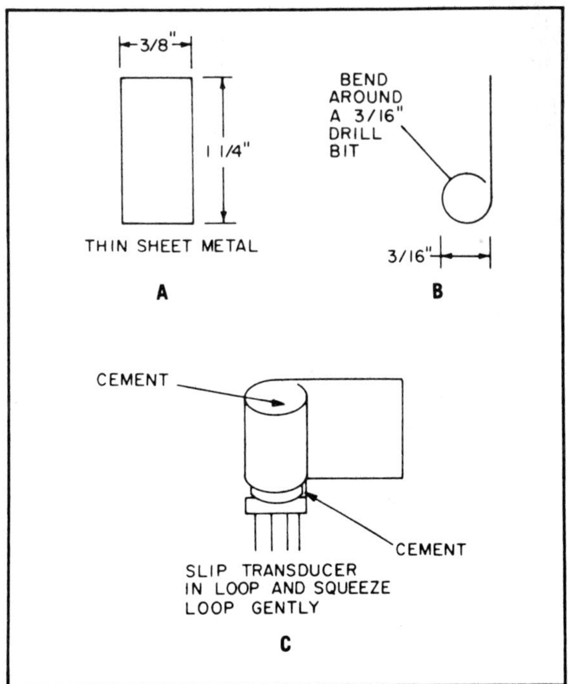

Fig. 8-17. How to construct a simple heatsink for the LX5700 temperature-sensing transducer. No heatsink is required if the PET will be used in water or to measure only cold situations (below 32° F, or 0°C).

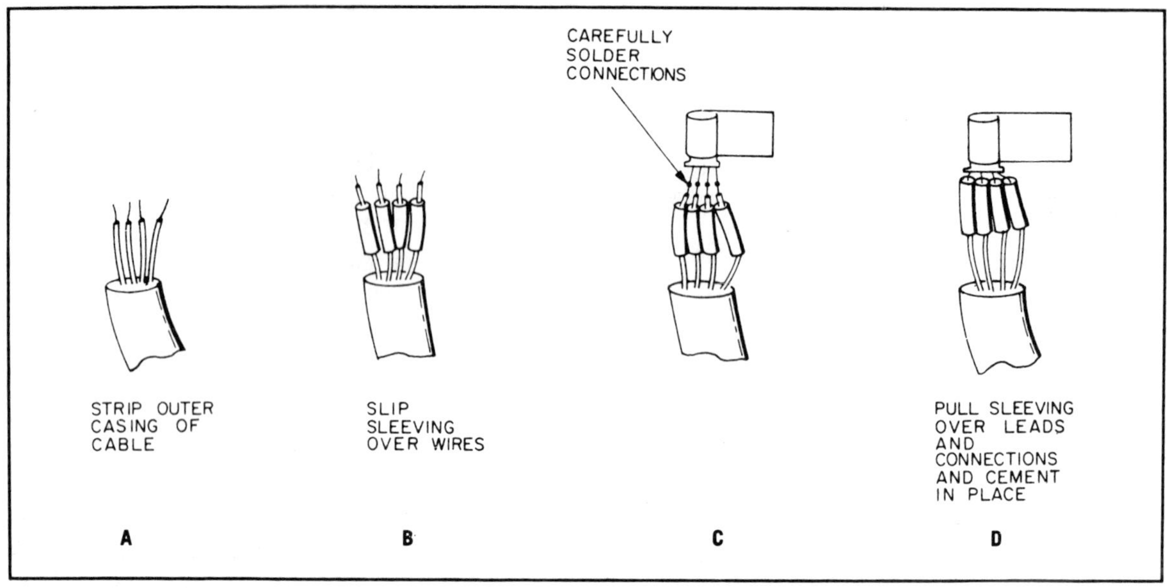

Fig. 8-18. Step-by-step drawings show how to connect a four-conductor cable to the temperature-sensing unit. Four separate wires can be used with pieces of tape every foot or so to keep them bundled together.

The exact values of R1, R2, and R3 are chosen so that at the minimum temperature we want to measure, $I = 0$, and at the maximum temperature we want to get a full scale deflection of our meter. (In the thermometer described in this article, $I\text{max} = 1$ mA).

In the actual circuit, R7 and the transducer's own zener diode form a voltage regulating system. R6 is an external load resistor for the transducer's output transistor. R8 and C1 increase the circuit's stability, and R4 and R5 form the calibration circuit.

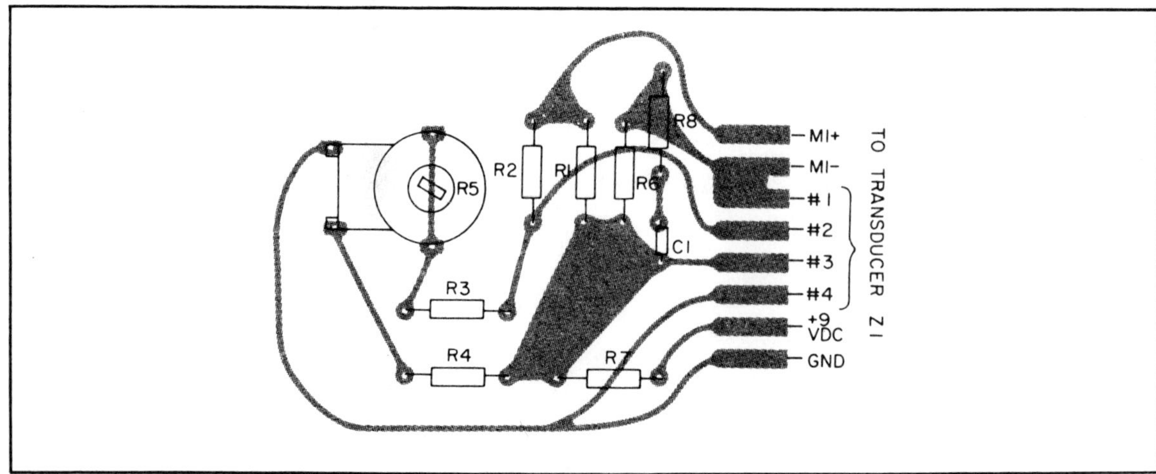

Fig. 8-19. Parts placement on the components board of PET. Component side shown.

Putting it Together. Although the circuit can easily be constructed with perfboard techniques, a printed circuit board foil layout and component guide is given for your convenience. See Fig. 8-20. If a printed circuit board is used, R5 should be of the printed circuit type. By far, the easiest way to make simple printed circuit boards is to draw directly on the copper clad board with a felt tip resist pen.

The only critical components in addition to the transducer are resistors R1, R2 and R3. For precision readings these three resistors should be the one-percent tolerance kind. However you can use the five percent resistors shown in the tables, although some adjustment of R1 may be required. See the section on Testing and Adjusting. The exact values of these three resistors depends upon the temperature range you want the thermometer to measure. The tables give the values. Notice that extra room is available on the foil layout of the printed circuit board for R1, R2, and R3 so that two resistors could be tied together in series, if necessary.

Although tiny, the power used by the transducer does raise its temperature slightly. If the transducer is to be operated in still air, a small heatsink should be used. Most other applications such as medicine, fishing, etc. do not require a heatsink. If small errors (up to about two degrees) can be tolerated, or if you take your temperature readings quickly, you don't even need a heatsink for measuring still air temperature. If you won't need the heatsink, disregard Fig. 8-17. and the next two paragraphs here.

The heatsink can be constructed from a ⅜ × 1¼-inch piece of aluminum or copper sheet metal as shown in Fig. 8-17. Place the transducer in the loop made out of the sheet metal and give the loop a

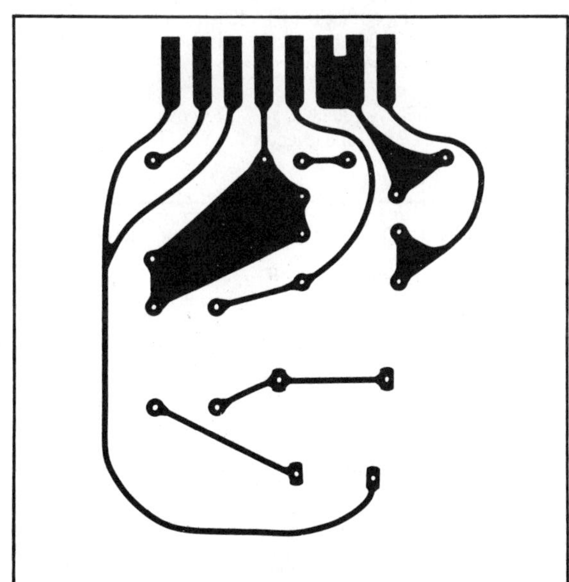

Fig. 8-20. Printed circuit board pattern shown is the same size as the actual foil for the components board of PET. Foil side down.

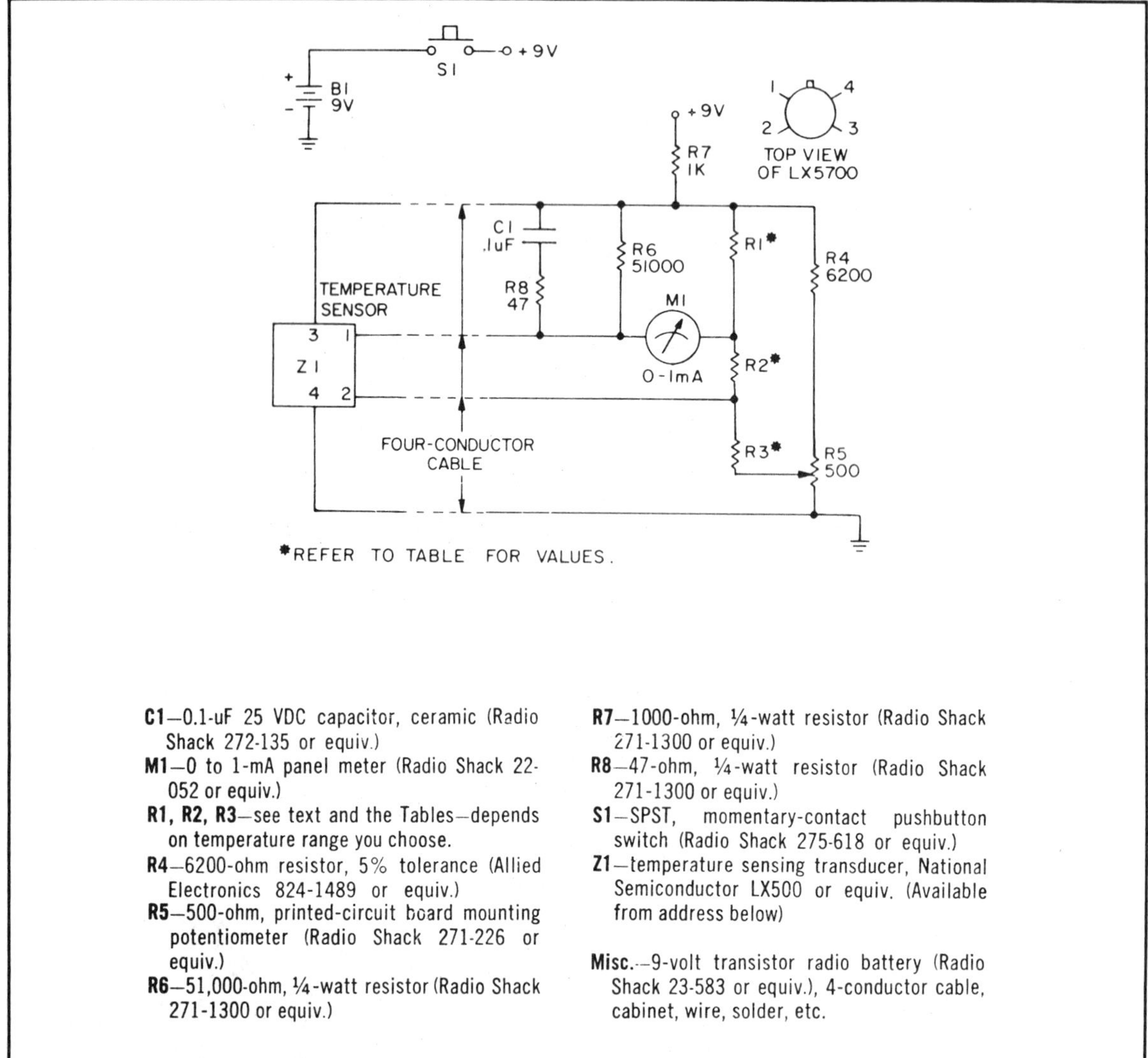

C1—0.1-uF 25 VDC capacitor, ceramic (Radio Shack 272-135 or equiv.)
M1—0 to 1-mA panel meter (Radio Shack 22-052 or equiv.)
R1, R2, R3—see text and the Tables—depends on temperature range you choose.
R4—6200-ohm resistor, 5% tolerance (Allied Electronics 824-1489 or equiv.)
R5—500-ohm, printed-circuit board mounting potentiometer (Radio Shack 271-226 or equiv.)
R6—51,000-ohm, ¼-watt resistor (Radio Shack 271-1300 or equiv.)
R7—1000-ohm, ¼-watt resistor (Radio Shack 271-1300 or equiv.)
R8—47-ohm, ¼-watt resistor (Radio Shack 271-1300 or equiv.)
S1—SPST, momentary-contact pushbutton switch (Radio Shack 275-618 or equiv.)
Z1—temperature sensing transducer, National Semiconductor LX500 or equiv. (Available from address below)

Misc.—9-volt transistor radio battery (Radio Shack 23-583 or equiv.), 4-conductor cable, cabinet, wire, solder, etc.

Fig. 8-21. Precision Electronic Thermometer schematic and parts list.

very gentle squeeze. See Fig. 8-17C. Using a good epoxy, cement the transducer permanently to the heatsink.

As shown in Fig. 8-17C, strip the outer casing of the four-conductor cable and slip plastic tubing or sleeving over the leads. This sleeving will be used shortly to insulate the connections and bare leads of the transducer.

Using a 25-watt or smaller soldering iron, solder the leads of the transducer to the end of the four-conductor cable you have prepared. Be sure to use a small alligator clip or long-nose pliers as a temporary heatsink between the soldering iron and the transducer. After soldering the leads, pull the sleeving over the connections and bare leads and cement the sleeving in place.

Next, spray all the exposed wires and connections with several coats of a clear sealant. Let the assembly dry between each coat. After the unit is completely dry, immerse it in the sealant. Let the assembly dry overnight and recoat if desired. To avoid accidentally leaving the thermometer on for long periods of time and depleting the battery, the power switch should be the momentary-contact pushbutton type.

If you decided to build the thermometer that measures temperatures between 0 and 100°F, you've just won the bench warmer prize—you won't have to touch the meter's dial at all! Just let 0 mA stand for 0°F, .10 mA for 10°F, .20 mA for 20°F, .30 mA for 30°F, .32 mA for 32°F etc. If you chose the −50 to + 50 Celsius (Centigrade) thermometer, your job isn't much harder. Just add the temperature markings shown in the chart to your meter's dial.

If you are building the same thermometer as the one which I built, which measures temperatures between −20°F and +105°F, you have a decision to make on how you mark the dial on the meter. The easy way out is to mark it according to Table 8-2, which labels every tenth of a milliamp with a temperature reading. Each division stands for 12.5°F.

If you find a thermometer that reads 5 or 10°F. Our prototype was labeled every 10°F. However, to have the thermometer show every 5°F you must place a label at .04 milliamperes intervals (or .08 mA every 10°F). While this can be easily done on some of the larger and better quality meters, most meters will require careful work and constant checking with drafting dividers (an instrument similar to a compass).

To label the temperatures on the meter dials, carefully remove the plastic or glass panel that protects the meter's movement and needle. This panel can often be pulled off with your fingers, but some meters have machine screws holding it in place.

Unless you are an artist or are just naturally great at lettering, use commercial dry transfer lettering. It's worth that extra little effort to do a nice job in building electronic projects, especially this thermometer, which you will use every day, for many years.

Testing and Adjusting. Connect the four-conductor cable to the thermometer. With the potentiometer, R5, set to its midpoint, depress the power switch. The needle should move upscale to the vicinity of the transducer's temperature. If the transducer is in still air, and has no heatsink, the needle will continue to slowly move upscale 1 to 2°F more.

Before attempting the following calibration, make sure all bare wires in the transducer probe assembly are completely insulated with one of the waterproofing materials.

Fill a small plastic pail half full of small ice cubes, bits of ice, or, if

	1% (ohms)	5% (ohms)
R1	1100	1100
R2	23.7K	24K
R3	43.2K + 470	43K + 680

For Farenheit Thermometer
0 to 100 degrees

1% (ohms)	5% (ohms)
900	910
25K	24K + 1100
42K	43K

For Celsius Thermometer
−50 to +50 degrees

	1% (ohms)	5% (ohms)
R1	1.5K	1.5K
R2	21.5K	22K
R3	46K	43K + 2.7K

Note that the plus sign between resistance means the two resistors should be placed in series.

Table 8-1. For Fahrenheit Thermometer.

available, clean, compacted snow. Pour enough cold water into the pail to fill it approximately two-thirds full.

Place the transducer probe assembly into the icy concoction in the plastic pail. Wait several minutes. Press the power switch and adjust the pot, R5, so the thermometer shows 32°F (0°C). That's about all there is to the calibration, easy huh? What's that! You didn't use the precision resistors recommended, and you have an *accurate* thermometer (surgical) available! Well then, you should test your new thermometer at another temperature. To do this obtain luke warm

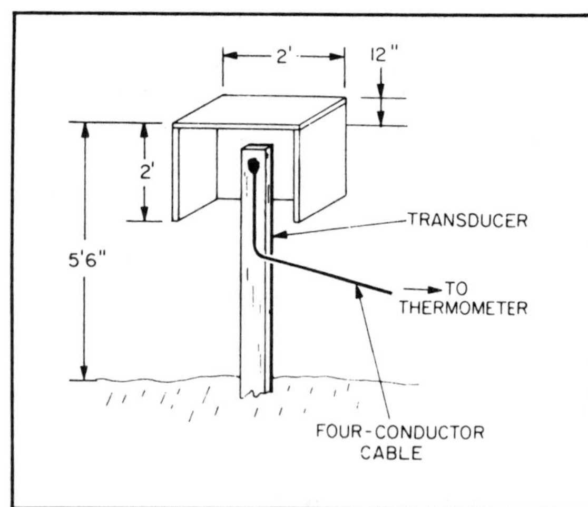

Fig. 8-22. If the PET's temperature sensor will be outdoors in sunlight, a simple cover should be built for it. Dimensions are approximate. If not sheltered, the readings might be off a degree or two.

Table 8-2. For Fahrenheit Thermometer −20 to + 105 degrees.

For Celcius Thermometer −50 to +50 degrees		For Celcius Thermometer −20 to +105 degrees	
mA	Degrees	mA	Degrees
0.0	−50	0.0	−20
.10	−40	.10	−7.5
.20	−30	.20	+5
.30	−20	.30	17.5
.40	−10	.40	30
.50	0	.416	32
.60	+10	.50	42.5
.70	20	.60	55
.80	30	.70	67.5
.90	40	.80	80
1.00	50	.90	92.5
		1.00	105

water from your tap. The temperature of the water should be near the maximum temperature your thermometer can measure. Place both the transducer probe assembly from your electronic thermometer and your accurate (store-bought) thermometer in the warm water. Stir the water every few seconds or so. After about five minutes take readings from both thermometers. If the readings differ by 2°F (1°C) or more, you can increase the thermometer's accuracy by slightly changing the value of R1. If the thermometer reads low, place a 100K resistor in parallel with R1. If it reads high, add a 10- or 15-ohm resistor in series with R1. Or you can replace fixed resistor R1 with a 500-ohm potentiometer in series with an 820- or 680-ohm resistor. However, you should do this approach *only* if a laboratory-grade thermometer is available for use as your standard.

Using PET. The real advantage of an electronic thermometer over the common garden variety is that it is the only kind of thermometer which can accurately measure temperature at a distance. Because of its accuracy and quick response, it can be used as a fast responding medical thermometer if the probe is placed under one's arm. Underarm temperature averages around 97.5°F (36.4°C) in healthy people—about 1°F lower than oral temperatures. PET can also be used as a super-accurate fishing thermometer to get the big ones, or in your car to measure the outside air temperature. When used in an automobile, it is most accurate when the car is moving. The only modification necessary to run PET directly off a car or boat's 12 Vdc electrical system is that the resistor R7 must be increased to 330 ohms. Because of its low power consumption, no power switch is necessary when a storage battery is used for power.

One of the most popular uses for electronic thermometers is the measurement of the outside air temperature from the comfort of one's home. However, even with such an accurate electronic thermometer

−20 to +105 degrees

mA	Degrees
0.0	−20
.04	−15
.08	−10
.12	−5
.16	0
.20	+5
.24	10
.28	15
.32	20
.36	25
.40	30
.416	32
.44	35
.48	40
.52	45
.56	50
.60	55
.64	60
.68	65
.72	70
.76	75
.80	80
.84	85
.88	90
.92	95
.96	100
1.00	105

Table 8-3. For Fahrenheit Thermometer.

as described here, this isn't as easy as it first may appear. Although just about everybody knows, you must keep the thermometer out of sight of that 11,000°F Ball of Fire in the sky to measure the true air temperature, few realize that a clear night sky has an effective temperature that would make a penguin shiver. Even on a clear, July night, the effective temperature of the sky is around −100°F!

A thermometer without a shelter, placed in the shade of the north side of a building, may read correctly during the day or on a dark and foggy night, but on a crystal-clear night it will read several degrees colder than the true air temperature. Moisture or frost will also lower a thermometer's temperature below that of the air on account of cooling by vaporization. Like it or not, you need some sort of thermometer shelter if you are to accurately measure the outside air temperature. The National Weather Service uses quite a sophisticated shelter, but for most purposes the shelter shown on page 324, is sufficient, especially if located in the shady area on the north side of a building. This shelter can be constructed out of an old apple box or built from scrap lumber. The transducer probe assembly should be mounted near the top of the inside of the enclosure and approximately 5 feet from the ground. All dimensions shown in the drawing are approximate and are given only as a guide. The shelter should be painted, inside and out, with a white, or better yet, aluminum exterior paint. Notice the enclosure has no bottom. This is to provide good air circulation for the transducer.

The Components Board. In this project the parts layout is entirely uncritical so the parts may be mounted on a printed circuit board, using the board pattern shown, or a similarly-sized piece of perfboard, with flea clips inserted into holes at roughly the same locations as are indicated in the components location diagram. If you're a beginner at constructing electronic projects, you'll find as you progress to more complex projects that making up a printed circuit board saves construction time and improves the project by making it neater and much more reliable (rigid mechanically) than other construction methods. However, you may like many other beginners be hesitant about making such a board the first time. If so, making a fairly simple printed circuit board such as the one shown for this project is a good idea. Although not necessary for this project, it is good practice and training for much more complex projects, where use of a printed circuit components board is virtually mandatory.

There are numerous other practical applications for PET, such as measuring your freezer and refrigerator's temperature. (Consumer's Union recommends a temperature of 37°F in the center of the fresh food space and 0°F in the freezer.) PET's variety of use is limited only by your needs, and by your imagination.

ELECTRONIC RAIN GAUGE

Who knows how much rain has fallen during any period of time? The Electronic Rain Gauge knows, and will tell you at a glance! This neat little unit will quickly tell you how much rain has fallen during any period of time, overnight for example. You won't have to wait for some giggly weatherman to tell you the precipitation over a 24-hour period either. The Electronic Rain Gauge will tell you with its eight-LED readout.

Unlike old-fashioned rain gauges, assembly is fast. The entire project cost is about $5 if you shop wisely, and best of all, you won't get soaked while using it either!

How the System Works. The project consists of two parts: A cabinet that houses all the circuitry including the LED display, and a water retaining bottle (with a funnel) that will hold the rain to be measured.

Refer to Figs. 8-23 and 8-24.

A ground wire, mounted on the bottom of the bottle, represents a negative potential that can turn on (bias) each of the transistors, Q1-Q8, through their respective sensing wires. When rain water bridges the gap, the ground wire and, for example, the 0.1 inch sensing wire (connected to the base of Q1) is turned on, and LED1 glows, showing that 0.1 inches of rain has fallen. The resistance of water is very high here, (20,000 to 100,000 ohms, typically) but it takes little negative potential to turn each of the eight transistors on.

Note that each sensing wire connects to the base of a transistor, which it controls. The length of the sensing wires can be of any convenient length. For example, you can have the cabinet with the circuitry and LED readout inside of your house, and the water retaining bottle outside for a real custom installation!

Circuit Assembly. To house the circuitry, choose a cabinet small enough to make the project compact, yet large enough to house all of the circuitry and LEDs. Since there are few parts, a neat layout on a small piece of perfboard would be your best bet. Point-to-point wiring, using short lengths of wire could be best with the perfboard arrangement.

Mount the LEDs on the front of the cabinet where they will be easily visible. If you mount the LEDs on a metal face, make sure that the LED leads do not short on the metal facing. Drill a hole in the top or side of the cabinet large enough to allow the eight sensing wires and the ground wire to pass through. The sensing wires can be of any gauge, but use solid, not stranded, wire. It would be a good idea to use

DIAMETER OF FUNNEL							
12	144	64	36	23.04	16	11.76	9.0
11	121	53.77	30.25	19.36	13.44	9.88	7.56
10	100	44.44	25	16	11.1	8.16	6.25
9	81	36	20.25	12.96	9.0	6.61	5.06
8	64	28.44	16	10.24	7.111	5.22	4.0
7	49	21.78	12.25	7.84	5.44	4.0	3.06
6	36	16	9,0	5.76	4,0	2.94	2.25
5	25	11.1	6.25	4.0	2.77	2.04	1.56
4	16	7.11	4.0	2.6	1.78	1.3	1.0
3	9.0	4.0	2.25	1.44	1.0	0.73	0.56
DIAMETER OF TUBE	1	1.5	2	2.5	3	3.5	4

Fig. 8-23. The numbers up the left side represent the diameter of the funnel in inches, and the numbers along the bottom represent the diameter of the retaining tube in inches. A six-inch funnel on a three-inch tube will have a height ratio of 4.0 where four inches of water in the retaining tube would represent one inch of rain. If a one-inch tube had a 12-inch funnel, that tube would need to be 144 inches high to measure one inch of rain collection.

color coded wire for the sensing wires so you can tell more easily where each wire will go when you install them into the water retaining bottle. The sensing wires can be any length that you wish, but if you use a really appreciable length, then use the thickest gauge of wire practical.

Rain Collector Setup. The water retainer bottle will catch rain directed into it and hold it for continuous measurement. You will need to drill eight holes for the sensing wires to go into at their respective levels. One additional hole, flush with the bottom of the bottle, will also be necessary for the ground wire.

The water retainer can be almost any plastic or metal container but it should have straight sides (the same diameter from top to bottom). You will also need a funnel to help collect the rain drops. You could use the container just as it is but little rain would get into the relatively small opening and it would be rare that the water depth ever exceeded a small fraction of an inch.

If you use a funnel that is twice the diameter of the container you will collect four inches of water in the tube for every inch that falls. This is because doubling the diameter of a circle increases its area by a factor of four, and four times as much rain is collected. You need to determine the ratio of the funnel area to the area of a cross-section of the measuring tube.

To do this you can use the attached chart or use the following formula:

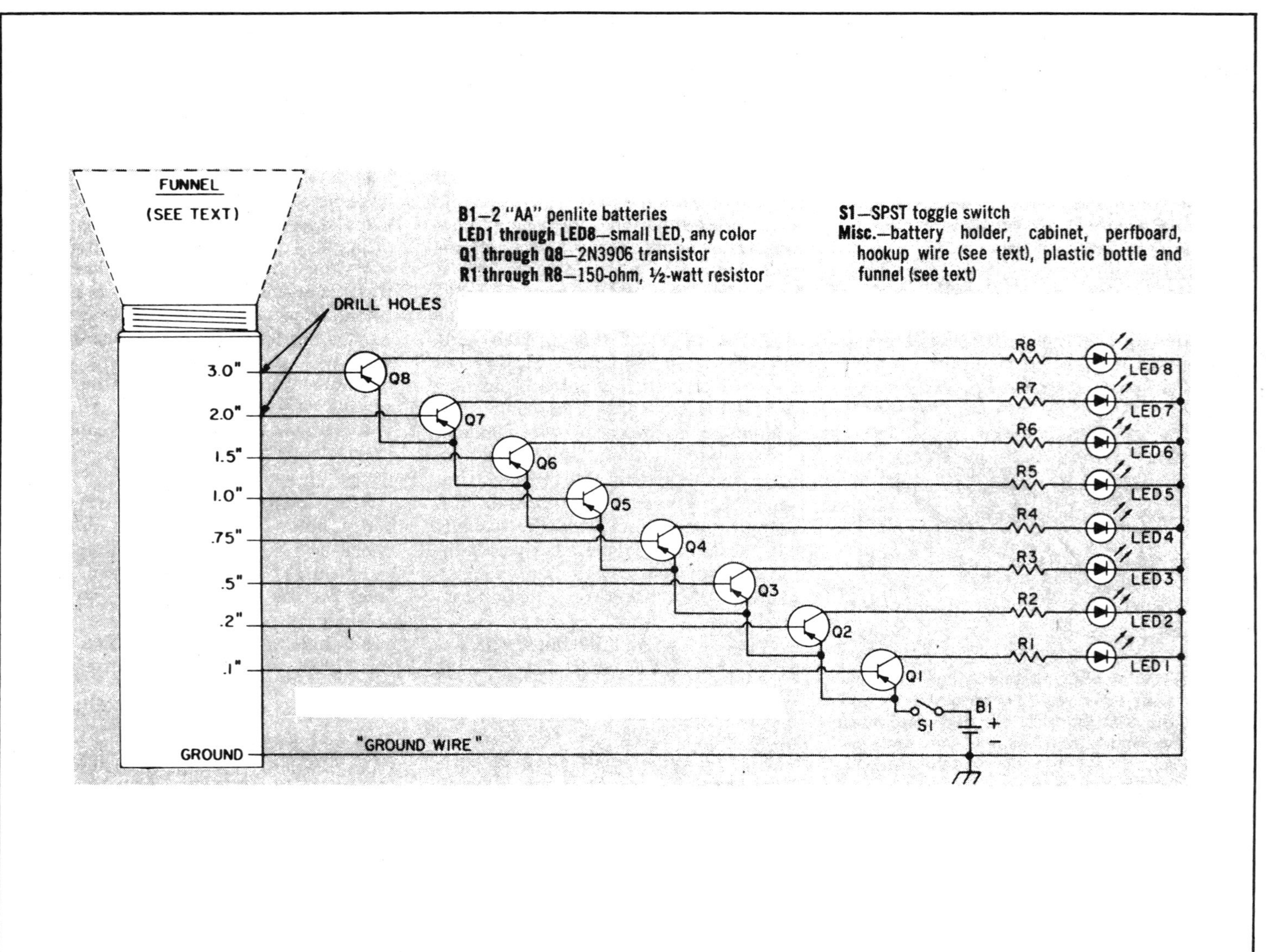

B1—2 "AA" penlite batteries
LED1 through LED8—small LED, any color
Q1 through Q8—2N3906 transistor
R1 through R8—150-ohm, ½-watt resistor
S1—SPST toggle switch
Misc.—battery holder, cabinet, perfboard, hookup wire (see text), plastic bottle and funnel (see text)

Fig. 8-24. Electronic Rain Gauge schematic.

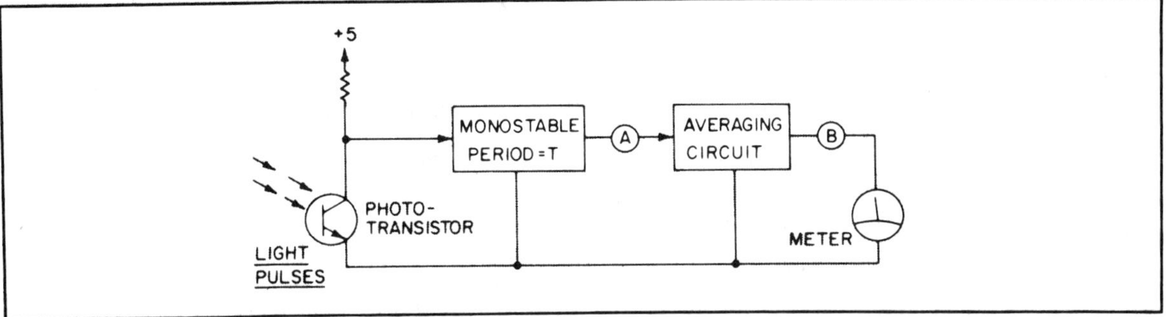

Fig. 8-38. When a light pulse falls on the phototransistor, the transistor conducts and triggers the monostable circuit which follows. When triggered, the output of this circuit goes high and then returns low. The output is fed into an averaging circuit and the meter.

tachometer; the choice of method depends on whether or not the rotating object can chop a beam of light. Consider first those devices which can chop a light beam, such as fans, propellers, pulley spokes, and even drive chains. With these you simply place the rotating object between the light source and photo-probe, thus allowing the propeller or whatever to chop the light source and photo-probe, thus allowing the propeller or whatever to chop the light that falls on Q1. Start with a distance of about six feet between your light source (100-watt lamp plus reflector) and the photo-probe. Decrease the spacing until you obtain a steady indication on M1. Further decrease in distance will not affect this reading. Make note of this working distance for future reference. Of course, if you are using sunlight, these directions don't apply. *Note that if the propeller has two blades, your reading on M1 will be twice the actual speed of rotation.* Likewise, four blades yield a reading that is four times too high, and so on. Do *not* use any backlighting (light coming from probe side of what you're measuring).

The other mode of operation relies on reflection to supply light pulses to Q1. We have diagrammed a dark-colored wheel, to which a small piece of aluminum foil has been attached. Once every cycle, the foil is in a position that enable it to reflect light from the source onto the photo-probe. Measurements by reflection may tend to be tricky, since you have to set up the angles just right. Nevertheless, a little experimentation is usually all that's necessary to get things working. The total light path—from source to reflector to photo-probe—should be less than or equal to the working distance you determined for the previous case with the propeller. Sometimes stray reflections in this mode can be troublesome, since they may prevent Q1 from cutting off (i.e., ceasing to conduct). A careful elimination of all extraneous sources of reflection will solve this problem.

As a final observation, note that Mack the Tach is a very flexible measuring instrument; its applications are limited only by your own ingenuity. So when you come upon a measuring task that has not been described here, be willing to experiment.

Index

$$\left(\frac{\text{Diameter of Funnel}}{\text{Diameter of Tube}}\right)^2 = \text{Height Ratio}$$

For example: a 5.75-inch funnel feeding a 1.66-inch tube would have a height ratio of:

$$\left(\frac{5.75}{1.66}\right)^2 = 11.9983 \text{ or } 12$$

So 12 inches of water in the tube will equal one inch of rainfall, 6 inches will equal ½-inch and 3 inches will equal ¼-inch of rain.

Drill holes in the bottle at the appropriate locations. Note that if you use the same funnel and tube as mentioned above, that tube will need to be three feet long to measure three inches of rain.

Strip at least ½ inch of insulation off the ends of the sensing wires and push them through their proper holes so all of the non-insulated wire sticks inside the bottle level with its hole. Put a drop of super glue right where the insulation of each wire butts up against its proper hole. This will hold the wire snug and also prevent water from leaking out. Be careful not to get any of the super glue on the non-insulated part of the wire.

Strip at least 1½ inches of insulation off the ground wire, and install it like the others in its hole at the bottom of the bottle. Make sure it sits flat on the bottom of the bottle.

Operation. When the glue is dry on the water retaining bottle, fill it with water to make sure that it doesn't leak. If it's watertight, you can then test the whole system out. If there aren't any dark clouds around to help you, you can fill the bottle slowly with tap water. Run water in slowly and as each sensing wire is submerged, its respective LED should glow.

If no water is in the water retaining bottle, none of the LEDs will glow, and thus no current from the batteries is drawn. Leaving the unit on with the bottle empty does not drain your batteries. So there you have it. You'll be nice and dry whenever the next time arrives that you need to know, or are just curious, how much rain has fallen in the next storm or drizzle.

TESLA COIL

Science teachers, amateur scientists, magicians, and many electronics hobbyists will find the *Tesla Coil* an interesting project both to build and to perform experiments with. Readily demonstrating the principles of radio and electricity, the "giant coil" may also lead to an interest in Dr. Nikola Tesla, whose work is little known by most beginning science students. Few, in fact, have heard of Tesla, or of the work he has done. But many experts believe Tesla was as great as Edison—or even greater—in his contributions to science.

Refer to Figs. 8-25 through 8-31.

Tesla's Contributions. During Dr. Tesla's time, in the late 19th century, only direct current (dc) was used commercially, though there was some knowledge of alternating current (ac). Tesla was the first person who investigated the practical uses of alternating currents to any great extent.

A few years after he came to the United States (1884), Tesla worked out several possible systems of alternating-current machinery, dynamos, motors, transformers, and distribution systems for which he was granted 45 patents between 1887 and 1893. In 1891, George Westinghouse purchased some of Tesla's systems and used them at the 1892 Chicago World's Fair to supply all the electrical power needed.

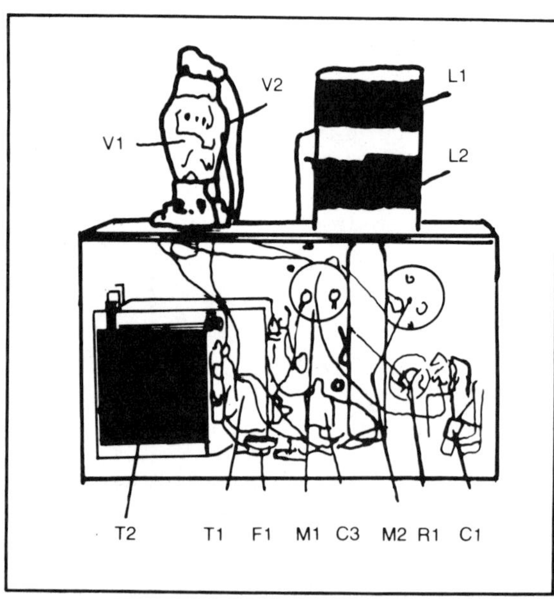

Fig. 8-25. Rear view of Tesla's Coil shows the location of the major components. Layout is not critical, but all leads to the coil, vacuum tubes, and bypass capacitors should be kept as short as possible, as done with rf wiring.

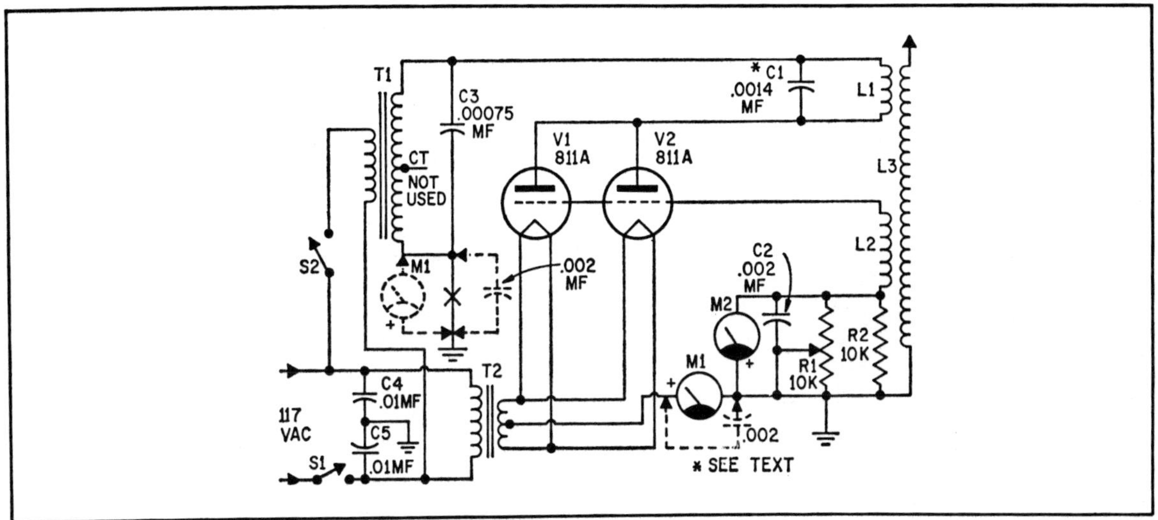

Fig. 8-26. Schematic diagram of the Tesla Coil shows alternate wiring connections in dotted lines. Circuit is quite flexible.

Dr. Tesla proposed a 7500-kilowatt transmitting plant producing rf currents at 100-million volts. One of the chief uses for this transmitted energy would be to illuminate isolated homes. The energy would be collected by a terminal a little above the roof of each house.

He actually built a generator similar to the one he proposed and lit 300 electric bulbs 25 miles from his laboratory. He started to build a transmitting station—probably the world's largest Tesla Coil—on a site near New York City before World War I, but he ran out of funds.

A Tesla Coil. This high-voltage rf generator is more commonly known as a Tesla Coil, although the actual construction hardly resembles the original apparatus of Dr. Tesla. The Tesla Coil is a dynamic demonstrator of electrical principles, and a crowd pleaser as a science fair project. It can be built for much less than the estimated cost. All the the materials used in building the unit pictured here were from war-surplus components. (The parts list gives stock, over-the-counter parts.)

The Tesla Coil is very easily built and it gives good results. Sparks several inches long may be obtained from the coil at voltages on the order of 50,000 volts.

Circuitry. The apparatus consists basically of a plate-tuned oscillator, as can be seen in the schematic diagram. Practically any transmitter-type triode vacuum tube will serve in the oscillator.

One of the things considered in selecting a tube was that it have a plate voltage rating of about 1500 volts. The higher the voltage, the more spectacular the experiments that may be performed with the apparatus. The second factor considered was that it have a fairly-high power output (of about 150 watts or more). Two 811A transmitting triodes met the specifications. Both of them are used, in parallel, to

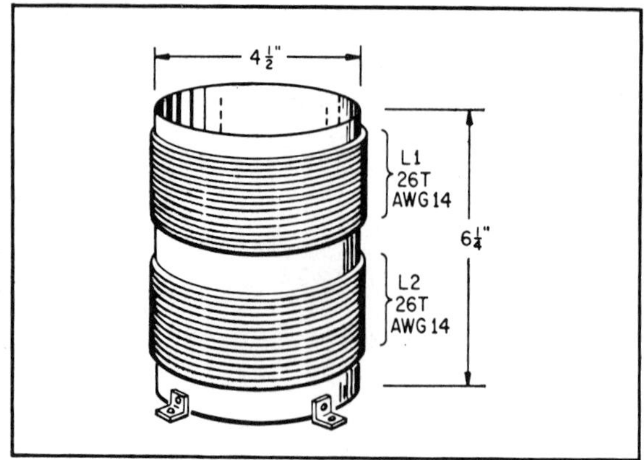

Fig. 8-27. Grid and plate coils are wound with same size wire. Large-diameter wire increases Q of grid inductance and that of plate tank L1-C1. Low-resistance coils have less internal losses and higher output—brush discharge is greater because of better efficiency.

deliver a total output of about 300 watts of rf power.

Supply Transformers. A high-voltage-secondary transformer (delivering about 1840 volts at about 250 milliamperes) is used as the ac plate-voltage supply. In the schematic diagram, a 0.00075-μf bypass capacitor is placed across the high-voltage secondary to prevent the high-frequency current from going through the transformer and out the ac power line.

The filament-transformer secondary supplies 6.3 volts, and 4 amperes are needed for each tube. Since the cathodes in these tubes are directly heated, the current from the plates must flow through the filament transformer to the filament. This is best accomplished by connecting the B– to the centertap of the filament winding, making a complete circuit as shown.

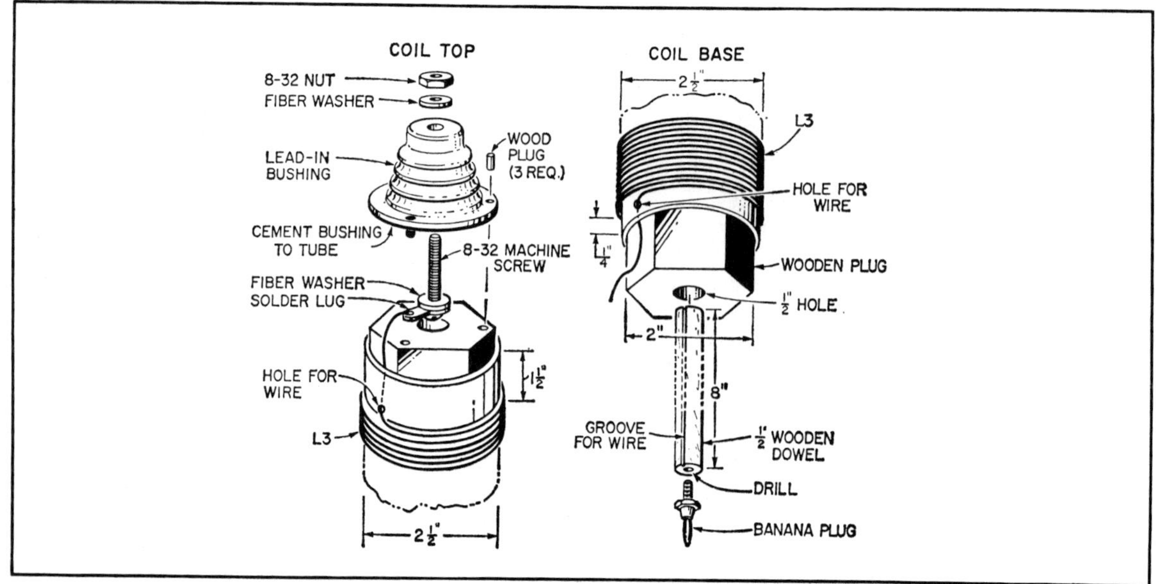

Fig. 8-28. Details of coil construction are important to simplify fabrication and winding. Plug-in coil makes storage easier.

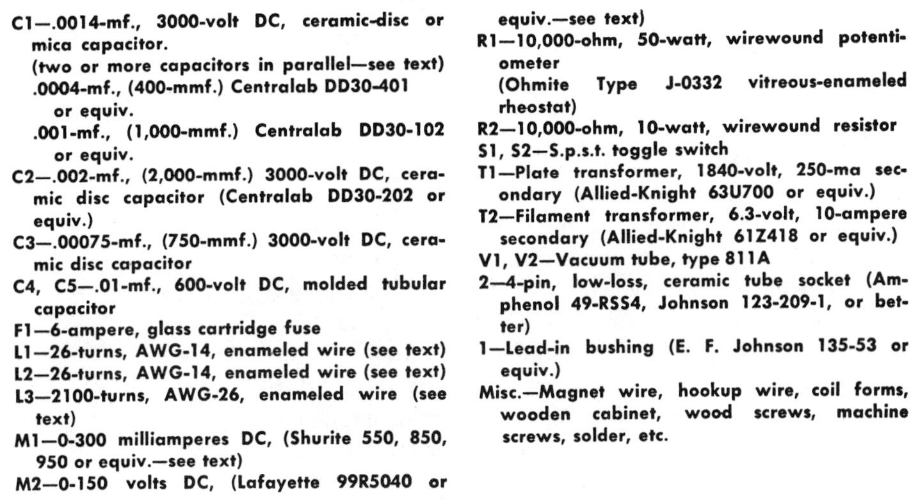

C1—.0014-mf., 3000-volt DC, ceramic-disc or mica capacitor.
(two or more capacitors in parallel—see text)
.0004-mf., (400-mmf.) Centralab DD30-401 or equiv.
.001-mf., (1,000-mmf.) Centralab DD30-102 or equiv.
C2—.002-mf., (2,000-mmf.) 3000-volt DC, ceramic disc capacitor (Centralab DD30-202 or equiv.)
C3—.00075-mf., (750-mmf.) 3000-volt DC, ceramic disc capacitor
C4, C5—.01-mf., 600-volt DC, molded tubular capacitor
F1—6-ampere, glass cartridge fuse
L1—26-turns, AWG-14, enameled wire (see text)
L2—26-turns, AWG-14, enameled wire (see text)
L3—2100-turns, AWG-26, enameled wire (see text)
M1—0-300 milliamperes DC, (Shurite 550, 850, 950 or equiv.—see text)
M2—0-150 volts DC, (Lafayette 99R5040 or equiv.—see text)
R1—10,000-ohm, 50-watt, wirewound potentiometer
(Ohmite Type J-0332 vitreous-enameled rheostat)
R2—10,000-ohm, 10-watt, wirewound resistor
S1, S2—S.p.s.t. toggle switch
T1—Plate transformer, 1840-volt, 250-ma secondary (Allied-Knight 63U700 or equiv.)
T2—Filament transformer, 6.3-volt, 10-ampere secondary (Allied-Knight 61Z418 or equiv.)
V1, V2—Vacuum tube, type 811A
2—4-pin, low-loss, ceramic tube socket (Amphenol 49-RSS4, Johnson 123-209-1, or better)
1—Lead-in bushing (E. F. Johnson 135-53 or equiv.)
Misc.—Magnet wire, hookup wire, coil forms, wooden cabinet, wood screws, machine screws, solder, etc.

Fig. 8-31. Tesla's Coil parts list.

spectacular demonstrations which can be performed—such as lighting fluorescent and neon tubes without wire connections. In fact, the energy of the discharge is sufficient to ignite candles and paper, or to drive a pin-wheel balanced on the discharge point. Even the smallest Tesla coil never fails to astound and interest onlookers with feats such as these.

MACK THE TACH

Like most people, you probably own a variety of motor-driven appliances and devices; autos, boats, washing machines, lawn mowers, power tools, model airplanes, movie projectors, tape recorders, and so forth. Again like most people, you probably never give a thought to the proper maintenance of these items until they break down. One of the surest indications of an upcoming breakdown is improper motor speed, and for a few dollars, you can make a tachometer to measure it.

Refer to Figs. 8-32 through 8-38.

A tachometer is an absolute necessity for the proper maintenance and tuning of motor-driven devices, and we here present *Mack the Tach*, every bit the equal of commercial units costing approximately $200. Motor speed can be measured on four ranges from 1000 rpm full-scale to 30,000 rpm full-scale. Accuracy is an excellent ±2½ percent of full-scale on the 10,000 rpm range, and ±3½ percent of full-scale on all other ranges. Furthermore, because this tachometer is optically coupled, no extra load is placed on a rotating device while it is

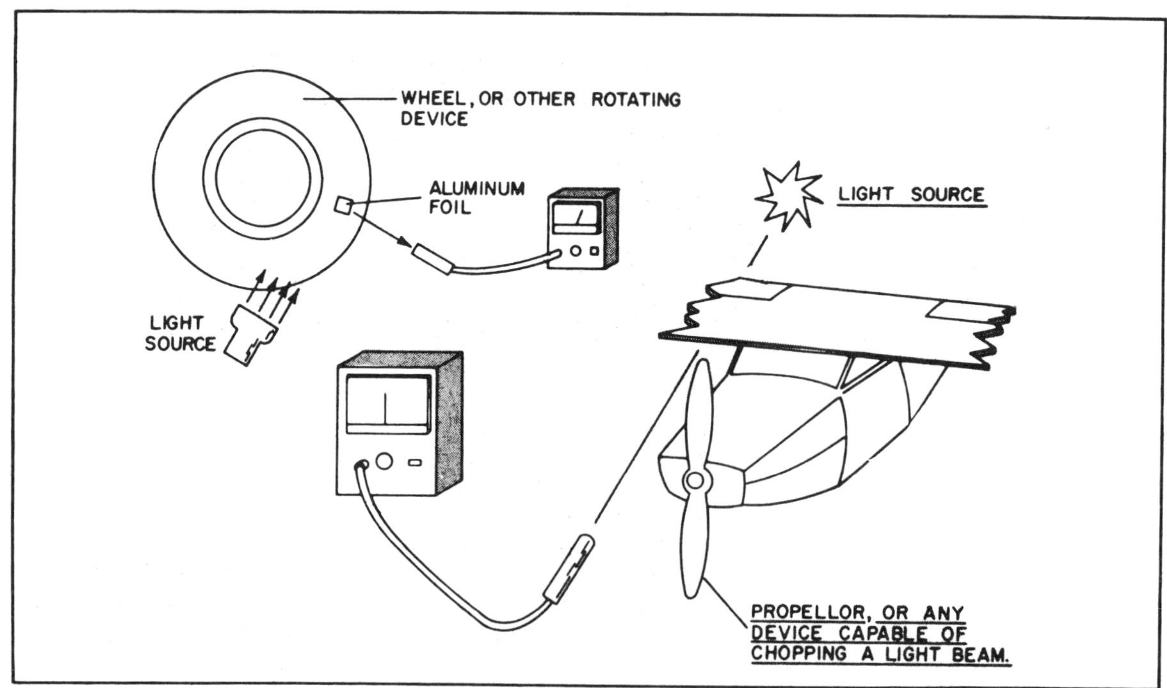

Fig. 8-32. Mack the Tach can measure the rotational speed of almost any object that can "chop" a light beam. Simply use it as shown here. If you wish to measure the rotational speed of something such as a wheel, use aluminum foil as shown, reflecting the light off of it into Mack's home-built, light-sensitive probe.

being tested. The result is better accuracy, especially with small, light-duty motors.

Seeing the Light. As you probably know, rpm measurements are just frequency measurements. In order to obtain an rpm reading on an analog meter, we need a frequency-to-voltage converter circuit. Such a circuit is detailed in the block diagram. Assume that we have arranged things so that every time the motor rotates, one pulse of light falls on the photo-transistor. This causes the photo-transistor to conduct and trigger the monostable circuit that follows. Each time the monostable is triggered, its output (point A) rises to a high potential for

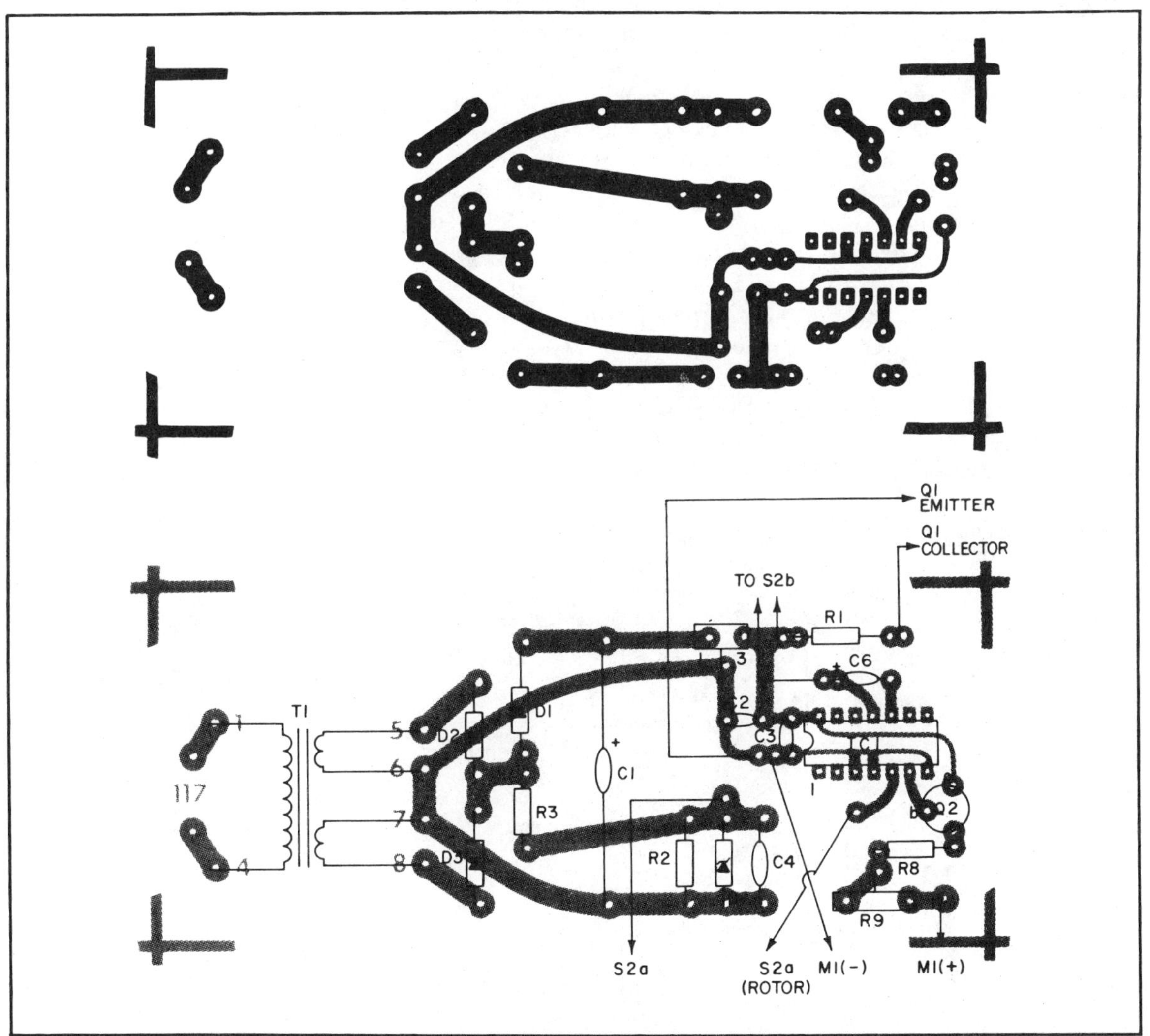

Fig. 8-33. At the top is a foil-side pattern for making your own Mack the Tach PC board. This circuit is simple enough to allow use of a resist pen to make the board if you do not have photographic means at your disposal. Beneath is the component-side view of Mack.

a fixed time interval T, then drops low again, remaining low until retriggered. The monostable output next feeds into an averaging circuit, whose output (point B) is ideally a dc voltage that drives a meter.

Operation of Mack the Tach is explained more clearly by the voltage diagrams. At low rpm, the monostable's output pulses are spaced fairly well apart. Consequently, the average value of the output is low, as indicated by the dc level dashed-in on the diagram. Since the average value is low, the meter will only deflect a little bit. Now, at higher rpm, the monostable gets triggered more frequently. The monostable's output spends proportionately more time at a high potential than at low potential. The average value of the monostable's output is now higher, and this results in a correspondingly higher reading on the meter. In both the high-rpm and low-rpm cases, the monostable's output rises to the same high potential for the same time interval (T); higher rpm decreases the time between pulses, and this alone results in a higher average voltage at the monostable's output.

Let's next examine Mack's schematic diagram. Diodes D2 and D3 full-wave rectify the ac voltage from transformer T1. This rectified current splits two ways: to D1 and to R3. Consider first the path through R3, D4, C4, and R2. The purpose of this four-component network is to provide a 120 Hz, clipped, full-wave-rectified sine wave, available whenever S2 is fully clockwise. This signal is used to trigger monostable IC1 during calibration.

Now let's consider the alternative path of the rectified current through D1 to C1 and IC2. C1 smooths out the rectified ac, while D1

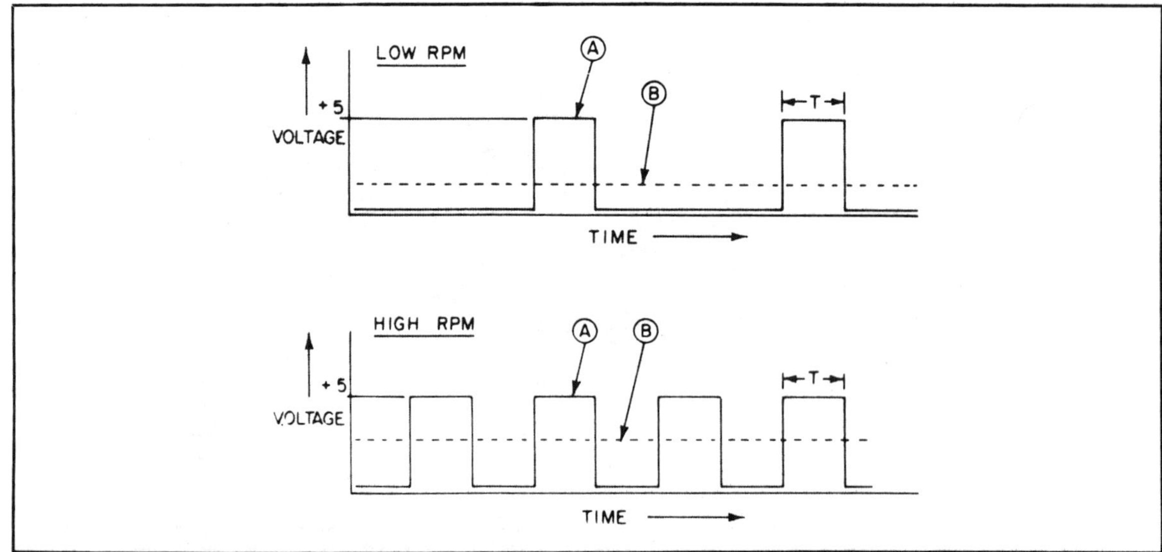

Fig. 8-34. At low rpm, the voltage pulses from the monostable output are far apart in time and this lowers the average voltage level. As the rpm increase, these pulses become closer and the voltage rises as read on the meter.

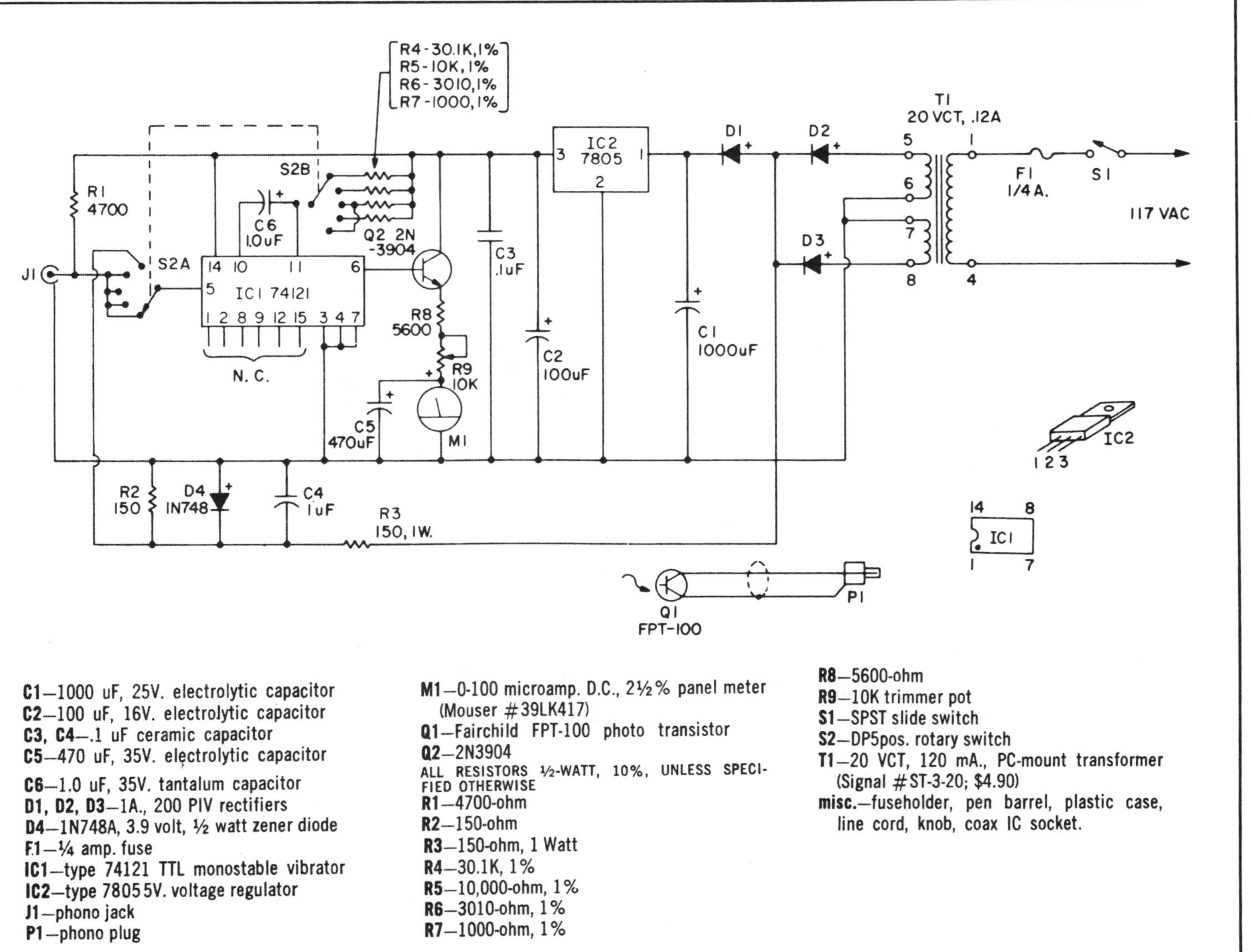

C1—1000 uF, 25V. electrolytic capacitor
C2—100 uF, 16V. electrolytic capacitor
C3, C4—.1 uF ceramic capacitor
C5—470 uF, 35V. electrolytic capacitor
C6—1.0 uF, 35V. tantalum capacitor
D1, D2, D3—1A., 200 PIV rectifiers
D4—1N748A, 3.9 volt, ½ watt zener diode
F1—¼ amp. fuse
IC1—type 74121 TTL monostable vibrator
IC2—type 7805 5V. voltage regulator
J1—phono jack
P1—phono plug

M1—0-100 microamp. D.C., 2½% panel meter (Mouser #39LK417)
Q1—Fairchild FPT-100 photo transistor
Q2—2N3904
ALL RESISTORS ½-WATT, 10%, UNLESS SPECIFIED OTHERWISE
R1—4700-ohm
R2—150-ohm
R3—150-ohm, 1 Watt
R4—30.1K, 1%
R5—10,000-ohm, 1%
R6—3010-ohm, 1%
R7—1000-ohm, 1%

R8—5600-ohm
R9—10K trimmer pot
S1—SPST slide switch
S2—DP5pos. rotary switch
T1—20 VCT, 120 mA., PC-mount transformer (Signal #ST-3-20; $4.90)
misc.—fuseholder, pen barrel, plastic case, line cord, knob, coax IC socket.

Fig. 8-35. Mack the Tach schematic and parts list.

isolates the R3-C4-D4-R2 network from the smoothing action of C1. Voltage regulator IC2 transforms the unregulated dc voltage across C1 into a regulated 5-volt potential at its output (pin 3). Capacitors C2 and C3 bypass the 5-volt supply and stabilize the circuit.

Transistor Q1 is the photo-transistor, and it connects to the rest of the circuit through a piece of coaxial cable terminated in P1. Plug P1 connects to jack J1.

So long as *Range Selector* S2 is not in the *Cal.* position (extreme clockwise), the trigger input (pin 5) of IC1 will be connected to photo-transistor Q1's collector through S2a.

Changes in the intensity of the light incident on Q1 produce changes in Q1's collector potential, thus triggering IC1. The duration of the output pulse available from monostable IC1 is controlled by C6 together with either R4, R5, R6, or R7. Switch S2b selects the resistor appropriate to the rpm range in use.

IC1's output (pin 6) drives transistor Q2, which then drives meter M1 through R8 and R9. Q2 provides some current gain, and it also ensures that all current to the meter gets cut off when pin 6 drops to its low level (a few tenths of a volt). The averaging in this circuit is performed to some extent by meter M1 itself because the inertia of the meter's needle causes the deflection to be proportional to the average current.

At low rpm, however, the meter needle would vibrate perceptibly, so capacitor C5 assists in averaging the pulses. Even so, you may still notice a little vibration when reading very low rpm; this is normal and not a cause for concern.

Light and Easy. Construction of the tachometer is particularly simple. Though a printed circuit board is optional, it does make construction even easier—transformer T1 is especially made for PC mounting. Instead of the usual solder lugs, this transformer's lead wires are brought out as pins, which then are soldered directly to the circuit board.

Parts layout within our Mack the Tach is not critical, so you may use any convenient arrangement. When drilling your cabinet, provide an access hole for R9; this will allow you to calibrate the circuit without removing the front cover every time.

Meter M1 is a 0-100 mA dc unit. If you happen to have a 0-100 mA meter at hand, you can use it, but remember that the accuracy of the meter will determine the overall accuracy of your tachometer.

A few further comments on some of the other components are in order. Note that resistor R3 should be a 150-ohm, 1-watt unit. If you don't have a 1-watt resistor, two 330-ohm, ½-watt resistors in parallel can be used instead. Almost any phototransistor can be used for Q1. Fairchild FPT-110s and FPT-110s (Radio Shack #276-130) were used with success. As noted in the parts list, resistors R4, R5, R6, and R7 were 1 percent units in the prototype. You can get by with 5 percent units which will leave the accuracy at about ±2½ percent of full-scale on the 10,000 rpm range; however, accuracy on all other ranges will now be less than or equal to ±7½ percent of full-scale.

Timing capacitor C6 is a 1.0-μF electrolytic, but be sure to use a tantalum device, not an aluminum capacitor. With this capacitor, you must be sure to get the proper orientation when installing it into the circuit. The same caution applies to all the semiconductors, meter M1, and capacitors C1, C2, and C5. As an added precaution, use a socket for IC1. In this way that IC can easily be removed if by chance it should turn out to be defective. Finally, note that voltage regulator IC2 is simply soldered to the circuit board. No heatsinking of this IC is required because only a small amount of supply current is consumed by the circuit.

In order to protect Q1, a photo-probe assembly will have to be constructed, which we have illustrated. Start by threading one end of a small-diameter coaxial cable (Belden 8417 or the equivalent) completely through the plastic barrel from an old pen. Now, grasp phototransistor Q1, and cut its base lead completely off. Solder the central conductor of the coax cable to Q1's collector, and then solder the coax's shield to Q1's emitter. Pull on the coax so as to retract Q1 far enough into the pen barrel so that its light-sensitive face is recessed one-half inch. Carefully secure Q1 and the coax to the pen barrel using epoxy cement. Finish up by attaching plug P1 to the far end of the coaxial cable.

Once you've completed construction of Mack, only calibration remains. Adjust R9 so that its resistance is maximum. Turn on the power, and put S2 into its Cal. position. Now adjust R9 for a reading of exactly 72 on M1. This completes the calibration. In the future you may re-check the calibration simply by repeating the above procedure. For most applications Mack the Tach as originally designed has adequate sensitivity. In fact, it is desirable for a photo-tachometer to have a minimum practical amount of sensitivity; in this way, ambient lighting conditions rarely affect a measurement. If added sensitivity is desired,

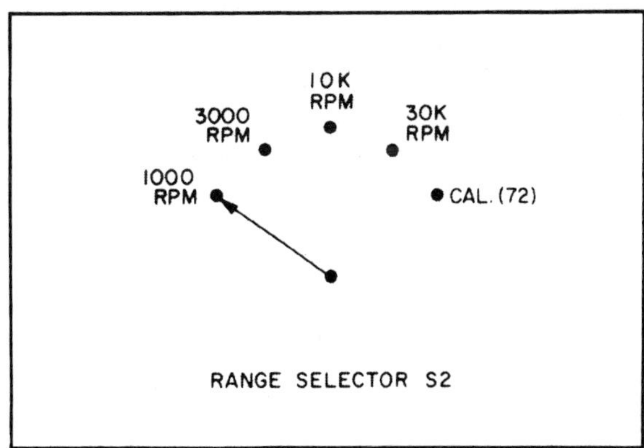

Fig. 8-36. Design your front panel as shown here. The rpm full-scale readings can be calculated using the table of RPM Full Scale vs. Mult. Factor, given in the text.

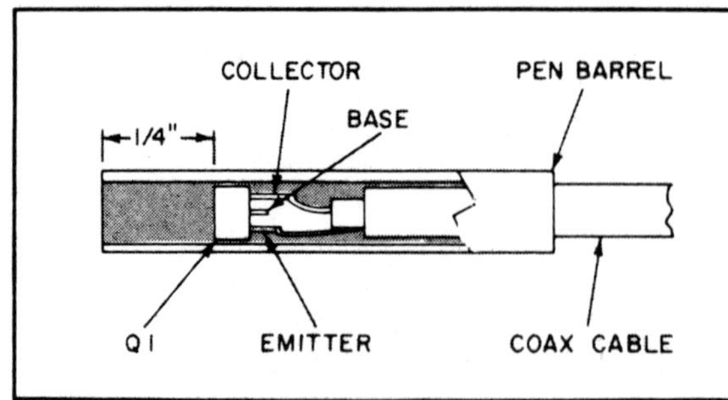

Fig. 8-37. Not only is the pen mightier than the sword, an old pen barrel can also show you the light! Build Mack's unique, light-sensitive probe—a pen to house the phototransistor.

however, the easiest course (besides going to a more powerful light source) is to replace Q1 with a photo-Darlington transistor, which must be NPN. One good choice is a type 2N5777, available from Poly-Paks as stock number 92CU2649. Other types may be used as well. Hookup is identical to that of a standard photo-transistor.

Let's now discuss the use of the tach. To begin with, you should place range switch S2 so that full-scale deflection is well above the motor's estimated rpm. After the first reading, drop down to a lower range if necessary. You'll notice in the photos of the prototype that meter M1's scale was left with its original markings: 0 to 100. It's then easy to apply a multiplication factor appropriate to the given range, as shown below.

RPM FULL SCALE	MULT. FACTOR
1000	X10
3000	X30
10,000	X100
30,000	X300

When setting up a measurement, it is important that the ratio of maximum to minimum illumination of Q1 be as high as possible. So far as the maximum illumination is concerned, a 100-watt bulb can saturate (fully turn on) Q1 from as far away as five feet, approximately. This assumes that the bulb is in a suitable reflector. If not, Q1 will have to be nearer to the lamp (about a foot away). The minimum light intensity on Q1 should be as low as possible. Recessing the phototransistor helps in this respect, since stray illumination is thus eliminated. *Try to avoid fluorescent lights* as sources of illumination for your Mack the Tach; incandescent bulbs and sunlight are best. The trouble with fluorescent lights is that their light output is intensity-modulated at 120 Hz., which is equivalent to 7200 rpm. Depending upon the exact characteristics of the fluorescent lamp and its distance from the photo-probe, erroneous readings can result from the use of such sources.

There are basically two different ways of using a photo-

Fig. 8-29. Wood block forms socket for plug-in coil L3. Six-inch deep hole accepts dowel which takes strain off jack and plug. Auger bits have necessary length for hole.

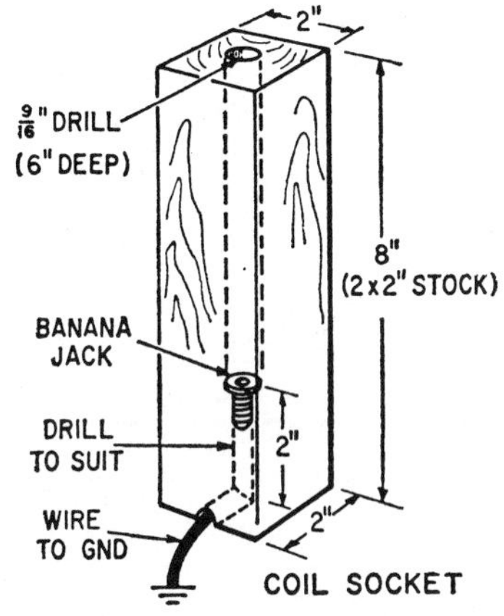

Coil Wiring. Plate coil L1 is 26 turns of No. 14 enameled magnet wire on a thoroughly wax-impregnated, 4¼-inch diameter cardboard form (phenolic or plastic can also be used). The grid coil (L2) is wound above L1, on the same form, in the same direction, and with the same number of turns of No. 14 wire as L1. Both coils must be given two coats of good insulating varnish. Brackets are put on the bottom of the form to mount it on the wooden cabinet.

To complete the resonant circuit (L1-C1), the value of C1 was calculated so the circuit oscillated at about 500 kHz. From experimentation it was found that the circuit used here oscillates best with a capacitance of about 0.0014-μF (.001-μF and .004-μF in parallel) with a corresponding frequency of about 500 kHz.

The output coil (L3) consists of 2100 closely-wound turns of No. 26 enameled wire on a thoroughly wax-impregnated 2¼-inch diameter cardboard form. Three coats of insulating varnish were put on the windings. One end of the winding of the coil is connected to the terminal of a cone-shaped standoff insulator. A wooden plug is fitted to the inside of the bottom portion of the coil with a portion of the coil with a dowel inserted in a hole drilled in the plug to mount the coil and make the ground-end connection. Shellac the wood to keep out moisture.

A box, with the back open, should then be constructed. Both triode tubes are mounted on top of the box along with L3—mounted concentrically with L1 and L2. Both transformers (T1 and T2) are mounted inside the box at one side.

Ac Input. Two .001-μF capacitors, connected from each side of the ac-line input to ground, prevent any high-frequency oscillations from escaping to the power lines. The transformers are protected with a 6-ampere fuse as shown in the schematic diagram—a circuit breaker can be used.

Power Switching. Two switches are mounted on the front panel

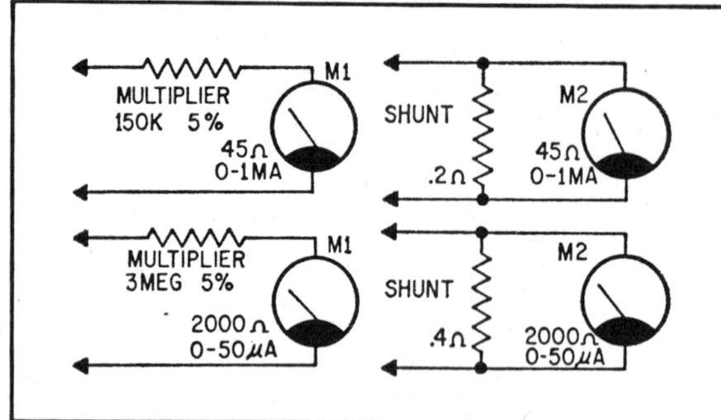

Fig. 8-30. Even 0-1 mA or 0-50 mA meters can be used to measure plate current and grid voltage in the Tesla Coil circuit if the proper multiplier or shunt is added to convert the meter to indicate proper range.

of the box. Plate transformer (T1) can not be turned *on* with S2 unless the filament transformer (T2) is *on*. This protects the tubes. For your own safety, it is wise to ground the apparatus to a waterpipe on some other ground.

In order to set R1 correctly, a dc voltmeter (M2) is wired across it to measure the grid bias. A dc milliammeter (M1) is also connected between the center tap of the filament transformer and ground to indicate the total plate and grid currents.

With the circuit completely wired—and double-checked for wiring mistakes—S1 may be turned *on* for filament warmup. After approximately 30 seconds, the plate-owner transformer T1) is turned *on* and the apparatus is in operation.

Control R1 is adjusted for a grid bias of about 115 volts—the plate-current meter should read about 200 mA. With the proper control adjustments some sort of brush discharge from the insulator terminal should be clearly visible.

With a variety of capacitors on hand, C1 can be changed in total capacitance until the brush discharge is maximum. With this condition maximum oscillation is obtained. (Maximum oscillation will vary somewhat with coil construction.) It was noticed that whenever the incorrect values of either R1 or C1 are used, the brush discharge was not as great.

The plates of the tubes must be watched at all times to make certain that their color does not get brighter than their normal dull-red.

Because each builder will want to use parts he has available, it is inadvisable to give specific instructions. Even substitutions for meters M1 and M2 may be made by using their proper shunt or multiplier. Though cardboard forms, shellac, and ac-fed oscillators may seem haywire today, the use of more deluxe components and circuitry is not justified for the average experimenter.

Of course you can build a lower-power version of the Tesla coil for a lot less money. But with less power available, the experiments will be limited and many may be impossible to perform. Lower power will give less brush discharge.

Experiments with the unit will reveal the many fascinating and